SOCIAL P 97/98

Twenty-Fifth Edition

Editor

Harold A. Widdison
Northern Arizona University

Harold A. Widdison, professor of sociology at Northern Arizona University in Flagstaff, holds degrees in Sociology and Business Administration from Brigham Young University and Case-Western Reserve University. Employed as an education specialist with the U.S. Atomic Energy Commission, he was awarded a Sustained Superior Performance Award. As a medical sociologist, Dr. Widdison is actively involved in his community with the local medical center's neonatal committee, a founding member of Compassionate Friends, a member of the board of directors of the Hozhoni Foundation for the mentally handicapped, and a consultant on death, dying, and bereavement.

A Library of Information from the Public Press
Dushkin/McGraw·Hill
Sluice Dock, Guilford, Connecticut 06437

Visit us on the Internet—http://www.dushkin.com

The Annual Editions Series

ANNUAL EDITIONS is a series of over 65 volumes designed to provide the reader with convenient, low-cost access to a wide range of current, carefully selected articles from some of the most important magazines, newspapers, and journals published today. ANNUAL EDITIONS are updated on an annual basis through a continuous monitoring of over 300 periodical sources. All ANNUAL EDITIONS have a number of features that are designed to make them particularly useful, including topic guides, annotated tables of contents, unit overviews, and indexes. For the teacher using ANNUAL EDITIONS in the classroom, an Instructor's Resource Guide with test questions is available for each volume.

VOLUMES AVAILABLE

- Abnormal Psychology
- Adolescent Psychology
- Africa
- Aging
- American Foreign Policy
- American Government
- American History, Pre-Civil War
- American History, Post-Civil War
- American Public Policy
- Anthropology
- Archaeology
- Biopsychology
- Business Ethics
- Child Growth and Development
- China
- Comparative Politics
- Computers in Education
- Computers in Society
- Criminal Justice
- Criminology
- Developing World
- Deviant Behavior
- Drugs, Society, and Behavior
- Dying, Death, and Bereavement
- Early Childhood Education
- Economics
- Educating Exceptional Children
- Education
- Educational Psychology
- Environment
- Geography
- Global Issues
- Health
- Human Development
- Human Resources
- Human Sexuality
- India and South Asia
- International Business
- Japan and the Pacific Rim
- Latin America
- Life Management
- Macroeconomics
- Management
- Marketing
- Marriage and Family
- Mass Media
- Microeconomics
- Middle East and the Islamic World
- Multicultural Education
- Nutrition
- Personal Growth and Behavior
- Physical Anthropology
- Psychology
- Public Administration
- Race and Ethnic Relations
- Russia, the Eurasian Republics, and Central/Eastern Europe
- Social Problems
- Social Psychology
- Sociology
- State and Local Government
- Urban Society
- Western Civilization, Pre-Reformation
- Western Civilization, Post-Reformation
- Western Europe
- World History, Pre-Modern
- World History, Modern
- World Politics

Cataloging in Publication Data
Main entry under title: Annual Editions: Social problems. 1997/98.
    1. United States—Social conditions—1960.—Periodicals. I. Widdison, Harold A., *comp.* II. Title: Social problems.
309'.1'73'092'05   73-78577   ISBN 0-697-37356-8
HN51.A78

© 1997 by Dushkin/McGraw-Hill, Guilford, CT 06437, A Division of The McGraw-Hill Companies.

Copyright law prohibits the reproduction, storage, or transmission in any form by any means of any portion of this publication without the express written permission of Dushkin/McGraw-Hill, and of the copyright holder (if different) of the part of the publication to be reproduced. The Guidelines for Classroom Copying endorsed by Congress explicitly state that unauthorized copying may not be used to create, to replace, or to substitute for anthologies, compilations, or collective works.

Annual Editions® is a Registered Trademark of Dushkin/McGraw-Hill,
A Division of The McGraw-Hill Companies.

Twenty-Fifth Edition

Cover image © 1996 PhotoDisc, Inc.

Printed in the United States of America   Printed on Recycled Paper

Editors/Advisory Board

Members of the Advisory Board are instrumental in the final selection of articles for each edition of ANNUAL EDITIONS. Their review of articles for content, level, currentness, and appropriateness provides critical direction to the editor and staff. We think that you will find their careful consideration well reflected in this volume.

EDITOR

Harold A. Widdison
Northern Arizona University

ADVISORY BOARD

Thomas E. Arcaro
Elon College

Sylven Seid Beck
George Washington University

Mamie Bridgeforth
Essex County College

T. Jesse Dent
Johnson C. Smith University

Roger G. Dunham
University of Miami

Nancy Federman
San Diego Mesa College

Mona Field
Glendale Community College

Kurt Finsterbusch
University of Maryland, College Park

Lorraine Greaves
Fanshawe College of
Applied Arts & Sciences

Ray A. Helgemoe
University of New Hampshire

Brian J. Jones
Villanova University

Fumie Kumagai
Kyorin University

Aline M. Kuntz
University of New Hampshire

Bruce D. LeBlanc
Black Hawk College

John Lynxwiler
University of Central Florida

Kathryn S. Mueller
Baylor University

Robert G. Newby
Central Michigan University

Marie Richmond-Abbott
Eastern Michigan University

Dean G. Rojek
University of Georgia

Larry Rosenberg
Millersville University

Donald F. Smith
George Mason University

Joseph L. Victor
Mercy College

Staff

Ian A. Nielsen, Publisher

EDITORIAL STAFF

Roberta Monaco, Developmental Editor
Addie Raucci, Administrative Editor
Cheryl Greenleaf, Permissions Editor
Deanna Herrschaft, Permissions Assistant
Diane Barker, Proofreader
Lisa Holmes-Doebrick, Program Coordinator
Joseph Offredi, Photo Coordinator

PRODUCTION STAFF

Brenda S. Filley, Production Manager
Charles Vitelli, Designer
Shawn Callahan, Graphics
Lara M. Johnson, Graphics
Laura Levine, Graphics
Mike Campbell, Graphics
Juliana Arbo, Typesetting Supervisor
Jane Jaegersen, Typesetter
Marie Lazauskas, Word Processor
Kathleen D'Amico, Word Processor
Larry Killian, Copier Coordinator

To the Reader

In publishing ANNUAL EDITIONS we recognize the enormous role played by the magazines, newspapers, and journals of the *public press* in providing current, first-rate educational information in a broad spectrum of interest areas. Many of these articles are appropriate for students, researchers, and professionals seeking accurate, current material to help bridge the gap between principles and theories and the real world. These articles, however, become more useful for study when those of lasting value are carefully *collected, organized, indexed,* and *reproduced* in a *low-cost format,* which provides easy and permanent access when the material is needed. That is the role played by ANNUAL EDITIONS. Under the direction of each volume's *academic editor,* who is an expert in the subject area, and with the guidance of an *Advisory Board,* each year we seek to provide in each ANNUAL EDITION a current, well-balanced, carefully selected collection of the best of the public press for your study and enjoyment. We think that you will find this volume useful, and we hope that you will take a moment to let us know what you think.

Welcome to *Annual Editions: Social Problems 97/98.* The year 1996 held great promise. Both political parties made promises that could not be realized, resulting in frustration in the electorate. The "Contract with America" proved to be more rhetoric than substance, and President William J. Clinton became more concerned with being reelected than with fulfilling campaign promises. Little of substance was accomplished in 1996 to correct and/or elevate major societal problems. Gains in one area seemed to be balanced by declines in others. Unemployment declined, but underemployment increased. Inflation declined, but the trade deficit increased. Serious violent crimes in the cities declined, but drug use among the youth increased. Election-year politics coupled with major political scandals have soured many individuals. Nevertheless, life goes on and with it, social problems. In this 1997/98 edition of *Annual Editions: Social Problems,* I have attempted to address concerns and issues raised by students and readers of the 1996/97 edition.

Hundreds of articles were reviewed in preparing this edition. In some cases, it was very difficult to select which among many well-written and well-researched pieces to include. In other cases, we had to search for quality materials to be included, sometimes unsuccessfully. I wish to thank those individuals who went out of their way to send materials for consideration. Your efforts made my task much easier. The criteria used in selecting the readings for this edition were timeliness, quality, content, relatedness to other articles, and readability. Some very good articles were excluded because the technical nature of the subjects made them too difficult to read and understand. Others were so long that to include them would necessitate extensive revisions, which could erode their content. As in the past, we discovered a superabundance of articles dealing with welfare reform, poverty, family and parenting issues, crime, drugs, terrorism, and the quality of life. What we were not able to find were good, readable, current, and nontechnical articles in the areas of aging, sex (other than AIDS), religion, and global concerns. As a result, some areas may not be covered as well as they should be, and your help in locating newer materials in these areas would be appreciated. To make room for newer materials, some excellent articles that had become dated had to be deleted. We believe that their replacements are of comparable quality.

This edition begins with two introductory essays examining various theories of the existence and persistence of social problems. Following this introductory section are seven units. Unit 1 clusters articles concerning the basic unit of society—the family—including the changes it is experiencing and the implications these changes have for the individual and society. Unit 2 looks at the causes and impact of crime, delinquency, and violence on the American society, and what, if anything, can be done to control them. This unit has been expanded to include materials on terrorism and the significance it is gaining locally as well as internationally. Unit 3 examines problems associated with access to, and the quality of, health care. Included are writings dealing with aging and the implications for the aged and for society of a rapidly growing number of older citizens. Unit 4 discusses issues, trends, and public policies impacting on poverty and inequality. Unit 5 explores the implications of mass immigration on American society, the desirability of cultural pluralism, and the advisability of, and continued need for, affirmative action programs. Unit 6 looks at some of the major problems facing cities in the 1990s, including their ability to provide their residents with high-quality life. Unit 7, the final unit, examines global issues that transcend international boundaries and have the potential to threaten global stability.

To assist the reader in identifying topics or issues covered in the articles, the *topic guide* lists various subject categories in alphabetical order as well as the articles in which they are discussed. A reader doing research on a specific topic is advised to check this guide first.

Most of the authors of the readings herein express serious concern about the troubled state of America's cities, families, economy, and deteriorating position as a world power, as well as concern about the condition of Earth's environment. However, they have not given up. They also suggest strategies to help the family, reduce crime, make cities safer, improve the environment, and so forth.

If you have suggestions for articles or topics to be included in future editions of this series, please let us know. You are also invited to use the postage-paid *article rating form* provided on the last page of this book. Your ideas and input would be appreciated.

Harold A. Widdison
Editor

Contents

To the Reader iv
Topic Guide 2

Introduction

Two introductory articles summarize the three major theoretical approaches to studying social problems: symbolic interactionism, functionalism, and conflict.

Overview 4

1. **Social Problems: Definitions, Theories, and Analysis,** Harold A. Widdison and H. Richard Delaney, *Dushkin Publishing Group/Brown & Benchmark Publishers,* 1995. 5
   This essay, written specifically for this volume, explores the complexities associated with defining, studying, and attempting to resolve "social" problems. The three major theoretical approaches—*symbolic interactionism, functionalism, and conflict*—are summarized.

2. **How Social Problems Are Born,** Nathan Glazer, *Public Interest,* Spring 1994. 14
   Nathan Glazer examines various strategies, techniques, and processes through which *some social concerns become defined as "social problems"* while others do not.

UNIT 1

Parenting and Family Issues

Five selections examine how the socially stabilizing force of the family has been assaulted by the dynamics of economic pressure and unemployment.

Overview 20

3. **Fount of Virtue, Spring of Wealth: How the Strong Family Sustains a Prosperous Society,** Charmaine Crouse Yoest, *The World & I,* August 1994. 22
   Cross-cultural anthropological studies reveal that *strong, stable families* reduce the incidence of violence, poverty, drug abuse, sickness, mental illness, and dropping out of school.

4. **Things That Go Bump in the Home,** John Leo, *U.S. News & World Report,* May 13, 1996. 34
   Is *domestic violence* primarily a male-initiated problem? That might have been the case in the past, but domestic violence is *no longer* a purely *gender issue.*

5. **Growing Up against the Odds,** Robert Royal, *The World & I,* July 1995. 36
   Robert Royal documents a direct link between the *breakup* of the traditional family and the increase of every type of *social pathology.*

6. **Why Leave Children with Bad Parents?** Michele Ingrassia and John McCormick, *Newsweek,* April 25, 1994. 40
   Under the pressure to preserve families, *social service agencies* bend over backwards to *return abused children to their parents.* As a result, thousands of abused kids' lives are placed in jeopardy each year.

7. **The Three R's Spell Success,** Sharon Darling, *State Government News,* March 1996. 45
   Family literacy programs not only increase children's chances of doing well in school but also help parents obtain skills for new jobs, which get them off welfare.

The concepts in bold italics are developed in the article. For further expansion please refer to the Topic Guide and the Index.

UNIT 2

Crime, Terrorism, and Violence

Five articles discuss the extent and significance of crime and delinquency in today's society.

Overview   48

8. **Terrorism in America,** Orrin Hatch and Doug Bandow, *The World & I,* August 1995.   50
   Senator Orrin Hatch, chair of the Senate Judiciary Committee, and Doug Bandow, a senior fellow at the Cato Institute, debate the *necessity of new legislation to combat terrorism*. Action and overreaction hang in the balance.

9. **Enemies of the State,** Jill Smolowe, *Time,* May 8, 1995.   56
   Secretive, paranoid, and obsessed, American "patriots" warn the American government to keep its hands off their *lands, wallets, and guns*—or else.

10. **How Nation's Largest Gang Runs Its Drug Enterprise,** Ann Scott Tyson, *Christian Science Monitor,* July 15, 1996.   61
    Well-organized, financed, and armed, the nation's *largest gang* defies attempts by the state and federal governments to bring it down.

11. **Forgiving the Unforgivable,** Jean Callahan, *New Age Journal,* September/October 1993.   64
    Survivors of crime and abuse are learning an *unlikely method of freeing themselves from their anguish*—forgiveness. Victims have discovered that this is a difficult but necessary process if they are to be able to move on with their lives.

12. **Fearsome Security: The Role of Nuclear Weapons,** Michael M. May, *The Brookings Review,* Summer 1995.   68
    Michael May argues that nuclear deterrence is a *fact of modern life* and that peacetime is the right time to maintain this deterrence and to formulate strategies on how to deal with it.

UNIT 3

Aging, Health, and Health Care Issues

Five articles address society's aging, health, and health care issues.

Overview   72

13. **Will America Grow Up before It Grows Old?** Peter G. Peterson, *The Atlantic Monthly,* May 1996.   74
    The long gray wave of *baby boomers* retiring could lead to an *all-engulfing economic crisis,* Peter Peterson contends—unless we balance the budget, reign in senior entitlements, raise retirement ages, and boost individual and pension savings.

14. **The Cruelest Choice,** Sharman Stein, *Chicago Tribune,* December 11, 1994.   91
    What role should parents and physicians play in deciding if, or how long, *severely premature babies* should be treated? Medicine is increasingly effective in treating premature babies, but the economic and social costs are horrific.

The concepts in bold italics are developed in the article. For further expansion please refer to the Topic Guide and the Index.

15. **A New Look at Health Care Reform,** Murray Weidenbaum, *Vital Speeches of the Day,* April 1, 1995. 96
    Murray Weidenbaum believes that any attempt to reform health care requires developing a ***sensible and sensitive*** means of balancing the demand for health care with the ability to supply it. ***Difficult choices*** must be made among imperfect alternatives.

16. **Guns, Money & Medicine,** *U.S. News & World Report,* July 1, 1996. 100
    The proliferation of powerful new weapons has sent the cost of crime spiraling, not just in the direct costs of the crime but in ***the enormous expenses associated with gunshot injuries.***

17. **Mental Illness Is Still a Myth,** Thomas Szasz, *Society,* May/June 1994. 104
    Psychiatrist Thomas Szasz, author of the classic book *The Myth of Mental Illness,* argues that ***psychiatry is a branch of the law*** or of secular religion, but it is not a science.

UNIT 4

Poverty and Inequality

Eight selections examine how inequality affects society and the institutions of education, women's rights, the economy, and welfare.

Overview 110

18. **'Glass Ceiling' Still Too Hard to Crack, U.S. Panel Finds,** Robert A. Rosenblatt, *Los Angeles Times,* March 16, 1995. 112
    Robert Rosenblatt documents the fact that while the doors to many lower-level jobs have been opened to minorities, ***positions at the top*** of the corporate structure have not.

19. **Death of the Middle Class,** John Cassidy, *New Internationalist,* July 1996. 114
    John Cassidy argues that ***the middle class is dead*** or mortally wounded, with significant and troubling implications for the "American dream."

20. **On the Battlefields of Business, Millions of Casualties,** Louis Uchitelle and N. R. Kleinfield, *New York Times,* March 3, 1996. 116
    To maintain or improve their corporate profits and competitive edge, many corporations have ***downsized,*** creating millions of ***human casualties.***

21. **Social Change One on One: The New Mentoring Movement,** Gary Walker and Marc Freedman, *The American Prospect,* July/August 1996. 122
    While many experts argue that nothing works to help poor kids stay out of trouble, the active involvement of kids in ***Big Brother or Big Sister*** programs suggests otherwise.

The concepts in bold italics are developed in the article. For further expansion please refer to the Topic Guide and the Index.

22. **Dismantling the Welfare State: Is It the Answer to America's Social Problems?** D. Stanley Eitzen, *Vital Speeches of the Day,* June 15, 1996. — 129
Is the *dismantling of the welfare state* the answer to America's social problems? Stanley Eitzen believes it is not and explains why.

23. **Welfare Fixers,** Adam Wolfson, *Commentary,* April 1996. — 134
There is no question but that the existing welfare system is not working. Adam Wolfson examines *three proposed solutions* and concludes that *separately* they will not work, but together they provide a coherent solution.

24. **It's Not Working: Why Many Single Mothers Can't Work Their Way Out of Poverty,** Chris Tilly and Randy Albelda, *Dollars and Sense,* November/December 1994. — 138
Chris Tilly and Randy Albelda show how and why many *single mothers* cannot work their way out of poverty no matter how hard they try. It appears that supplemental support is essential. Seven types of needed reform are suggested.

25. **Aid to Dependent Corporations: Exposing Federal Handouts to the Wealthy,** Chuck Collins, *Dollars and Sense,* May/June 1995. — 141
It is not just the poor who are receiving handouts from the government. Many major corporations receive handouts, or what Chuck Collins calls *"wealthfare."*

UNIT 5

Cultural Pluralism and Affirmative Action

Five selections discuss various aspects of cultural pluralism: that it has been minimized, that diverse differences promote unity, and that diversity is the sign of social maturity.

Overview — 144

26. **Reclaiming the Vision: What Should We Do after Affirmative Action?** Constance Horner, *The Brookings Review,* Summer 1995. — 146
Constance Horner accepts the fact that *affirmative action programs* will be modified significantly and suggests six approaches that would help align minority interests with ascendant and long-standing American values.

27. **Balancing Budgets on Women's Backs: The World Bank and the 104th U.S. Congress,** Shea Cunningham and Betsy Reed, *Dollars and Sense,* November/December 1995. — 151
Shea Cunningham and Betsy Reed show how the efforts of nations around the world to *stimulate their respective economies* have succeeded—but at a great cost for women and children.

28. **Black America's Moment of Truth,** Dinesh D'Souza, *The American Spectator,* October 1995. — 155
The problem facing *black America* today is not white racism. The problem is a culture that embraces violence and celebrates ignorance—and a leadership that does not have the courage to say so, contends Dinesh D'Souza.

The concepts in bold italics are developed in the article. For further expansion please refer to the Topic Guide and the Index.

29. **A Twofer's Lament,** Yolanda Cruz, *The New Republic,* October 17, 1994. — 166

Yolanda Cruz relates from her own experience how affirmative action can *damage self-esteem*. She reflects on the difference between opportunity and privilege.

30. **Crisis of Community: Make America Work for Americans,** William Raspberry, *Vital Speeches of the Day,* June 1, 1995. — 168

William Raspberry believes that the *struggle for group advantage* has virtually destroyed a sense of community in the United States.

Overview — 172

A. CITIES

31. **Can We Stop the Decline of Our Cities?** Stephen Moore and Dean Stansel, *USA Today Magazine (Society for the Advancement of Education),* March 1994. — 174

Investing massive amounts of federal money in major U.S. cities has not stemmed urban decline. The authors suggest that the federal aid approach can never meet its urban objectives until *city government abandons its self-serving agenda* in favor of a "people-first" commitment.

32. **Can Churches Save America?** *U.S. News & World Report,* September 9, 1996. — 179

Are *small community-based groups* such as churches the solution for many social problems? Some politicians are beginning to believe that they are.

B. QUALITY OF LIFE

33. **The Projects Come Down,** Rob Gurwitt, *Governing,* August 1995. — 183

Because *centralized public housing projects* have not worked, cities are now experimenting with decentralized programs, with considerable success.

34. **Time Out,** John Marks, *U.S. News & World Report,* December 11, 1995. — 187

Plagued by stress and inability to spend quality time with their families, a growing number of people are *reevaluating what they really want out of life*.

UNIT 6

Cities, Urban Growth, and the Quality of Life

Eight articles examine the current state of cities in the United States.

The concepts in bold italics are developed in the article. For further expansion please refer to the Topic Guide and the Index.

C. EDUCATION

35. **Can the Schools Be Saved?** Chester E. Finn Jr., *Commentary*, September 1996. 191
   Chester Finn believes that public schools can and must be saved. Saving them is not easy because it means bucking *entrenched special interests and bureaucratic inertia*.

36. **Off Course**, Karen Lehrman, *Mother Jones*, September/October 1993. 196
   What is or should be the role for any academic program? Karen Lehrman looks at *women's studies programs* across the nation and concludes that many emphasize subjectivity over objectivity, feelings over facts, and instinct over logic.

D. DRUGS

37. **A Society of Suspects: The War on Drugs and Civil Liberties,** Steven Wisotsky, *USA Today Magazine (Society for the Advancement of Education)*, July 1993. 202
   "A decade after President Reagan launched *the war on drugs*, all we have to show for it are city streets ruled by gangs, a doubled prison population, and a substantial erosion of constitutional protections," writes Steven Wisotsky.

38. **It's Drugs, Alcohol, and Tobacco, Stupid!** Joseph A. Califano Jr., *Center on Addiction and Substance Abuse at Columbia University,* Annual Report, 1994. 208
   Is *drug abuse* the primary culprit undermining the effectiveness of welfare, health care, and criminal justice systems? Joseph Califano says yes and shows how.

Overview 214

39. **U.S. Competitiveness: "Resurgence" versus Reality,** Robert H. Hayes, *Challenge,* March/April 1996. 216
   Robert Hayes argues that if the United States has not *lost its competitive edge* in the global economy, it soon will. He explains how and why.

40. **The Civil Rights Issue of the '90s,** Nancie G. Marzulla, *The World & I,* October 1994. 225
   Just how far can state and federal governments go in implementing laws designed to protect the environment? *What rights do private property holders have* to use their own lands?

UNIT 7

Global Issues

Five articles discuss common social problems faced by people worldwide. Topics include the environment, drugs, and cultural misunderstanding.

The concepts in bold italics are developed in the article. For further expansion please refer to the Topic Guide and the Index.

41. **Global Reach: The Threat of International Drug Trafficking,** Rensselaer W. Lee III, *Current History,* May 1995. 229
    The *international trade in drugs* has become an increasingly important issue in global security. It is a problem, however, that *falls outside traditional national security concerns,* even though it threatens the political stability of many states.

42. **A Growing Global Crisis,** Ann Marie Kimball, *The World & I,* June 1996. 234
    By far "the most significant global *pandemic of disease* in the latter half of the twentieth century" is *AIDS,* says Ann Marie Kimball. How it became so and its implications for the future are discussed.

43. **Earth Is Running Out of Room,** Lester R. Brown, *USA Today Magazine (Society for the Advancement of Education),* January 1995. 238
    A great number of the planet's leading scientists (including 102 Nobel Prize winners) have noted that the continuation of destructive human activities may so alter the living world that it will be *unable to sustain life as we know it.*

Index	241
Article Review Form	244
Article Rating Form	245

The concepts in bold italics are developed in the article. For further expansion please refer to the Topic Guide and the Index.

Topic Guide

This topic guide suggests how the selections in this book relate to topics of traditional concern to students and professionals involved with the study of social problems. It is useful for locating articles that relate to each other for reading and research. The guide is arranged alphabetically according to topic. Articles may, of course, treat topics that do not appear in the topic guide. In turn, entries in the topic guide do not necessarily constitute a comprehensive listing of all the contents of each selection.

TOPIC AREA	TREATED IN	TOPIC AREA	TREATED IN
Abuse	4. Things That Go Bump in the Home 6. Why Leave Children with Bad Parents? 11. Forgiving the Unforgivable	Drinking	2. How Social Problems Are Born 38. It's Drugs, Alcohol, and Tobacco, Stupid!
Affirmative Action	18. 'Glass Ceiling' Still Too Hard to Crack	Drugs	2. How Social Problems Are Born 5. Growing Up against the Odds 10. How Nation's Largest Gang Runs Its Drug Enterprise 21. Social Change One on One 33. Projects Come Down 37. Society of Suspects 38. It's Drugs, Alcohol, and Tobacco, Stupid! 41. Global Reach
African Americans	26. Reclaiming the Vision 28. Black America's Moment of Truth		
Aging	13. Will America Grow Up before It Grows Old?		
Aid to Families with Dependent Children (AFDC)	24. It's Not Working		
		Economy	5. Growing Up against the Odds 20. On the Battlefields of Business, Millions of Casualties 22. Dismantling the Welfare State 29. Twofer's Lament 36. Off Course 39. U.S. Competitiveness
AIDS	42. Growing Global Crisis		
Antigovernment	9. Enemies of the State		
Big Brother/ Big Sister	21. Social Change One on One		
		Education	7. The Three R's Spell Success 21. Social Change One on One 35. Can the Schools Be Saved?
Children	5. Growing Up against the Odds 6. Why Leave Children with Bad Parents? 10. How Nation's Largest Gang Runs Its Drug Enterprise 14. Cruelest Choice 27. Balancing Budgets on Women's Backs		
		Entitlements	13. Will America Grow Up before It Grows Old?
		Environment	40. Civil Rights Issue of the '90s
Cities	30. Crisis of Community 31. Can We Stop the Decline of Our Cities? 33. Projects Come Down	Ethics	14. Cruelest Choice
		Family	3. Fount of Virtue, Spring of Wealth 4. Things That Go Bump in the Home 5. Growing Up against the Odds 6. Why Leave Children with Bad Parents? 7. Three R's Spell Success 20. On the Battlefields of Business, Millions of Casualties 22. Dismantling the Welfare State 23. Welfare Fixers 34. Time Out
Civil Liberties	37. Society of Suspects 40. Civil Rights Issue of the '90s 42. Growing Global Crisis		
Crime and Delinquency	5. Growing Up against the Odds 11. Forgiving the Unforgivable 16. Guns, Money & Medicine 21. Social Change One on One 22. Dismantling the Welfare State 32. Can Churches Save America? 37. Society of Suspects 38. It's Drugs, Alcohol, and Tobacco, Stupid! 41. Global Reach		
		Fathering	5. Growing Up against the Odds
		Feminism	4. Things That Go Bump in the Home 36. Off Course
		Foreign Policy	41. Global Reach
Disabilities	14. Cruelest Choice		
Divorce	5. Growing Up against the Odds	Gangs	10. How Nation's Largest Gang Runs Its Drug Enterprise 16. Guns, Money & Medicine 33. Projects Come Down

TOPIC AREA	TREATED IN	TOPIC AREA	TREATED IN
Guns	1. Social Problems 2. How Social Problems Are Born 9. Enemies of the State 16. Guns, Money & Medicine	Sexism	18. 'Glass Ceiling' Still Too Hard to Crack
		Single Parents	24. It's Not Working
Health	13. Will America Grow Up before It Grows Old? 14. Cruelest Choice 15. New Look at Health Care Reform 16. Guns, Money & Medicine 34. Time Out 38. It's Drugs, Alcohol, and Tobacco, Stupid! 42. Growing Global Crisis 43. Earth Is Running Out of Room	Social Security	13. Will America Grow Up before It Grows Old?
		Sociological Theories	1. Social Problems
		Stress	34. Time Out
		Suicide	5. Growing Up against the Odds
		Taxes	13. Will America Grow Up before It Grows Old? 15. New Look at Health Care Reform 22. Dismantling the Welfare State 25. Aid to Dependent Corporations 31. Can We Stop the Decline of Our Cities? 32. Can Churches Save America?
Inequality	18. 'Glass Ceiling' Still Too Hard to Crack 19. Death of the Middle Class 22. Dismantling the Welfare State		
Law	8. Terrorism in America 17. Mental Illness Is Still a Myth 26. Reclaiming the Vision 38. It's Drugs, Alcohol, and Tobacco, Stupid!	Terrorism	8. Terrorism in America
		Trade Deficit	39. U.S. Competitiveness
		Underemployment	20. On the Battlefields of Business, Millions of Casualties
Medicine	15. New Look at Health Care Reform 17. Mental Illness Is Still a Myth	Unemployment	20. On the Battlefields of Business, Millions of Casualties 22. Dismantling the Welfare State
Mental Health	3. Fount of Virtue, Spring of Wealth 11. Forgiving the Unforgivable 17. Mental Illness Is Still a Myth		
Middle Class	19. Death of the Middle Class	Unions	31. Can We Stop the Decline of Our Cities? 35. Can the Schools Be Saved?
Militia	9. Enemies of the State	Victims	11. Forgiving the Unforgivable
Minorities	35. Can the Schools Be Saved?	Violence	3. Fount of Virtue, Spring of Wealth 4. Things That Go Bump in the Home 6. Why Leave Children with Bad Parents? 7. Three R's Spell Success 8. Terrorism in America 9. Enemies of the State 16. Guns, Money & Medicine 21. Social Change One on One 28. Black America's Moment of Truth
Multicultural/ Cultural Pluralism	30. Crisis of Community		
Parenting	14. Cruelest Choice		
Population	43. Earth Is Running Out of Room		
Poverty	22. Dismantling the Welfare State 23. Welfare Fixers 24. It's Not Working 27. Balancing Budgets on Women's Backs 28. Black America's Moment of Truth 32. Can Churches Save America? 33. Projects Come Down 38. It's Drugs, Alcohol, and Tobacco, Stupid!	Volunteerism	21. Social Change One on One
		War	12. Fearsome Security
		Wealth	19. Death of the Middle Class 25. Aid to Dependent Corporations
		Welfare	5. Growing Up against the Odds 7. Three R's Spell Success 22. Dismantling the Welfare State 23. Welfare Fixers 24. It's Not Working 27. Balancing Budgets on Women's Backs 28. Black America's Moment of Truth 32. Can Churches Save America?
Psychiatry	17. Mental Illness Is Still a Myth		
Race and Ethnic Issues	18. 'Glass Ceiling' Still Too Hard to Crack 28. Black America's Moment of Truth 29. Twofer's Lament 30. Crisis of Community		
Retirement	13. Will America Grow Up before It Grows Old?		

Introduction

Before initiating any analysis of social problems, it is always useful to agree on what it is that is being talked about. Things that are symbolic or seem to represent a serious social problem to one group might be seen by others as a symptom of a much larger problem, or even as no problem at all.

Two articles are included in this section that explore the complexities of social problems. While some individuals take a very simplistic black-and-white approach in defining social problems and, in turn, what must be done to eliminate them, sociologists realize how complex and intertwined social problems are in all aspects of social life. But even sociologists do not agree as to the best approach to take in the study of social issues.

Harold Widdison and H. Richard Delaney, in the first article, introduce the reader to sociology's three dominant theoretical positions and give examples of how those espousing each theory would look at specific issues. The three theories—symbolic interactionism, functionalism, and conflict—represent three radically different approaches to the study of social problems and their implications for individuals and societies. The perceived etiology of problems and their possible resolutions reflect the specific orientations of those studying them. As you peruse the subsequent articles, try to determine which of the three theoretical positions the various authors seem to be utilizing. Widdison and Delaney conclude this article by suggesting several approaches students may wish to consider in defining conditions as "social" problems and how they can and should be analyzed.

The second article explores how social problems are born, that is, why some issues, actions, or behaviors become defined as significant social problems while others do not. Nathan Glazer believes that the logic underlying the symbolic language is the critical factor in determining if an individual concern will eventually evolve into a societal-level social problem.

The introduction of this book of readings with a discussion of this type is desirable in order to help readers understand the social and sociological aspects of problems and issues plaguing modern society. The other articles included in this edition range from living in single-parent families to the pending destruction of the world's environment. The reader should ask why the issue addressed in each article is a social problem. Is it a case of rights in conflict, a case of conflicting values, or a consequence of conflicting harms? To find out, the reader might first skim over each article to get a general idea of where the author is coming from—that is, the author's theoretical position—and then reread the article to see just what the author indicates is the cause of the problem and what can or should be done to resolve or eliminate it.

SOCIAL PROBLEMS:

Definitions, Theories, and Analysis

Harold A. Widdison and H. Richard Delaney

INTRODUCTION AND OVERVIEW

When asked, "What are the major social problems facing humanity today?" college students' responses tend to mirror those highlighted by the mass media—particularly AIDS, child abuse, poverty, war, famine, racism, sexism, crime, riots, the state of the economy, the environment, abortion, euthanasia, homosexuality, and affirmative action. These are all valid subjects for study in a social problems class, but some give rise to very great differences of opinion and even controversy. Dr. Jack Kevorkian in Michigan and his killing machine is one example that comes to mind. To some he evokes images of Nazi Germany with its policy of murdering the infirm and helpless. Others see Kevorkian's work as a merciful alternative to the slow and agonizing death of individuals with terminal illnesses. In the latter light, Kevorkian is not symbolic of a potentially devastating social issue, but of a solution to an escalating social problem.

The same controversy exists at the other end of life—specifically, what obligations do pregnant women have to themselves as opposed to the unborn? Some individuals see abortion as a solution to the problems of population, child abuse, disruption of careers, dangers to the physical and emotional health of women, as well as the prevention of the birth of damaged fetuses, and they regard it as a right to self-determination. Others look at abortion as attacking the sanctity of life, abrogating the rights of a whole category of people, and violating every sense of moral and ethical responsibility.

Affirmative action is another issue that can be viewed as both a problem and a solution. As a solution, affirmative action attempts to reverse the effects of hundreds of years of discrimination. Doors that have been closed to specific categories of people for many generations are, it is hoped, forced open; individuals, regardless of race, ethnicity, and gender, are able to get into professional schools, and secure good jobs, with the assurance of promotion. On the other hand, affirmative action forces employers, recruiting officers, and housing officials to give certain categories of individuals a preferred status. While affirmative action is promoted by some as a necessary policy to compensate for centuries of exclusion and discrimination, others claim that it is discrimination simply disguised under a new label but with different groups being discriminated against. If race, sex, age, ethnicity, or any other characteristic other than merit is used as the primary criterion for selection or promotion, then discrimination is occurring. Discrimination hurts both sides. William Wilson, an African American social scientist, argues that it is very damaging to the self-esteem of black individuals to know that the primary reason they were hired was to fill quotas.

Both sides to the debate of whether these issues themselves reflect a social problem or are solutions to a larger societal problem have valid facts and use societal-level values to support their claims. Robin William Jr. in 1970 identified a list of 15 dominant value orientations that represent the concept of the good life to many Americans:

1. Achievement and success as major personal goals.
2. Activity and work favored above leisure and laziness.
3. Moral orientation—that is, absolute judgments of good/bad, right/wrong.
4. Humanitarian motives as shown in charity and crisis aid.
5. Efficiency and practicality: a preference for the quickest and shortest way to achieve a goal at the least cost.
6. Process and progress: a belief that technology can solve all problems and that the future will be better than the past.
7. Material comfort as the "American dream."
8. Equality as an abstract ideal.
9. Freedom as a person's right against the state.
10. External conformity: the ideal of going along, joining, and not rocking the boat.
11. Science and rationality as the means of mastering the environment and securing more material comforts.
12. Nationalism: a belief that American values and institutions represent the best on earth.
13. Democracy based on personal equality and freedom.
14. Individualism, emphasizing personal rights and responsibilities.

INTRODUCTION

15. Racism and group-superiority themes that periodically lead to prejudice and discrimination against those who are racially, religiously, and culturally different from the white northern Europeans who first settled the continent.

This list combines some political, economic, and personal traits that actually conflict with one another. This coexistence of opposing values helps explain why individuals hold contradictory views of the same behavior and why some issues generate such intensity of feelings. It is the intent of this article and the readings included in this book to attempt to help students see the complex nature of a social problem and the impact that various values, beliefs, and actions can have on them.

In the next segment of this article, the authors will look at specific examples of values in conflict and the problems created by this conflict. Subsequently the authors will look at the three major theoretical positions that sociologists use to study social problems. The article will conclude with an examination of various strategies and techniques used to identify, understand, and resolve various types of social problems and their implications for those involved.

As noted above, contemporary American society is typified by values that both complement and contradict each other. For example, the capitalistic free enterprise system of the United States stresses rugged individualism, self-actualization, individual rights, and self expression. This economic philosophy meshes well with Christian theology, particularly that typified by many Protestant denominations. This fact was the basis of German sociologist Max Weber's "The Protestant Ethic and The Spirit of Capitalism" (1864). He showed that the concepts of grace (salvation is a gift—not something you can earn), predestination (the fact that some people have this gift while others do not), and a desire to know if the individual has grace gave rise to a new idea of what constitutes success. Whereas, with the communitarian emphasis of Catholicism where material success was seen as leading to selfishness and spiritual condemnation, Protestantism viewed material success as a sign of grace. In addition, it was each individual's efforts that resulted in both the economic success and the spiritual salvation of the individual. This religious philosophy also implied that the poor are poor because they lack the proper motivation, values, and beliefs (what is known as the "culture of poverty") and are therefore reaping the results of their own inadequacies. Attempts to reduce poverty have frequently included taking children from "impoverished" cultural environments and placing them in "enriched" environments to minimize the potentially negative effects parents and a bad environment could have on their children. These enrichment programs attempt to produce attitudes and behaviors that assure success in the world but, in the process, cut children off from their parents. Children are forced to abandon the culture of their parents if they are to "succeed." Examples of this practice include the nurseries of the kibbutz in Israel and the Head Start programs in America. This practice is seen by some social scientists as a type of "cultural genocide." Entire cultures were targeted (sometimes explicitly, although often not intentionally) for extinction in this way.

This fact upsets a number of social scientists. They feel it is desirable to establish a pluralistic society where ethnic, racial, and cultural diversity exist and flourish. To them attempts to "Americanize" everyone are indicative of racism, bigotry, and prejudice. Others point to the lack of strong ethnic or racial identities as the unifying strength of the American system. When immigrants came to America, they put ethnic differences behind them, they learned the English language and democratic values, and they were assimilated into American life. In nations where immigrants have maintained their ethnic identities and held to unique cultural beliefs, their first loyalty is to their ethnic group. Examples of the destructive impact of strong ethnic loyalties can be seen in the conflict and fragmentation now occurring in the former Soviet Union, Czechoslovakia, and Yugoslavia.

James Q. Wilson (1994:54–55) noted in this regard:

> We have always been a nation of immigrants, but now the level of immigration has reached the point where we have become acutely conscious, to a degree not seen, I think, since the turn of the century, that we are a nation of many cultures. I believe that the vast majority of those who have come to this country came because they, too, want to share in the American Dream. But their presence here, and the unavoidable tensions that attend upon even well-intentioned efforts at mutual coexistence, makes some people—and alas, especially some intellectuals in our universities—question the American Dream, challenge the legitimacy of Western standards of life and politics, and demand that everybody be defined in terms of his or her group membership. The motto of this nation—*E pluribus unum,* out of the many, one—is in danger of being rewritten to read, *Ex uno plures*—out of the one, many."

THEORETICAL EXPLANATIONS: SYMBOLIC INTERACTION, FUNCTIONALISM, CONFLICT

In their attempts to understand social phenomena, researchers look for recurring patterns, relationships between observable acts, and unifying themes. The particular way in which researchers look at the world reflects not only their personal views and experiences, but their professional perspective as well. Sociologists focus on interactions between individuals, between individuals and groups, between groups, and between groups and the larger society in which they are located. They try to identify those things that facilitate or hinder interaction, and the consequences of each. But not all sociologists agree as to the most effective/appropriate approach to take, and they tend to divide into three major theoretical camps: symbolic interactionism, functionalism, and conflict theory. These three approaches are not mutually exclusive, but they do represent radically different perspectives of the nature of social reality and how it should be studied.

Symbolic Interaction

This theoretical perspective argues that no social condition, however unbearable it may seem to some, is inherently or objectively a social problem until a significant number of politically powerful people agree that it is contrary to the public good. Scientists, social philosophers, religious leaders, and medical people may "know" that a specific action or condition has or will eventually have a devastating effect on society or a specific group in society, but until they can convince those who are in a position to control and perhaps correct the condition, it is not considered a social problem. Therefore it is not the social condition, but how the condition is defined and by whom, that determines if it is or will become a social problem. The social process whereby a specific condition moves from the level of an individual concern to a societal-level issue can be long and arduous or very short. An example of the latter occurred in the 1960s when some physicians noticed a significant increase in infants born with severe physical deformities. Medical researchers looking into the cause made a connection between the deformities and the drug thalidomide. Pregnant women suffering from severe nausea and health-threatening dehydration were prescribed this drug, which dramatically eliminated the nausea and appeared to have no bad side effects. But their babies were born with terrible deformities. Once the medical researchers discovered the connection, they presented their findings to their colleagues. When the data were reviewed and found to be scientifically valid, the drug was banned immediately. Thus a small group's assessment of an issue as a serious problem quickly was legitimized by those in power as a societal-level social problem and measures were taken to eliminate it.

Most situations are not this clear-cut. In the mid-1960s various individuals began to question the real reason(s) why the United States was involved in the war in Southeast Asia. They discovered data indicating that the war was not about protecting the democratic rights of the Vietnamese. Those in power either ignored or rejected such claims as politically motivated and as militarily naive. Reports from the Vietcong about purported U.S. military atrocities were collected and used as supportive evidence. These claims were summarily dismissed by American authorities as Communistic propaganda. Convinced of the validity and importance of their cause, the protesters regrouped and collected still more evidence including data collected by the French government. This new information was difficult for the U.S. government to ignore. Nevertheless, these new claims were rejected as being somewhat self-serving since the Vietcong had defeated the French in Indochina and presumably the French government could justify its own failure if the United States also failed.

Over the years the amount of data continued to accumulate augmented by new information collected from disenchanted veterans. This growing pool of evidence began to bother legislators who demanded an accounting from the U.S. government and the Department of Defense, but none was forthcoming. More and more students joined the antiwar movement, but their protests were seen as unpatriotic and self-serving—that is, an attempt to avoid military service. The increasing numbers of protesters caused some legislators to look more closely at the claims of the antiwar faction. As the magnitude of the war and the numbers of American servicemen involved grew, the numbers of people affected by the war grew as well. Returning veterans' reports of the state of the war, questionable military practices (such as the wholesale destruction of entire villages), complaints of incompetent leadership in the military, and corrupt Vietnamese politicians gave greater credibility to the antiwar movement's earlier claims and convinced additional senators and representatives to support the stop-the-war movement, even though those in power still refused to acknowledge the legitimacy of the movement.

Unable to work within the system and convinced of the legitimacy of their cause, protesters resorted to unconventional and often illegal actions, such as burning their draft cards, refusing to register for the draft, seeking refuge in other countries, attacking ROTC (Reserve Officers' Training Corps) buildings on college campuses, and even bombing military research facilities. These actions were initially interpreted by government officials as criminal activities of self-serving individuals or activities inspired by those sympathetic with the Communist cause. The government engaged in increasingly repressive efforts to contain the movement. But public disaffection with the war was fueled by rising American casualties; this, coupled with the discontent within the ranks of the military, eventually forced those in power to acquiesce and accept the claims that the war was the problem and not the solution to the problem. Reaching this point took nearly 15 years.

For the symbolic interactionist, the fact that socially harmful conditions are thought to exist is not the criterion for what constitutes a social problem. Rather the real issue is to understand what goes into the assessment of a specific condition as being a social problem. To the symbolic interactionist, the appropriate questions are, (a) How is it that some conditions become defined as a social problem while others do not? (b) Who, in any society, can legitimate the designation of a condition as a social problem? (c) What solutions evolve and how do they evolve for specific social problems? (d) What factors exist in any specific society that inhibit or facilitate resolution of social problems?

In summary, symbolic interactionists stress that social problems do not exist independently of how people define their world. Social problems are socially constructed as people debate whether or not some social condition is a social problem and decide what to do about it. The focus is on the meanings the problem has for those who are affected by it and not on the impact it is having on them.

Functionalism

A second major theory sociologists use to study social problems is functionalism. Functionalists argue that society is a social system consisting of various integrated parts. Each of these parts fulfills a specific role that contributes to the overall functioning of society. In well-integrated systems, each part contributes to the stability of the whole. Functionalists examine each part in an attempt to determine the role it plays in the operation of the system as a whole. When any part fails, this creates a problem for the whole. These failures (dysfunctions) upset the equilibrium of the system and become social problems. To functionalists, anything that impedes the system's ability to achieve its goals is, by definition, a social problem. Unlike the symbolic interactionists, the functionalists argue that

INTRODUCTION

a social problem is not contingent on someone's assessment that it is a problem. Serious social problems may exist without anyone being aware of the detrimental effects they are having on various members of society or on society itself. Functionalists examine the stated objectives, values, and goals of a group; observe the behaviors of the members of the group; and assess how their behaviors impact on the abilities of the group to achieve its goals.

Many times the stated objectives (what sociologists call the "manifest functions") produce results that were not desired nor intended (what sociologists call "latent functions") and are in fact working against the group's abilities to accomplish its goals. Sociologists attempt to make the members of the group aware of the consequences of specific behaviors. For example, in an attempt to help single mothers of infants to provide for their children adequately, the American government created Aid to Families with Dependent Children (AFDC), a program designed to provide single women with enough money to feed, clothe, and house their children. This program was motivated by the Judeo-Christian philosophy that society is obligated to care for those who cannot care for themselves. In this regard the program has been a success. But it also has had a dark side in that it has discouraged the establishment of stable households with a father present, since two-parent families do not qualify for aid. If a woman marries or is known to be living with a man, she loses her eligibility and her benefits. As a result, males have been pushed to the periphery of the family. Many lower-class, unskilled males cannot earn as much as a single mother under AFDC. It does not mean that men are not around, only that they are discouraged from becoming permanent fixtures in these families. As a result, a program designed to help families ended up altering the structures of families and in the process created a whole new social problem.

Functionalists also argue that if a behavior or social institution persists, it must be meeting some need within the society. Merely defining a behavior as a problem does not assure its demise. To eliminate any behavior researchers/society must first find out what functions it is serving and then make the behavior dysfunctional, which in turn will cause the behavior to disappear. As poverty, crime, and inequality exist and persist in all societies, the task of the social scientist is to discover how and why. In this regard most individuals would argue that poverty is not desirable and should, if possible, be eliminated. Yet, as discussed by Herbert J. Gans in his article "The Uses of Poverty: The Poor Pay All," poverty benefits a significant portion of society. The incentive to eradicate poverty is neutralized by specific benefits to the nonpoor. Five of the 13 functions Gans identified are as follows:

1. Poverty ensures that society's "dirty work" will be done. Poverty functions to provide a low-wage labor pool that is willing, or rather unable to be unwilling, to perform dirty work at low cost.
2. Because the poor are required to work at low wages, they subsidize a variety of economic activities that benefit the affluent.
3. Poverty creates jobs for a number of occupations and professions that serve or "service" the poor, or protect the rest of society from them, such as social workers, police, and prison staff.
4. The poor buy goods others do not want, thus prolonging the economic usefulness of such goods—that is, day-old bread, fruit, and vegetables that would otherwise have to be thrown out, secondhand clothes, and deteriorating automobiles and buildings.
5. The poor, being powerless, can be made to absorb the costs of change and growth in American society. Urban renewal and expressways, for example, have typically been located in poor neighborhoods.

Although not explicitly stated by Gans, the poor cannot afford the ever-spiraling costs of health care and become those upon which the fledgling physician can practice his or her profession. As part of the learning process, mistakes are common and the poor are thus likely to have a lower level of medical expertise. Many medical, dental, and nursing schools are located within the inner city. In exchange for free or greatly reduced fees, poor people become guinea pigs to help student nurses, doctors, and dentists become experienced enough to practice on the more affluent.

The functionalists examine conditions, behaviors, and institutions in an attempt to try to understand the functions being met by these specific phenomena. To eliminate any of these problems the associated behaviors have to become dysfunctional. But because many of the functional alternatives to each problem would be dysfunctional for the affluent and powerful members of society, there is an incentive for the behavior to persist.

In summary, functionalists emphasize the interrelationship of the various parts of a system and believe that changes in one part will have significant implications for other parts. Any particular social problem is only a part of a larger whole. This means that in order to understand a social problem, one must place it in a broader context. A social problem is a consequence of the way a social system is put together.

Conflict Theory

Those social philosophers adhering to the conflict perspective view life and all social interaction as a struggle for power and privilege. They see every person and every group as being in competition for scarce and valued resources. They believe that even though people occasionally may have to cooperate with each other or even form alliances, they are still essentially in conflict. As soon as the alliance is no longer beneficial, conflict will often ensue. Unlike the functionalists who see the elements of a society as harmoniously working together and contributing to the whole, conflict theorists view all the parts as being in competition with each other. They see the guiding principle of social life as disequilibrium and change, not equilibrium and harmony. But, like the functionalists, they argue that social problems can and do exist independently of people's assessments of them. They argue that whether people are aware of it or not, they are enmeshed in a basic struggle for power and survival. Each group in society is attempting to achieve gains for itself that must necessarily be at the expense of other groups. It is this consistent conflict over limited resources that threatens societal peace and order.

Whereas the functionalists try to understand how different positions of power came into existence (Davis & Moore 1945), the conflictists show how those in power attempt to stay in power (Mills 1956). The conflict theorists see social problems as the natural and inevitable consequences of groups in society struggling to survive and gain control over those things that can affect their ability to survive. Those groups that are successful then attempt to use whatever means they must to control their environment and consolidate their position, thus increasing their chances of surviving. According to conflict theorists, those in power exploit their position and create poverty, discrimination, oppression, and crime in the process. The impact of these conditions on the exploited produces other pathological conditions such as alienation, alcoholism, drug abuse, mental illness, stress, health problems, and suicide. On occasions, such as that which occurred in Los Angeles in the summer of 1991 when policemen were found innocent of the use of excessive force in the beating of Rodney King, the feelings of helplessness and hopelessness can erupt as rage against the system in the form of violence and riots or as in Eastern Europe as rebellion and revolution against repressive governments.

The conflict theorists argue that drug abuse, mental illness, various criminal behaviors, and suicide are symptoms of a much larger societal malaise. To understand and eliminate these problems, society needs to understand the basic conflicts that are producing them. The real problems stem from the implications of being exploited. Being manipulated by the powerful and denied a sense of control tends (a) to produce a loss of control over one's life (powerlessness), (b) to lead to an inability to place one's productive efforts into some meaningful context (meaninglessness), (c) not to being involved in the process of change but only in experiencing the impact resulting from the changes (normlessness), and (d) to cause one to find oneself isolated from one's colleagues on the job (self-isolation). Conflictists see all of these problems as the product of a capitalistic system that alienates the worker from himself and from his or her fellow workers (Seeman 1959).

To protect their positions of power, privilege, prestige, and possessions, those in power use their wealth and influence to control organizations. For example, they manipulate the system to get key individuals into positions where they can influence legislation and decisions that are designed to protect their power and possessions. They might serve on or appoint others to school boards to assure that the skills and values needed by the economy are taught. They also assure that the laws are enforced internally (the police) or externally (the military) to protect their holdings. The war in the Persian Gulf is seen by many conflict theorists as having been fought for oil rather than for Kuwait's liberation.

When the exploited attempt to do something about their condition by organizing, protesting, and rebelling, they threaten those in power. For example, they may go on a strike that might disrupt the entire nation. Under the pretext that it is for the best good of society, the government may step in and stop the strike. Examples are the air-traffic-controllers strike of 1987 and the railroad strike in 1991. In retaliation the workers may engage in work slow-down, stoppage, and even sabotage. They may stage protests and public demonstrations and cast protest votes at the ballot boxes. If these do not work, rebellions and revolutions may result. Those in power can respond very repressively as was the case in Tiananmen Square in China in 1989, threaten military force as the Soviet Union did with the Baltic countries in 1990, or back down completely as when the Berlin Wall came down. Thus reactions to exploitation may produce change but inevitably lead to other social problems. In Eastern Europe and the former Soviet Union, democracy has resulted in massive unemployment, spiraling inflation, hunger, crime, and homelessness.

Sometimes those in power make concessions to maintain power. Conflict theorists look for concessions and how they placate the poor while still protecting the privileged and powerful. The rich are viewed as sharing power only if forced to do so and only to the extent absolutely necessary.

Robert Michels (1949), a French social philosopher, looked at the inevitable process whereby the members of any group voluntarily give their rights, prerogatives, and power to a select few who then dominate the group. It may not be the conscious decision of those who end up in positions of power to dominate the group, but, in time, conscious decisions may be made to do whatever is necessary to stay in control of the group. The power, privilege, and wealth they acquire as part of the position alters their self-images. To give up the position would necessitate a complete revision of who they are, what they can do, and with whom they associate. Their "selves" have become fused/confused with the position they occupy, and in an attempt to protect their "selves," they resist efforts designed to undermine their control. They consider threats to themselves as threats to the organization and therefore feel justified in their vigorous resistance. According to Michels, no matter how democratic an organization starts out to be it will always become dominated and controlled by a few. The process whereby this occurs he labeled the "Iron Law of Oligarchy." For example, hospitals that were created to save lives, cure the sick, and provide for the chronically ill, now use the threat of closure to justify rate increases. The hospital gets its rate increase, the cost of health goes up, and the number of individuals able to afford health care declines, with the ultimate result being an increase in health problems for the community. Although not explicitly stated, the survival of the organization (and its administrators) becomes more important than the health of the community.

In summary, the conflict theoretical model stresses the fact that key resources such as power and privilege are limited and distributed unequally among the members/groups in a society. Conflict is therefore a natural and inevitable result of various groups pursuing their interests and values. To study the basis of social problems, researchers must look at the distribution of power and privilege because these two factors are always at the center of conflicting interests and values. Moreover, whenever social change occurs, social problems inevitably follow.

Conflict and Functionalism: A Synthesis
While conflict theorists' and functionalists' explanations of what constitutes the roots of social problems appear to be completely

INTRODUCTION

contradictory, Dahrendorf (1959) sees them as complementary. "Functionalism explains how highly talented people are motivated to spend twenty-five years of study to become surgeons; conflict theory explains how surgeons utilize their monopoly on their vital skills to obtain rewards that greatly exceed that necessary to ensure an adequate supply of talent." (See also Ossowski 1963; van de Berghe 1963; Williams 1966; Horowitz 1962; and Lenski 1966 for other attempts at a synthesis between these two theoretical models.)

SOCIAL PROBLEMS: DEFINITION AND ANALYSIS

Value Conflicts

It is convenient to characterize a social problem as a conflict of values, a conflict of values and duties, a conflict of rights (Hook, 1974), or a social condition that leads to or is thought to lead to harmful consequences. Harm may be defined as (a) the loss to a group, community, or society of something to which it is thought to be entitled, (b) an offense perceived to be an affront to our moral sensibilities, or (c) an impoverishment of the collective good or welfare. It is also convenient to define values as individual or collective desires that become attached to social objects. Private property, for example, is a valued social object for some while others disavow or reject its desirability; because of the public disagreement over its value, it presents a conflict of values. A conflict of values is also found in the current controversy surrounding abortion. Where pro-life supporters tend to see life itself as the ultimate value, supporters of pro-choice may, as some have, invoke the Fourteenth Amendment's right-to-privacy clause as the compelling value.

Values-versus-Duties Conflicts

A second format that students should be aware of in the analysis of social problems is the conflict between values and obligations or duties. This approach calls our attention to those situations in which a person, group, or community must pursue or realize a certain duty even though those participating may be convinced that doing so will not achieve the greater good. For example, educators, policemen, bureaucrats, and environmentalists may occupy organizational or social roles in which they are required to formulate policies and follow rules that, according to their understanding, will not contribute to the greater good of students, citizens, or the likelihood of a clean environment. On the other hand, there are situations in which, we, as a individuals, groups, or communities, do things that would not seem to be right in our pursuit of what we consider to be the higher value. Here students of social problems are faced with the familiar problem of using questionable, illogical, or immoral means to achieve what is perhaps generally recognized as a value of a higher order. Police officers, for example, are sometimes accused of employing questionable, immoral, or deceptive means (stings, scams, undercover operations) to achieve what are thought to be socially helpful ends and values such as removing a drug pusher from the streets. Familiar questions for this particular format are, Do the ends justify the means? Should ends be chosen according to the means available for their realization? What are the social processes by which means themselves become ends? These are questions to which students of social problems and social policy analysis should give attention since immoral, illegal, or deceptive means can themselves lead to harmful social consequences.

Max Weber anticipated and was quite skeptical of those modern bureaucratic processes whereby means are transformed into organizational ends and members of the bureaucracy become self-serving and lose sight of their original and earlier mission. The efforts of the Central Intelligence Agency (CIA) to maintain U.S. interests in Third World countries led to tolerance of various nations' involvement in illicit drugs. Thus the CIA actually contributed to the drug problem the police struggle to control. A second example is that of the American Association of Retired Persons (AARP). To help the elderly obtain affordable health care, life insurance, drugs, and so forth, the AARP established various organizations to provide or contract for services. But now the AARP seems to be more concerned about its corporate holdings than it is about the welfare of its elderly members.

RIGHTS IN CONFLICT

Finally, students of social problems should become aware of right-versus-right moral conflicts. With this particular format, one's attention is directed to the conflict of moral duties and obligations, the conflict of rights and, not least, the serious moral issue of divided loyalties. In divorce proceedings, for example, spouses must try to balance their personal lives and careers against the obligations and duties to each other and their children. Even those who sincerely want to meet their full obligations to both family and career often find this is not possible because of the real limits of time and means.

Wilson (1994:39,54) observes that from the era of "Enlightenment" and its associated freedoms arose the potential for significant social problems. We are seeing all about us in the entire Western world the working out of the defining experience of the West, the Enlightenment. The Age of Enlightenment was the extraordinary period in the eighteenth century when individuals were emancipated from old tyrannies—from dead custom, hereditary monarchs, religious persecution, and ancient superstition. It is the period that gave us science and human rights, that attacked human slavery and political absolutism, that made possible capitalism and progress. The principal figures of the Enlightenment remain icons of social reform: Adam Smith, David Hume, Thomas Jefferson, Immanuel Kant, Isaac Newton, James Madison.

The Enlightenment defined the West and set it apart from all of the other great cultures of the world. But in culture as in economics, there is no such thing as a free lunch. If you liberate a person from ancient tyrannies, you may also liberate him or her from familiar controls. If you enhance his or her freedom to create, you will enhance his or her freedom to destroy. If you cast out the dead hand of useless custom, you may also cast out the living hand of essential tradition. If you give an individual freedom of expression, he or she may write *The Marriage of Figaro* or he or she may sing "gangsta rap." If you enlarge the number of rights one has, you may shrink the number of responsibilities one feels.

There is a complex interaction between the rights an individual has and the consequences of exercising specific rights. For example, if an individual elects to exercise his or her right to consume alcoholic beverages, this act then nullifies many subsequent rights because of the potential harm that can occur. The right to drive, to engage in athletic events, or to work, is jeopardized by the debilitating effects of alcohol. Every citizen has rights assured him or her by membership in society. At the same time, rights can only be exercised to the degree to which they do not trample on the rights of other members of the group. If a woman elects to have a baby, must she abrogate her right to consume alcohol, smoke, consume caffeine, or take drugs? Because the effects of these substances on the developing fetus are potentially devastating, is it not reasonable to conclude that the rights of the child to a healthy body and mind are being threatened if the mother refuses to abstain during pregnancy? Fetal alcohol effect/syndrome, for instance, is the number-one cause of preventable mental retardation in the United States, and it could be completely eliminated if pregnant women never took an alcoholic drink. Caring for individuals with fetal alcohol effect/syndrome is taking increasingly greater resources that could well be directed toward other pressing issues.

Rights cannot be responsibly exercised without individuals' weighing their potential consequences. Thus a hierarchy of rights, consequences, and harms exists and the personal benefits resulting from any act must be weighed against the personal and social harms that could follow. The decision to use tobacco should be weighed against the possible consequences of a wide variety of harms such as personal health problems and the stress it places on society's resources to care for tobacco-related diseases. Tobacco-related diseases often have catastrophic consequences for their users that cannot be paid for by the individual, so the burden of payment is placed on society. Millions of dollars and countless health care personnel must be diverted away from other patients to care for these individuals with self-inflicted tobacco-related diseases. In addition to the costs in money, personnel, and medical resources, these diseases take tremendous emotional tolls on those closest to the diseased individuals. To focus only on one's rights without consideration of the consequences associated with those rights often deprives other individuals from exercising their rights.

The Constitution of the United States guarantees individuals rights without clearly specifying what the rights really entail. Logically one cannot have rights without others having corresponding obligations. But what obligations does each right assure and what limitations do these obligations and/or rights require? Rights for the collectivity are protected by limitations placed on each individual, but limits of collective rights are also mandated by laws assuring that individual rights are not infringed upon. Therefore, we have rights as a whole that often differ from those we have as individual members of that whole.

For example, the right to free speech may impinge in a number of ways on a specific community. To the members of a small Catholic community, having non-Catholic missionaries preaching on street corners and proselytizing door-to-door could be viewed as a social problem. Attempts to control their actions such as the enactment and enforcements of "Green River" ordnances (laws against active solicitation), could eliminate the community's problem but in so doing would trample on the individual's constitutional rights of religious expression. To protect individual rights, the community may have to put up with individuals pushing their personal theological ideas in public places. From the perspective of the Catholic community, aggressive non-Catholic missionaries are not only a nuisance but a social problem that should be banned. To the proselytizing churches, restrictions on their actions are violations of their civil rights and hence a serious social problem.

Currently another conflict of interests/rights is dividing many communities, and that is cigarette smoking. Smokers argue that their rights are being seriously threatened by aggressive legislation restricting smoking. They argue that society should not and cannot legislate morality. Smokers point out how attempts to legislate alcohol consumption during the Prohibition of the 1920s and 1930s was an abject failure and, in fact, created more problems than it eliminated. They believe that the exact same process is being attempted today and will prove to be just as unsuccessful. Those who smoke then go on to say that smoking is protected by the Constitution's freedom of expression and that no one has the right to force others to adhere his or her personal health policies, which are individual choices. They assert that if the "radicals" get away with imposing smoking restrictions, they can and will move on to other health-related behaviors such as overeating. Therefore, by protecting the constitutional rights of smokers, society is protecting the constitutional rights of everyone.

On the other hand, nonsmokers argue that their rights are being violated by smokers. They point to an increasing body of research data that shows that secondhand smoke leads to numerous health problems such as emphysema, heart disease, and throat and lung cancer. Not only do nonsmokers have a right not to have to breathe smoke-contaminated air, but society has an obligation to protect the health and well-being of its members from the known dangers of breathing smoke.

These are only a few examples of areas where rights come into conflict. Others include environmental issues, endangered species, forest management, enforcement of specific laws, homosexuality, mental illness, national health insurance, taxes, balance of trade, food labeling and packaging, genetic engineering, rape, sexual deviation, political corruption, riots, public protests, zero population growth, the state of the economy, and on and on.

It is notable that the degree to which any of these issues achieves widespread concern varies over time. Often, specific problems are given much fanfare by politicians and special interests groups for a time, and the media try to convince us that specific activities or behaviors have the greatest urgency and demand a total national commitment for a solution. However, after being in the limelight for a while, the importance of the

INTRODUCTION

problem seems to fade and new problems move into prominence. If you look back over previous editions of this book, you can see this trend. It would be useful to speculate why, in American society, some problems remain a national concern while others come and go.

The Consequences of Harm

To this point it had been argued that social problems can be defined and analyzed as (a) conflicts between values, (b) conflicts between values and duties, and (c) conflicts between rights. Consistent with the aims of this article, social problems can be further characterized and interpreted as social conditions that lead, or are generally thought to lead, to harmful consequences for the person, group, community, or society.

Harm—and here we follow Hyman Gross's (1979) conceptualization of the term—can be classified as (a) a loss, usually permanent, that deprives the person or group of a valued object or condition it is entitled to have, (b) offenses to sensibility—that is, harm that contributes to unpleasant experiences in the form of repugnance, embarrassment, disgust, alarm, or fear, and (c) impairment of the collective welfare—that is, violations of those values possessed by the group or society.

Harm can also be ranked as to the potential for good. Physicians, to help their patients, often have to harm them. The question they must ask is, "Will this specific procedure, drug, or operation, produce more good than the pain and suffering it causes?" For instance, will the additional time it affords the cancer patient be worth all the suffering associated with the chemotherapy? In Somalia, health care personnel are forced to make much harder decisions. They are surrounded by starvation, sickness, and death. If they treat one person, another cannot be treated and will die. They find themselves forced to allocate their time and resources, not according to who needs it the most, but according to who has the greatest chance of survival.

Judges must also balance the harms they are about to inflict on those they must sentence against the public good and the extent to which the sentence might help the individual reform. Justice must be served in that people must pay for their crimes, yet most judges also realize that prison time often does more harm than good. In times of recession employers must weigh harm when they are forced to cut back their workforce: Where should the cuts occur? Should they keep employees of long standing and cut those most recently hired (many of which are nonwhites hired through affirmative action programs)? Should they keep those with the most productive records, or those with the greatest need for employment? No matter what employers elect to do, harm will result to some. The harm produced by the need to reduce the workforce must be balanced by the potential good of the company's surviving and sustaining employment for the rest of the employees.

The notion of harm also figures into the public and social dialogue between those who are pro-choice and those who are pro-life. Most pro-lifers are inclined to see the greatest harm of abortion to be loss of life, while most pro-choicers argue that the compelling personal and social harm is the taking away of a value (the right to privacy) that everyone is entitled to. Further harmful consequences of abortion for most pro-lifers are that the value of life will be cheapened, the moral fabric of society will be weakened, and the taking of life could be extended to the elderly and disabled, for example. Most of those who are pro-choice, on the other hand, are inclined to argue that the necessary consequence of their position is that of keeping government out of their private lives and bedrooms. In a similar way this "conflict of values" format can be used to analyze, clarify, and enlarge our understanding of the competing values, harms, and consequences surrounding other social problems. We can, and should, search for the competing values underlying such social problems as, for example, income distribution, homelessness, divorce, education, and the environment.

Loss, then, as a societal harm consists in a rejection or violation of what a person or group feels entitled to have. American citizens, for example, tend to view life, freedom, equality, property, and physical security as ultimate values. Any rejection or violation of these values is thought to constitute a serious social problem since such a loss diminishes one's sense of personhood. Murder, violence, AIDS, homelessness, environmental degradation, the failure to provide adequate health care, and abortion can be conveniently classified as social problems within this class of harms.

Offenses to our sensibilities constitute a class of harm that, when serious enough, becomes a problem affecting moral issues and the common good of the members of a society. Issues surrounding pornography, prostitution, and the so-called victimless crimes are examples of behaviors that belong to this class of harm. Moreover some would argue that environmental degradation, the widening gap between the very rich and the very poor, and the condition of the homeless also should be considered within this class of harm.

A third class of harms—namely impairments to the collective welfare—is explained, in part, by Gross (1979:120) as follows:

> Social life, particularly in the complex forms of civilized societies, creates many dependencies among members of a community. The welfare of each member depends upon the exercise of restraint and precaution by others in the pursuit of their legitimate activities, as well as upon cooperation toward certain common objectives. These matters of collective welfare involve many kinds of interests that may be said to be possessed by the community.

In a pluralistic society, such as American society, matters of collective welfare are sometimes problematic in that there can be considerable conflict of values and rights between various segments of the society. There is likely to remain, however, a great deal of agreement that those social problems whose harmful consequences would involve impairments to the collective welfare would include poverty, poor education, mistreatment of the young and elderly, excessive disparities in income distribution, discrimination against ethnic and other minorities, drug abuse, health and medical care, the state of the economy, and environmental concerns.

BIBLIOGRAPHY

Dahrendorf, R. (1959). *Class and class conflict in industrial society.* Stanford, CA: Stanford University Press.

Davis, Kingsley, & Moore, Wilbert E. (1945). Some principles of stratification. *American Sociological Review,* 10, 242–249.

Gans, Herbert J. (1971). The uses of poverty: The poor pay all. *Social Policy.* New York: Social Policy Corporation.

Gross, Hyman. (1979). *A theory of criminal justice.* New York: Oxford University Press.

Hook, Sidney. (1974). *Pragmatism and the tragic sense of life.* New York: Basic Books.

Horowitz, M. A. (1962). Consensus, conflict, and cooperation. *Social Forces, 41,* 177–188.

Lenski, G. (1966). *Power and privilege.* New York: McGraw-Hill.

Michels, Robert. (1949). *Political parties: A sociological study of the oligarchical tendencies of modern democracy.* New York: Free Press.

Mills, C. Wright. (1956). *The power elite.* New York: Oxford University Press.

Ossowski, S. (1963). *Class structure in the social consciousness.* Translated by Sheila Patterson. New York: The Free Press.

Seeman, Melvin. (1959). On the meaning of alienation. *American Sociological Review, 24,* 783–791.

Van den Berghe, P. (1963). Dialectic and functionalism: Toward a theoretical synthesis. *American Sociological Review, 28,* 695–705.

Weber, Max. (1964). *The protestant ethic and the spirit of capitalism.* Translated by Talcott Parsons. New York: Scribner's.

William, Robin Jr. (1970). *American society: A sociological interpretation,* 3rd. ed. New York: Alfred A. Knopf.

Williams, Robin. (1966). Some further comments on chronic controversies. *American Journal of Sociology, 71,* 717–721.

Wilson, James Q. (1994, August). The moral life." *Brigham Young Magazine,* pp. 37–55.

Wilson, William. (1978). *The declining significance of race.* Chicago: University of Chicago Press.

CHALLENGE TO THE READER

As you read the articles that follow, try to determine which of the three major theoretical positions each of the authors seems to be using. Whatever approach the writer uses in his or her discussion suggests what he or she thinks is the primary cause of the social problem/issue under consideration.

Also ask yourself as you read each article, (1) What values are at stake or in conflict? (2) What rights are at issue or in conflict? (3) What is the nature of the harm in each case, and who is being hurt? (4) What do the authors suggest as possible resolutions for each social problem?

ern sociologists like Herbert Blumer, Murray Edelman, and others.

How Social Problems Are Born

Nathan Glazer

Nathan Glazer is co-editor of The Public Interest.

How do we get more attention, more public action, for a problem we consider important? More important, how do we get the right kind of public attention and action, right in scale, and right in the kinds of solutions the public is willing to accept and fund?

Contemporary social scientists are skeptical about the possibilities of achieving such a rational ordering of things. Consider the following from the sociologist Joseph Gusfield:

> Human problems do not spring up, full-blown and announced, into the consciousness of bystanders. Even to recognize a situation as painful requires a system for categorizing and defining events.... "Objective" conditions are seldom so compelling and so clear in their form that they spontaneously generate a "true" consciousness. Those committed to one or another solution to a public problem see its genesis in the necessary consequences of events and processes; those in opposition often point to "agitators" who impose one or another definition of reality.

This passage is taken from Gusfield's *The Culture of Public Problems: Drinking-Driving and the Symbolic Order* (University of Chicago Press, 1981) and he exhibits in it a common approach in today's social sciences to the issue of how we make social problems out of social conditions, which may be crudely summarized as: It's all in the head. We need a system of defining and categorizing events before we know we have a problem. When most of us agree we do have a problem—when a social condition has been changed into a problem—we interpret this as a case of the problem having become worse, or a case of increasing empathy and sympathy on the part of the public for those suffering. Paradoxically, we often recognize that we have a problem when the condition we are responding to has improved. Recall how the problem of "poverty" burst upon us in the early days of the Kennedy administration. John Kenneth Galbraith had just published *The Affluent Society,* and Michael Harrington had published *The Other America,* but as we now know poverty had been declining all through the forties and fifties.

So our first explanation of how a condition has become a problem may not hold—the problem may not have become worse. Our second, that we have become wiser or more understanding or more sympathetic to the plight of others, is flattering to us, but Gusfield does not give us that credit. It is our categories, rather than reality, that have changed. As we look further into his study of drinking-driving in the book from which I have quoted, we find it is rife with discussions of symbolism, dramaturgy, rhetoric, metaphor, and the like. "The Fiction and Drama of Public Consciousness," one chapter title announces. "The Literary Art of Science: Drama and Pathos in Drinking-Driver Research," another reads.

This is not a case of individual idiosyncrasy. Much of the writing by leading social scientists on how we fix upon social problems, on how they get on the agenda of public attention, is skeptical as to the kind of simple and direct relation

Reprinted with permission of the author and *The Public Interest,* No. 115, Spring 1994, pp. 31-44. © 1994 by National Affairs, Inc.

we might imagine: the problem gets worse, or we become more sensitive to it. More likely, an interest group of some sort, an advocacy group, has taken it up and made it a matter of public concern. The arts of publicity are more relevant than the findings of science.

Thus, in Gusfield's *The Culture of Public Problems,* devoted to the problem of the drinking driver—one would think a serious enough issue to deserve direct attention—we will find rather more references to the literary critic Kenneth Burke than to any scientist or social scientist.

The issue for Gusfield is not only the social construction of public problems, which do indeed have many dimensions, among which the determination of fact, of the existing situation, is only one, but the social construction of science itself, a rather popular theme among social scientists and advanced literary critics these days.

THE ROLE OF RHETORIC

There is undoubtedly a degree of overkill in Gusfield's approach but there is something to learn from it too, as we consider how we get the right kind of public attention for an issue of importance. One problem to which Gusfield points is that we move very rapidly from the problem itself, which may be both undeniable and important, to the arts of publicity and attention-getting, and, as he argues, these arts also affect almost immediately the very facts that we use to get attention and that are the bedrock of our initial concern. Thus, if we examine the facts which we use as the basis to claim attention, public money, funds for research, we see that the facts themselves become shaped by the need to compete with other claims, other problems, for which the arts of publicity are also employed.

Half of the 50,000 deaths a year from automobile accidents are attributed regularly, we are told, to drink on the part of the driver. When we examine this oft-repeated statistic, according to Gusfield—and he goes into the source of the figure in detail—it turns out that it is hardly solidly based, that many questions can be raised about it. Similarly with the statistic on how many Americans have serious drinking problems—a common figure of 9 or 10 million was used when Gusfield was writing his book in the early 1980s; its sources are murky and uncertain, and of course depend on what we mean by drinking problems.

It is not only in the case of drinking-driving and alcoholism that the first necessary step in defining a problem—finding out just what the scale of the problem is—immediately gets mixed up with the necessary requirements of the next steps in getting attention for it, bringing it to the notice of necessary publics. And so we are familiar with disputes right now about the scale of date rape on campus, as well as with the prior question of just what date rape is. Similarly with child abuse, and many other public issues.

Is Gusfield only playing games when he asks just how do the authorities decide that an accident was based on drinking, and other questions which undermine the statistic that half of all automobile accident fatalities are owing to drinking? Of course drinking-driving is a serious problem, so why does he bother us with the figures used in making a case to congressional committees or attracting publicity or funds? But his approach does alert us to a number of things of importance. First, that rhetoric, drama, the arts of gaining access to the mass media or congressmen, are implicated at the very beginning in all our efforts to gain attention to social problems. Second, that there is no easy way of scaling social problems from the point of view of how "important" they are. In the passage I reprinted from his *The Culture of Public Problems,* he placed the words "objective" and "true" in quotation marks. Third, that because this is so there is the constant danger of overkill, over-dramatization, the constant possibility that those most gifted in the arts of publicity and drama will engross a larger share of funds and attention than *their* problem warrants. (I leave aside for the moment the question of whether we can decide on any objective basis how much money or attention one problem deserves as against another, or the methods by which we might decide. Whatever our answer, we would probably all agree that some problems seem to have gained an inordinate amount of attention compared to others of apparently similar or greater scale. This has been argued in the case of AIDS. Very likely the attention AIDS gets is in part related to the number of people in the arts and fashion and publicity who are affected.)

It is revealing that Gusfield titled an earlier book, on the temperance move-

2. How Social Problems Are Born

ment, *Symbolic Crusade,* and it is clear that the use of the term "crusade" is meant to suggest to us that a movement that tried to deal with a problem that was serious enough at the time, and that may be as serious today, was overdone, excessive, shrill, in some ways more than the problem called for. (After all, it led, astonishingly, to the passage of a constitutional amendment banning alcoholic drinks.) The use of the word "crusade" today implies that we are confronted with something that is rather too grand, too much, for its object, and the word tends to evoke skepticism of the cause to which it is attached, rather than inspiring greater commitment to the cause. What, after all, in our laid-back contemporary world, used to horrors of all sorts, deserves a "crusade," with its religious implications? (We seem happier with the word "war," as in war against poverty, and war against drugs, headed by a czar rather than a pope.)

THE KNOWLEDGE PROBLEM

Gusfield does I believe emphasize too much the social construction of problems rather than their objective realities (and I am not using quotation marks, as he did around "objective"), but his work draws our attention to a surprising fact about social scientists' examination of the question of how a social condition becomes a social problem: There is a considerable degree of skepticism of how we go about it, or indeed how we can go about it in a democracy in which the mass media inevitably shape public perception and knowledge.

One finds the same in another social scientist, the late Aaron Wildavsky, who devoted a considerable part of his enormous energy and great talents to the study of how society deals with risk. Wildavsky gave more credit to the objective realities than Gusfield does, but he doubted they could play a major role in determining how we allocate our resources in dealing with risk. While he dealt primarily with environmental risks, he would have said the same thing about social risks and social problems. But despite his much greater respect for science and scientists and their ability to determine the degree of danger from various environmental risks (had he been speaking about social problems he might well have approached Gusfield in skepti-

15

INTRODUCTION

cism), he also believed that any hope of matching our resources to our problems by taking account of the scale of the danger they posed was probably a vain one. The problems were primarily political, and when Wildavsky said political he meant also cultural, tied up with our interests, our values, our perceptions, which brings him not far from Gusfield.

"What would be needed," he and Mary Douglas asked in their book *Risk and Culture,*

> to make us able to understand the risks that face us?—Nothing short of total knowledge (a mad answer to an impossible question). The hundreds of thousands of chemicals about whose dangers so much is said are matched easily by the diversity of the causes of war or the afflictions of poverty or the horrors of religious and racial strife. Just trying to think of what categories of objects a person might be concerned about is alarming. Indeed, it might be better for mental health to limit rather than expand sources of concern. Since no one can attend to everything, some sort of priority must be established among dangers. . . . Ranking dangers . . . so as to know which ones to address and in what order, demands prior agreement on criteria. There is no mechanical way to produce a ranking.

Scientists may come together on this (less likely social scientists) but there is no way of making their agreement public policy: We are not a nation of philosophers and kings. The issue then becomes political, with everything involved in that term. Douglas and Wildavsky quote some other authorities on risk:

> Values and uncertainties are an integral part of every acceptable-risk problem. As a result, there are no value-free processes for choosing between risky alternatives. The search for an "objective method" [again, in quotation marks] is doomed to failure and may blind the searchers to the value-laden assumptions they are making.

Another quotation from a different source:

> Not only does each approach fail to give a definitive answer, but it is predisposed to representing particular interests and recommending particular decisions. Hence, choice of a method is a political decision with a distinct message about who should rule and what should matter.

Scientists may agree on what risks should be addressed, in what order, with what resources, but even that is not assured, and when it comes to social problems and social scientists agreement is even less likely. Popular passions will be aroused, they will affect what politicians and administrators do, and one can only hope that knowledge—authentic knowledge, solidly based, scientifically established, something I still believe in despite the assault on its possibility we have seen in the newer trends in the humanities and social sciences—will play some role in determining what legislators and administrators do.

So in almost all transitions from social condition to social problem we are in the grip of passions, interests, perceptions, values that are not going to be affected much by what the scientists tell us. Gusfield called the fight for temperance a "crusade," Aaron Wildavsky uses the term "sectarian," with its religious connotations, to describe those who devote themselves to getting the public and government to pay attention to what they conceive of as major risks, and he uses the term not as an epithet but as a carefully constructed concept which for him describes the character of the people and groups who have done so much to alert us to environmental risks.

Despite the fact that there is a great deal that we can learn from the work of Gusfield and Wildavsky on how social conditions become social problems, I will separate myself from the full scope of their arguments. I believe that there are objective ways of determining the scale of a problem, even if all our efforts are somewhat corrupted by our political attitudes, by human failings that affect even scientists, and other factors; and that while we have undoubtedly seen cases in which the attention to a problem and the resources devoted to it can properly arouse skepticism, there are indeed conditions which hardly need to be "socially constructed," which spring to our eyes and appeal to our human sympathies and simply demand attention, and for which there would appear to be only one central question to consider: What to do about it. Yet we must be alerted to the issues the social scientists raise when we consider pragmatically how to make a social condition a social problem.

THE CASE OF PROHIBITION

Our two authors have tended to concentrate on issues which have in some way been misconstrued owing to interests and passions, perceptions and values, they do not share. Thus, temperance was initially raised as a moral problem, a problem of making people better. This was in time joined by other considerations: temperance would fight poverty among workingmen, making them better workers, fathers, husbands. Eventually the method chosen to make them better was that of depriving them of the means for bad behavior. We consider Prohibition a great failure, but it did (according to the best authority) reduce the consumption of alcohol by half.

It is not clear how the problem of excessive alcohol consumption might have been better construed at the time; it could not easily have been construed differently from what it was in a largely rural and small-town, Protestant and evangelical America, fearful of the rising numbers of immigrants and Catholics, of the growth of the big cities with their wider range of acceptable behavior, their greater tolerance. We now see the crusade as mistaken, because we see the problem of alcohol consumption, when it becomes a problem, as one of mental health, and its incidence and impact have declined, at least in public perception, perhaps in reality. That decline has much to do with our viewing alcohol—as so much else—in the context of health rather than sin, as we have seen the decline of the theological ethic, and the rise of the therapeutic ethic. (The decline in alcoholism may also be connected with changes in taste and fashion, from hard liquor to wine, from wine to water.)

Just as Gusfield's temperance crusaders are seen as motivated by the fight against sin, evil, bad behavior, when they might have chosen (and in time their successors did) a more effective way of viewing the problem, Wildavsky's environmental crusaders are viewed as sectarians, impassioned, moving from one topic that arouses their indignation to another, incapable of placing in the balance the goods the technologies they oppose have brought, or comprehending the impact on lives and economies of the measures they demand. Religion has now been replaced by a suspicion of science and technology, a suspicion of big organizations, whether industrial firms or government, even though government is called upon to restore the ecological balance (but the distrust of government is such that it is primarily the courts and judges who are depended on to keep government in line in enforcing the rules and regulations).

2. How Social Problems Are Born

TOBACCO, A SUCCESS STORY

Despite the rather sour tone adopted by many of our best social scientists toward crusaders and sectarians and indeed toward the passion for reform in general (a tone we may trace, perhaps, to Richard Hofstadter's *The Age of Reform*), we live in a society afflicted with problems that are hardly imaginary, hardly the result of misperceptions inspired by the passions of crusaders and sectarians. And we have seen successes in transforming conditions into problems in which we do not sense that crusaders and sectarians are driving forces, but rather scientific understanding and pragmatic policymakers. Perhaps the largest success (partial it is true), is the decline in tobacco consumption, and the decline in its social acceptance. While the battle against smoking is not without its sectarians, the costs of smoking to health are undeniable and ever more solidly documented, and the efforts that have been devoted to reducing its incidence have been balanced and, in time, effective.

Interestingly enough, the effort to reduce the consumption of tobacco has not been conducted, as so many others have, by means of major federal legislation, giving responsibility to a major federal agency, new or old, operating under law and issuing detailed regulations. Nor do we have a single major reform organization devoted to the cause of eradicating smoking—there is no equivalent to the Women's Christian Temperance Union, the Anti-Saloon League, the major environmental organizations. There are anti-tobacco crusaders, but they are local rather than national, hardly organized, and their effectiveness has been in getting local restrictive legislation, and in getting large organizations to set rules limiting the areas where smoking is permitted. At the national level, there have been warnings rather than prohibitions, the voluntary—though under pressure—banning of advertising on TV, the local pressure against billboards.

The campaign for the reduction of smoking has thus been characterized first by the fact that it was initiated by almost unambiguous scientific findings, announced by high medical authority, rather than by mass pressure from a mass organization; second that it has been conducted more on the local level than on the national; third that it has been characterized more by voluntary concessions, as in the case of advertising, than by national prohibition.

I have been trying to understand just why the campaign against smoking has been successful to the degree it has, so much more successful than the campaign against drugs for example, and why its characteristics have been so instinctive, but it has not been easy to get light on this matter. Seeking for some understanding on the shelves of the Widener Library at Harvard, I discover to my surprise that there is very little on the subject. "Smoking" comes in the Library of Congress cataloguing system between "drink" and "drugs," and while one finds shelf upon shelf of material on these two subjects, on smoking, there is almost nothing, a few disparate volumes. I wonder whether this is because we prefer studying failure rather than success; or because the limited degree of national coercive action means there is less to study; or because tobacco is inherently less glamorous than drink and drugs. But it kills as many people and if we have an example of success, even partial success, it should be worthy of study.

I am sure that to the smoker my assessment of the moderate nature of the campaign against smoking, compared to the crusade against drink, or the war against drugs, is too benign. Yet it is my impression there is something to be learned from the smoking story as we try to understand the more effective ways in which we convert social conditions into social problems. One thing to be learned, for example, is that moderation in the campaign, the willingness to accept slow but steady progress, may prevent a major backlash. In the case of drinking, the backlash was the repeal of the Eighteenth Amendment. Much had been learned from Prohibition and one important thing that had been learned was that it was better to leave the matter to the states and the localities. As a result, drink disappeared from the national agenda, and an issue that had troubled American political life for generations was domesticated. Excessive drinking became a medical and health problem, not a moral or legal problem.

BACKLASHES

The war against drugs has not yet met such a backlash, but it may in the campaign for legalization. There are incredible problems around legalization, but there are incredible problems around our efforts to eradicate drug use, too, and we may well find in time that questions will be raised about why we spend billions ineffectively in trying to eliminate the sources of drugs and in trying to eradicate dealing in drugs, why so large a proportion of our criminal justice resources—in police, judges, prosecutors, courts, jails, prisons—is devoted to the attempt to wipe out drug use.

We may shortly find another case of backlash in the case of child abuse. The story of child abuse is also one of a case, as in smoking, in which reform starts from the top, rather than as a result of mass pressure from crusaders or sectarians, and in which doctors rather than movement people play the central role, at least at the beginning. The story is an interesting one. It begins with interest in sponsoring research on the issue of child abuse in the Children's Bureau in the then-Department of Health, Education, and Welfare. They funded the work on this subject of Dr. C. Henry Kempe, a pediatrician who specialized in immunology, and who noted that the interns and residents he supervised were more interested in diagnosing rare blood diseases in the children under their care than in noting physical injuries to the children. Dr. Kempe wanted to draw attention to this problem. In one of his first efforts to report on the physical abuse of children, he organized a seminar at an academic meeting on the physical abuse of children. We are told by Barbara J. Nelson, who has researched this story, that

> fellow members of the program committee suggested that a title such as physical abuse which emphasized legally liable and socially deviant behavior might make some members wary of attending the seminar. Kempe agreed, renaming the seminar the "Battered Child Syndrome." In one stroke he labeled the problem in a manner which downplayed the deviant aspects while highlighting the medical aspects. From an agenda-setting perspective the effect of the label cannot be overestimated.

In 1962 Kempe published an article under that title In the *Journal of the American Medical Association*. The rest is history. The mass media took up the phrase and the issue (scarcely a case of a situation that had become observably worse, but rather one for which the right label had been found) became a national

INTRODUCTION

one, a classic case of how a social condition becomes a social problem. The Children's Bureau proposed a model child abuse reporting statute in 1963, and by 1967 every state had passed a reporting law. Laws were revised over time to become ever stronger. To the physical abuse of children, attractive enough to the mass media, was added concern for their sexual abuse, and we may well have now reached a stage where considerably more is reported and even prosecuted than exists, and we may be on the verge of a backlash against the attention and resources devoted to the problem, a common stage in the history of such "victories."

Of course it is inevitable, what with changing cycles of attention and fashion, that at one point there will be great attention to a social problem, at a subsequent point much less, with very little change in the problem itself. But what most impresses when one considers the range of problems with which the mass media, scientists, social scientists, legislators, voluntary organizations, all deal, is that no problem is fully neglected. This is not surprising in a democracy where everything is open, every issue has its advocates, and the mass media are ever ready to exploit a problem which has been lying fallow and relatively neglected. There are entrepreneurs of problem-making, problem-enhancing, at all levels: professionals in given areas, who see a problem others do not, such as battered children; bureaucrats seeking to maintain old missions, expand into new missions, as in the case of the Children's Bureau; legislators who leap into an area in which there is no or little legislation and try to make it their own, as Congressman Mario Biaggi did in the case of child abuse; scientists, natural and social, who see opportunities for research; editors and journalists, seeking new and interesting topics; advocacy organizations, some single-mindedly dealing with one clientele, one issue, others seeking to expand into new issues as old ones lose interest and salience. We see the change in perception starting in some cases through some administrative or bureaucratic action at the top, in others being initiated by an outside advocacy group, in others seemingly launched on the public stage by a single book, e.g., Rachel Carson's *Silent Spring,* Ralph Nader's *Unsafe at Any Speed.* The number of enterprises of this sort that have failed (including I am sure many books as good as Nader's and Carson's and trying to draw attention to a problem as serious as the effect of pesticides on the environment or automobile safety) are far more numerous than the few that have achieved remarkable success.

There is one factor in the potential success of such enterprises that has not been much noted. I think the scale and steadiness of public response to the entrepreneurs of problem-making depends not only on the seriousness of the problem, on the degree to which it impinges directly on public perception, the degree to which it agitates and concerns the public, but on whether any effective action to deal with it is visible. Consider what is called the "urban crisis," the problem of the inner cities. One may argue with the term, it is not well-defined, and contains within itself a host of other problems. It was once high on the public agenda, in the 1960s and 1970s, and then declined, despite steady study, publications, popular books, advocacy groups, special academic programs, and I would conclude this was for one reason: There was nothing to be done, or at least nothing much to be done. Other problems present clear targets, things to be done, even if they will fail: prohibit drink, ban smoking in enclosed spaces, interdict drugs. We learn from the experience, even if slowly and poorly, and eventually we learn there is not much to be done: people will drink and take drugs, there will be cycles of greater or lesser use, and moderate impact is all we can hope for.

We are now at a moment in which great attention is being given in the mass media to a specific problem that can be considered part of the "urban crisis," gun violence, the use of guns in situations, whether of fights among youths or of robbery, in which they were until recently less available or less used, with an accompanying high rate of homicide particularly affecting young black males (who are also those who are using the guns), bystanders, shopkeepers, taxicab drivers. We are now seeing efforts by those in the field of public health to try to recast the problem as a public health one. Thus, it is likened to an "epidemic," which it certainly is, as the word is popularly used, but public health people wonder whether equating it to epidemics such as tuberculosis or AIDS will give us more insight into what is happening, or direct more attention and more effective attention to it, or suggest more tools with which to deal with it.

The model for the campaign against gun violence that is now developing is the successful reduction of smoking, or of driving under the influence of drink. The public health model calls for such actions as proper tracing of the incidence of the condition, relating it to other social conditions, developing campaigns of information and modes of treatment, devising techniques of education or publicity which change habits—which, for example, would make the possession and use of guns reprehensible. Undoubtedly we will be hearing and seeing a good deal more about this in the coming months, perhaps years. Any such effort, if we take as a model the case of driving under the influence of drink, or smoking, must be long-range. The campaign will take forms we cannot imagine now, and we cannot know whether we will achieve the relative success of our efforts to reduce smoking and drinking-driving, or the relative failure of our efforts to reduce the use of drugs.

But one contrast comes to mind and suggests a caution: smoking and drinking-driving were not matters in which incidence was concentrated in one social class, one ethnic or racial group. One could argue the same with gun violence, if one concentrates on the possession of guns, and the overall incidence of deaths from guns, and if one places in the same statistical category suicides from the use of an available gun, hunting accidents, domestic violence which climaxes with shooting. These kinds of deaths from guns are old matters, and there has not been, I think, any marked change in recent years. The gun violence that is now the subject of so many newspaper and TV stories is a different matter, concentrated in the inner city, affecting largely one major minority group—young blacks—even though it is also present generally and affects almost everybody.

Is anything gained by lumping this phenomenon into a general category of "gun violence," and considering it under the category of public health? I think not. It is one thing to change behavior, through publicity appeals, through campaigns that change popular attitudes toward a behavior from acceptance to disapproval, when that behavior is found throughout the society, and when those to whom one is appealing are representative of the society in general. It must be a different matter when one deals with behavior that is encapsulated in a specific social group. Consider one consequence of cre-

2. How Social Problems Are Born

ating a general category of "gun violence." One possible approach, the one most popular today in the mass media and among political leaders, is: Make it harder to get guns. This is likely to work among the middle classes and among groups who presently hold guns legally. It is likely to be completely ineffective among the groups whose very behavior has raised the issue to the high pitch of current concern, the young people who are killing each other in the urban ghettoes, and more occasionally (but often enough) killing others. It suggests to us that it is still important to consider just what the problem is, and we move in rather ineffective directions if we frame it improperly. Is the problem "gun violence," or is it rather the larger one of the complex of social problems in the inner cities that we still have no effective means of attacking? Can the problem of gun violence be effectively isolated from this larger complex and reduced? It seems to me doubtful, but many efforts are under way, from Jesse Jackson's exhortations to the attempt to reframe the issue as one of public health to the campaign of Jay Winston of the Harvard School of Public Health to get anti-gun messages into TV programs. And we will learn from these efforts whether gun violence is more like smoking or more like drugs.

Parenting and Family Issues

Throughout history the family has been the most effective and primary transmitter of values, beliefs, and behaviors, for good or ill. But the American family has been under increasing assault since the 1960s. Some individuals argue that it is not so much an assault as a restructuring of an antiquated social institution. Single parents, couples with no children, single individuals, unmarried people living together, homosexual couples—almost any and all combinations of persons living under a common roof are now classified by some as a "family."

Questions raised by the articles in this unit include (a) Just what is a family? (b) What impact is the "new" family structure having on its members, especially the children? (c) What impact are other social institutions having on the family? (d) How does what is happening in the family impact on other institutions? (e) What can and should be done to strengthen the family? (f) What must be done to reconnect fathers with the family?

The essay "Fount of Virtue, Spring of Wealth: How the Strong Family Sustains a Prosperous Society" examines the role that families play in society. Cross-cultural anthropological studies reveal that there are links between violence, poverty, drug abuse, health (both physical and mental), educational accomplishments, and family stability.

The short report "Things That Go Bump in the Home" claims that family violence is not just a problem of males abusing females and children. Increasingly, domestic violence is being displayed by both sexes, with the rate of dangerous assaults on males by their female companions being twice that of assaults on females by their male companions.

"Growing Up against the Odds" documents a significant link between the rate at which "traditional" families are breaking down and the increase in every type of social pathology. Robert Royal concludes that if children are to have a meaningful chance for success and for fulfilling lives, the traditional family must thrive, not merely survive.

"Why Leave Children with Bad Parents?" shows that programs that stress parental rights and/or family preservation over the welfare of children are extremely hazardous to the health and well-being of the children. In attempting to keep families intact, social welfare agencies aggressively encourage the return of abused children to their abusive parents, even when no evidence exists that the parents will not repeat the abuse.

Sharon Darling, in "The Three R's Spell Success," states that education can help families break out of poverty when parents as well as children become literate. She shows that literacy programs tend to stimulate active involvement of parents in the activities of their children both at home and at school.

Looking Ahead: Challenge Questions

What personal experiences have you had with problems in your family?

Is the current high divorce rate good or bad for society? Explain your answer.

How do the problems facing children who live with one parent differ from those of children who live with two parents?

To what degree is domestic violence a "male" problem?

Is it possible and/or desirable to try to preserve the traditional family? Explain your answer.

UNIT 1

Under what conditions should abused children be returned to their (abusive) parents?

In what major ways are the lives of children in peril, especially children in troubled families?

What are the consequences of being forced to survive in families living on welfare?'

How have literacy programs impacted on families?

In what ways would the approaches of symbolic interactionists, functionalists, or conflict theorists differ in the study of family issues?

What conflicts in rights, values, and duties seem to underlie each issue?

FOUNT OF VIRTUE, SPRING OF WEALTH

How the Strong Family Sustains a Prosperous Society

Charmaine Crouse Yoest

Charmaine Crouse Yoest is a Bradley fellow at the University of Virginia and a public policy consultant. She is former deputy director of policy at the Family Research Council, a Washington-based study institute.

"Daddy," I asked, "will you have to go to war?"

At that point, the Vietnam War, for me, consisted primarily of television footage of helicopters and soldiers in the jungle. I was terrified that my father would be called to go to war and die at any moment.

"No, honey," he replied, "the military doesn't draft men who are in school."

Still not understanding why men had to fight and die in wars, I pressed on. "But what if you did have to go? Would you really go?"

"Yes, I would. Sometimes men have to fight to protect the people they love," he explained. "If a burglar wanted to break into our house, I would fight to keep your mom and you and your brother safe."

As he talked, I understood: There are things that are worth dying for. For my dad, like many other people, the list begins with family.

That night my father taught me the integral connection between family and country—that there are times when the burglar is another country and the house our families live in is our country.

And so it is with most of the lessons children need to learn as they grow and mature—they are best taught and modeled in the context of family. Children are the future citizens of any nation. For this reason, societies have a stake in whether children are raised to become good citizens rather than bad ones.

President Theodore Roosevelt wrote: "Sins against pure and healthy family life are those which of all others are sure in the end to be visited most heavily upon the nation." Sociologist Urie Bronfenbrenner, in outlining the needs of children, provides a compelling rationale for the state's interest in preserving the family as the primary context for cultivating healthy children and seeing them develop into productive citizens.

> The informal education that takes place in the family is not merely a pleasant prelude, but rather a powerful prerequisite for success in formal education from the primary grades onward. This empowering experience reaches further still. As evidenced in longitudinal studies, it appears to provide a basis, while offering no guarantee, for the subsequent development of the capacity to function responsively and creatively as an adult in the realms of work, family life, and citizenship.[1]

Quite simply, the family lays the foundation. As increasing bodies of research attest, our families are the fertile ground from which children acquire the patterns, habits, lessons, and values that, in our increasingly interdependent society, affect us all.

From a societal viewpoint, the family is important because of its role in shaping good citizens. Fundamentally, the state

---
1. Urie Bronfenbrenner, "What Do Families Do?" *Family Affairs* (New York: Institute for American Values, Winter/Spring 1991), 4.

3. Fount of Virtue, Spring of Wealth

President Theodore Roosevelt wrote: "Sins against pure and healthy family life are those which of all others are sure in the end to be visited most heavily upon the nation."

needs stability, achievement, and loyalty from its citizens, and families foster these three qualities.

THE STATE'S NEED FOR STABILITY IN ITS CITIZENRY

There can be no more graphic testimony to a society's need for law and order than the smoldering images of riot-torn Los Angeles. Events like the L.A. riots, New York City's "wilding" episode, and the depredations of Jeffrey Dahmer overshadow the complex policy discussions over the merit of various government spending programs and actions, and they are a keen reminder that, above all else, society is meant to protect its citizens. Although we rarely do it consciously, each of us gives up a measure of individual rights and autonomy in order to live within the bounds of community. We do so with the expectation of greater stability and security than we would have on our own. Thus, while we chafe at the delay in stopping at intersections when the light is red, few of us begrudge that infringement on our personal freedom because of the safety and order it brings to our daily travels.

Some do, however; there are more and more individuals claiming their right to become a law unto themselves. While running red lights is a relatively minor breach of the community contract, the daily escalation of violence in our country attests to increasing instability and weakening of the social order.

Respect for authority and willingness to accept personal limits are character traits upon which social order is built. The lessons that build that kind of private discipline, so important to public stability, are best taught in the home. Kay James,

former assistant secretary at the U.S. Department of Health and Human Services (HHS), tells a true story about two young boys growing up in a poverty-stricken neighborhood.

One day while roaming the neighborhood, the two boys broke into the local school and stole several chickens from the cafeteria refrigerator. Both boys then proudly took their bounty home to their mothers for dinner that night.

The first mother cried out in delight: "Boy, I don't know where you got this, but we sure are going to eat good tonight!"

The second boy came home to a different response. "Son," his mother asked, "I know you don't have a job, and you don't have money. Where did you get those chickens?"

When she heard the details, she took the chickens by the feet and started pummeling her son with them. She backed him into a corner, and with one hand still holding the birds and the other pointed right in his face, she said, "Boy, I will starve before I let one of my children bring stolen food into this house!"

That was all she said before she turned and opened the back door and flung those chickens into the backyard. "If you want to help out around here," she declared, "you can get a job."[2]

Kay recounts that several years later the first mother was left grieving beside the casket of her son, shot to death in a drug deal. The second woman was Kay's own mother. Both she and her brother learned

---

2. Kay James, *Never Forget* (Grand Rapids, Mich.: Zondervan Publishing House, 1993), 46.

1. PARENTING AND FAMILY ISSUES

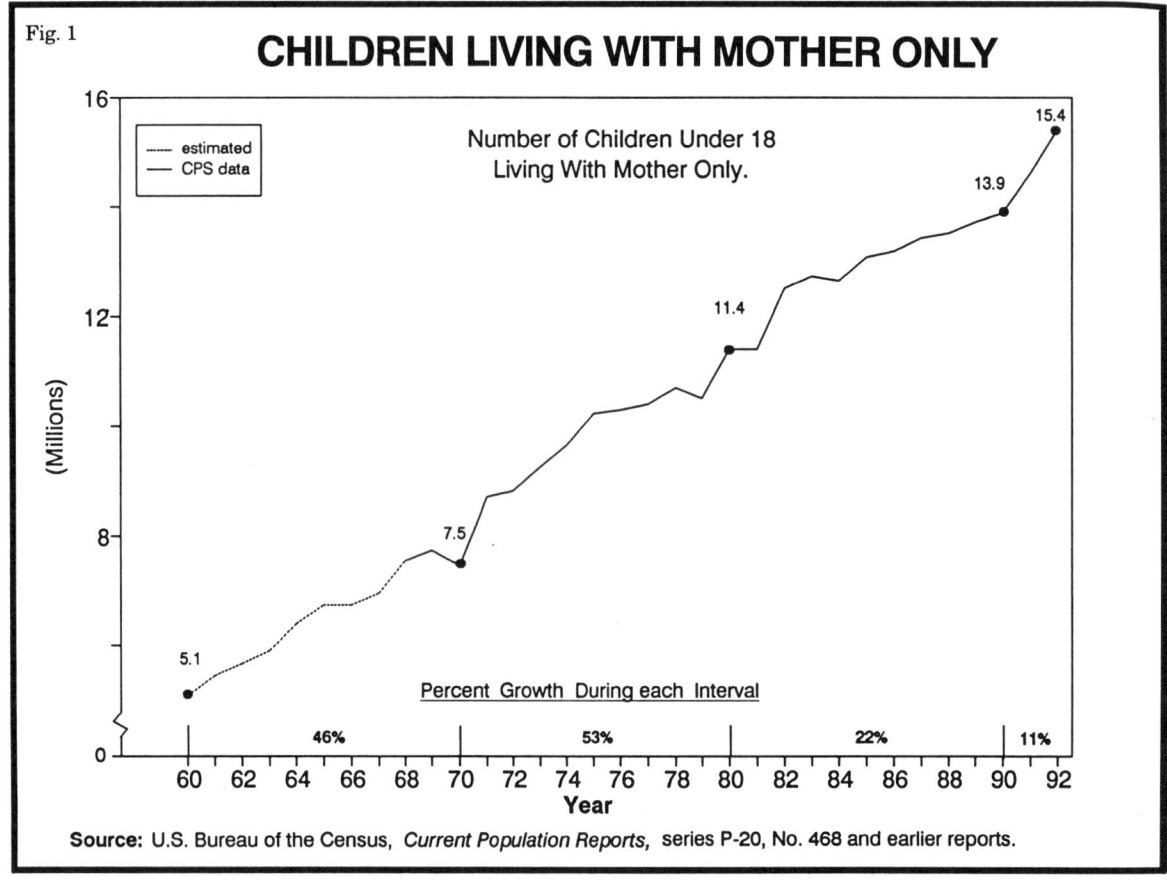

an enduring and valuable lesson in honesty and respect for authority that night.

Many children today are not being taught respect for any authority—in part because of the increase in father absence. The active involvement of fathers in the raising and disciplining of children, particularly young boys, is crucial. Myriam Miedzian, author of *Boys Will Be Boys: Breaking the Link Between Masculinity and Violence*, stated in testimony before the U.S. House of Representatives that "raising a son without a nurturant father in the home significantly increases the likelihood of the boy becoming violent."[3] She added that this is true even in other cultures:

> Cross-cultural anthropological studies indicate that violent behavior is often characteristic of male adolescents and adults whose fathers were absent or played a small role in their son's early rearing. For example, anthropologists Beatrice Whiting's and John Whiting's study of children in six cultures revealed that those tribes in which the father was most loosely connected with the family and had least to do with the rearing of children, were the most violent. These findings are corroborated by other studies.[4]

Is it any wonder, then, that violence is increasing in our society when the number of children in single-parent, female-headed homes has increased 202 percent since 1960? Today, nearly one-third—28 percent—of all children are born into single-parent homes (see figure 1). In the black community, it has reached as high as 68 percent.

Unfortunately, because of the disparity between the white and black communities in the percentages of single-parent

---

3. Myriam Miedzian, testimony to the Select Committee on Children, Youth, and Families, U.S. House of Representatives, "Babies and Briefcases: Creating a Family-Friendly Workplace for Fathers," June 11, 1991, 1.

4. Miedzian, "Babies and Briefcases," 3.

3. Fount of Virtue, Spring of Wealth

"Cross-cultural anthropological studies indicate that violent behavior is often characteristic of male adolescents and adults whose fathers were absent or played a small role in their son's early rearing."

homes, there has developed a mistaken perception that increased levels of crime in some communities is a racial issue rather than one of family structure. A study by Douglas A. Smith and G. Roger Jarjoura, published in 1988, disproved this fallacy:

> Many studies that find a significant association between racial composition and crime rates have failed to control for community family structure and may mistakenly attribute to racial composition an effect that is actually due to the association between race and family structure.[5]

The research points to a strong correlation between the increasing numbers of children growing up in single-parent homes and the rising levels of violence and crime in our country. One reason is the appalling number of adolescent boys who live on the streets and whose need for belonging, identity, camaraderie, and security that should come from family is filled by gangs. Leon Bing, author of *Do or Die*, talked with two fourteen-year-old gang members, "Sidewinder" and "Bopete," in a youth detention center:

> Bopete: "Sometimes I think about not goin' back to banging' when I get outta here. I play in sports a lot here, and I . . ."
>
> Sidewinder's laugh interrupts. "Sound like a regular ol' teenager, don't he? I sound like that, too, after the drive-by. I got shot twice in the leg . . . and when that happen I didn't want to bang no more, either. Makin' promises to God, all like that. But when it heal up . . ." He is silent for a moment; then, "I tell you somethin'—*I don't feel connected to any other kids in this city or in this country or in this world. I only feel comfortable in my 'hood. That's the only thing I'm connected to, that's my family. One big family—that's about it.*" (emphasis mine)
>
> "In my 'hood, in the Jungle, it ain't like a gang. It's more like a nation, everybody all together as one. Other kids, as long as they ain't my enemies, I can be cool with 'em." Bopete lapses into silence. "I'll tell you, though—if I didn't have no worst enemy to fight with, I'd probably find somebody."
>
> Sidewinder picks it up. "I'd find somebody. 'Cause if they ain't nobody to fight, it ain't no gangs. It ain't no life. I don't know . . . it ain't no . . ."
>
> "It ain't no fun."
>
> "Yeah! Ain't no fun just sittin' there. Anybody can just sit around, just drink, smoke a little Thai. But that ain't fun like shootin' guns and stabbin' people. *That's* fun."[6]

The family also promotes societal stability by providing men with a proper channel for their sexuality and providing appropriate role models for adolescent boys of a stable, healthy masculinity. Without the nuclear family of husbands and wives, mothers and fathers, to provide these crucial social functions, a vacuum of enormous and devastating proportions is developing. Miedzian quoted the classic work of sociologist Walter Miller:

> Miller pointed out that the extreme concern with toughness and the frequent

---

5. Douglas Smith and G. Roger Jarjoura, "Social Structure and Criminal Victimization," *Journal of Research in Crime and Delinquency* 25 (February 1986), 27–52.

6. Leon Bing, *Do or Die*, as quoted by William Tucker, "Is Police Brutality the Problem?" *Commentary* 95 (January 1993), 26.

1. PARENTING AND FAMILY ISSUES

> *"Anybody can just sit around, just drink, smoke a little Thai,"*
> *a gang member said. "But that ain't fun like shootin'*
> *guns and stabbin' people. That's fun."*

violence in lower-class culture probably originates in the fact that for a significant percentage of these boys there is no consistently present male figure whom they can identify with and model themselves on. Because of this they develop an "almost obsessive . . . concern with 'masculinity'" which Miller refers to as "hypermasculinity."[7]

Little boys, as they grow toward manhood, must make a break from the feminine role modeled for them by their mothers and establish their own masculine identity. With a father in the home, they can do this relatively painlessly by imitating their dads; when they become men, they can find masculine roles in becoming husbands and fathers themselves. But in our society today, far too many boys are growing up without that male role model and entering an adult society that has ceased to value and support marriage.

In his book *Men and Marriage*, George Gilder champions marriage as an indispensable social construct for the appropriate channeling of male sexual aggression into creativity. "Without a durable relationship with a woman," explains Gilder, "a man's sexual life is a series of brief and temporary exchanges, impelled by a desire to affirm his most rudimentary masculinity."[8]

He goes on to make the case that this male impulse, biologically based though it is, results in a destabilizing influence on society. And, ultimately, it is not fulfilling for men themselves. In the end, looking for meaning in life and given impetus through societal constructs, a man will marry: "The man's love . . . offers a promise of dignity and purpose. For he then has to create, by dint of his own effort . . . a life that a woman could choose. Thus are released and formed the energies of civilized society. He provides, and he does it for a lifetime, for a life."[9]

A civilized society, a stable society—the antithesis of the mayhem engendered by the new "evolving" family forms—is precisely the objective. Family, based on the fundamental marital union, is the foundation for a strong nation.

As society experiments and individuals reject the responsibilities of family, we all pay a price. Sen. Daniel Patrick Moynihan sounded this alarm in a 1965 article that was greeted with widespread approbation. Three decades later, daily headlines confirm his prescience:

> From the wild Irish slums of the nineteenth century eastern seaboard, to the riot-torn suburbs of Los Angeles, there is one unmistakable lesson in American history: a community that allows a large number of young men to grow up in broken families, dominated by women, never acquiring any stable relationship to male authority, never acquiring any set of rational expectations about the future—that community asks for and gets chaos. Crime, violence, unrest, disorder—most particularly the furious, unrestrained lashing out at the whole social structure—that is not only to be expected; it is very near to inevitable. And it is richly deserved.[10]

---

7. Miedzian, "Babies and Briefcases," 2.
8. George Gilder, *Men and Marriage* (Gretna, La.: Pelican Books, 1992), 14.
9. Gilder, *Men and Marriage*, 290.
10. Daniel Patrick Moynihan, *Family and Nation: The Godkin Lectures, Harvard University* (San Diego, New York, London: Harcourt, Brace, Jovanovich, 1986), 9.
11. Alan Carlson, *Family Questions: Reflections on the American Social Crisis* (New Brunswick and Oxford: Transaction Books, 1988), 7.

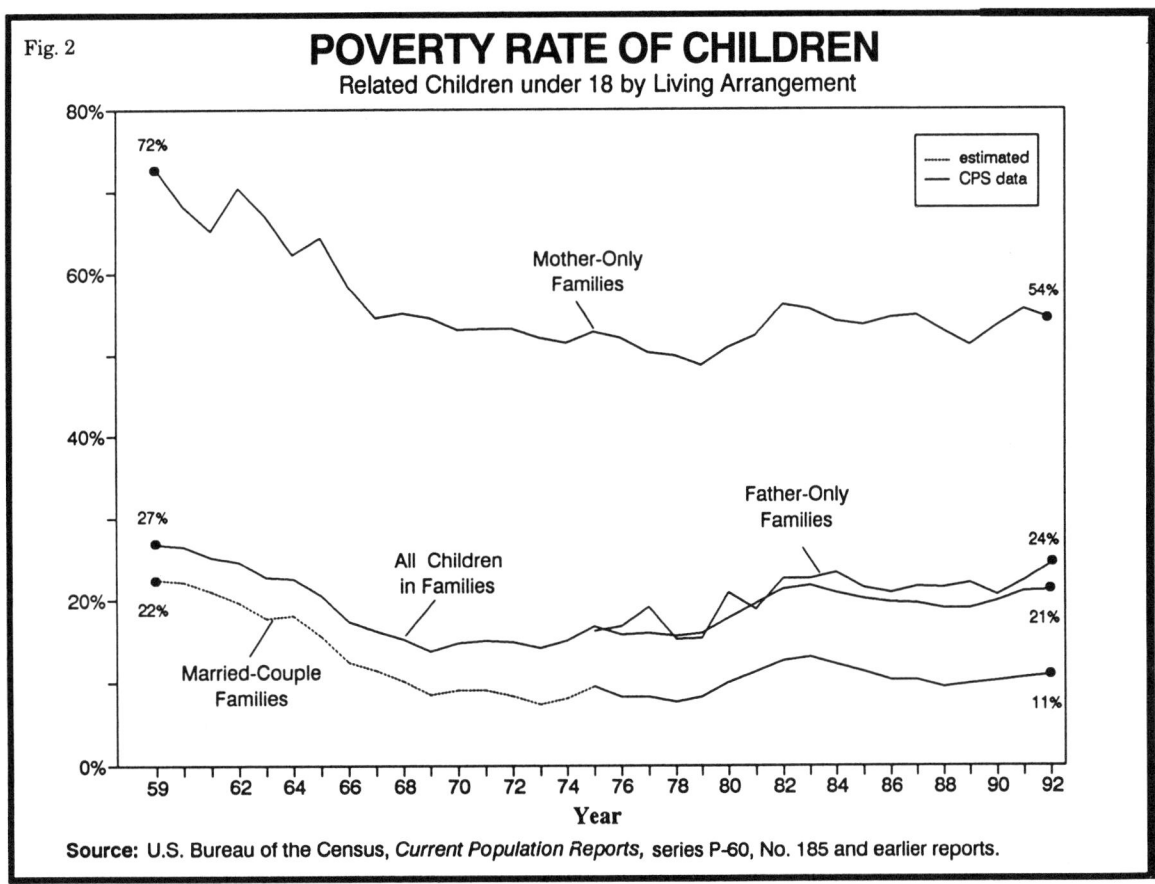

Source: U.S. Bureau of the Census, *Current Population Reports*, series P-60, No. 185 and earlier reports.

THE FAMILY AND ECONOMIC STABILITY

The family fosters stability through its role in character formation. Stability has other facets, however. Among these is economic stability.

The White House Conference on Children in 1970 came to the conclusion that "society has the ultimate responsibility for the well-being and optimum development of all children."[11] Like many ideas that at first glance pass muster, this one contains a kernel of truth: When a child has been abandoned or abused, someone must step in to care for that child. But it is essential to recognize that this is merely a remedial action; the family is the first line of defense and provision of our children. Nevertheless, it has become increasingly common to hear terminology that reflects the philosophy inherent in the White House Conference statement. Some refer to "society's children," but even more often we hear, "America's children." In many cases, this is an innocent turn of phrase; in others, it reflects a lack of recognition of the primacy of the parental responsibility for children.

But what relevance do parental rights and responsibilities have in the context of economic stability? Headlines scream, "Numbers of American Children in Poverty Rising," and some public opinion brokers draw the erroneous conclusion that economics is the major factor in producing poverty.

The truth is that not all American children have shared equally in the rising poverty rates. Children in single-parent homes are five times more likely to live in poverty than children in two-parent homes (see figure 2). Additionally, children in two-parent homes move more easily out of poverty through fluctuations in the economy; children in single-parent homes live in more persistent poverty despite improvements in economic conditions.

The strong correlation between family structure and poverty among children is consistently overlooked or downplayed

1. PARENTING AND FAMILY ISSUES

by policymakers. Of course, some children in two-parent homes live in poverty. However, with a differential of this magnitude—*five times*—the importance of the nuclear family in providing basic economic essentials for children should be unquestioned.

The inescapable difficulty of raising children alone is even more clear if we look at the poverty levels of children living with only their fathers: These children are less likely than children living with only their mothers to live in poverty, but they are *still twice as likely* to live in poverty as children in two-parent homes.

There is a striking contrast between the changes in the poverty levels of children living with only their mothers and children in two-parent homes. With improvements in the economy in the past, the poverty level of children in two-parent families has dramatically dropped; at the same time, the poverty level of children in single-parent homes has steadily increased with very few fluctuations. During the 1975 and 1982 recessions, more children became impoverished, with a subsequent drop occurring during the recovery in families with married parents.

By meeting children's and other family members' most basic economic needs so effectively and efficiently, the family functions as a stabilizing bulwark. Its absence leaves gaping holes. Just look at the last three decades in our country—welfare dependency has increased dramatically in the wake of unprecedented family breakdown. The majority of women receiving Aid to Families with Dependent Children (AFDC) are not married to the fathers of their children. Fifty-three percent of all AFDC recipients have "no marriage tie." The second most prevalent reason for AFDC payments, 38.5 percent, is divorce or separation. The group most likely to become long-term (ten years or more) AFDC recipients are women who are single mothers. Marriage remains the No. 1 escape route from the welfare rolls.

When parents shirk their responsibility to provide financially for their own children, everyone suffers. Even though the government has stepped into the breach to provide subsistence for children abandoned by their fathers, a bureaucratic safety net has proved a poor substitute for paternal provision. The fact remains, the family is society's best vehicle for the self-reliance of individuals and the care of children. Even the centrist-Democrat Progressive Policy Institute, in its *Mandate for Change* document, states: "It is no exaggeration to say that a stable, two-parent family is an American child's best protection against poverty."[12] Research clearly demonstrates this truth. The U.S. Department of Health and Human Services found that in 1985–86 "the poverty rate for married teens living with husband and children was 28% compared with 81% for unmarried teen mothers living alone with their children."[13]

THE ROLE OF THE FAMILY IN PHYSICAL AND MENTAL HEALTH

Another facet of societal stability is the general level of health of its population. In an era of high-tech medical care, health is an area where the family has a significant, but little recognized, effect. Yet research demonstrates another, slightly surprising, contribution the family makes to its members: Children in stable, two-parent homes are healthier than other children, and adults who are married are healthier than those who are not.

The National Commission on America's Urban Families released its *Families First* report in January 1993, saying that: "In sum, problems of psychological distress and poor mental health, which carry profound social as well as personal consequences, are among the most pervasive and most damaging consequences of current family fragmentation in the United States."[14]

There is increasing recognition that divorce is a prime contributor to health problems for both adults and children. However, skeptics have claimed that the ill effects of divorce on children can actually be attributed to the negative predivorce environment of the home and that

---

12. Elaine Ciulla Kamarck and William Galston, "A Progressive Family Policy for the 1990s," *Mandate for Change* (New York: Berkley Books, 1993), 157.
13. Gilbert Crouse and David Larson, "Cost of Teenage Childbearing: Current Trends," *ASPE Research Notes: Information for Decision Makers*, U.S. Department of Health and Human Services (August 1992).
14. *Families First*, Report of the National Commission on America's Urban Families (Washington, D.C.: January 1993), 29.

divorce is an improvement for those children. While it is certainly true that the nature of the marital relationship has a profound effect on children, one study cited by the Urban Commission indicates that divorce may not be the best solution for a contentious home. A 1991 study compared two sets of troubled boys, the first from divorced homes and the second from intact homes. After a five-year follow-up, the researchers found that the former group of boys had higher rates of both substance abuse and mental health problems than the latter.[15]

One of the most influential studies to be done in recent years was the 1991 report by Deborah Dawson of the National Center for Health Statistics based on the 1988 National Health Interview Survey on Child Health, which was a survey of seventeen thousand children nationwide. On the positive side, Dawson found "overall good health of the child population." However, she found that after controlling for social and demographic characteristics, "children who had experienced the separation of their natural parents . . . were more likely than other children to have had an accident, injury, or poisoning in the preceding year."[16]

But the most startling results came when Dawson turned to the emotional and psychological health of children: "Children living with formerly married mothers were more than three times as likely as those living with both biological parents to have received treatment for emotional or behavioral problems in the preceding twelve months."[17]

She found an elevated score for children living with stepparents and never-married mothers, as well. Finally, in looking at the "overall behavioral problem score," she found that children living with their biological parents once again scored better. "This pattern was repeated in the scores for antisocial behavior, anxiety or depression, headstrong behavior, hyperactivity, dependency, and peer conflict or social withdrawal," reported Dawson.

Adults as well as children are affected. Researchers at Yale and the University of California at Los Angeles have found that marriage is a significant buffer against the stresses of life for adults. Among both blacks and whites, men and women, being married correlates with a lower rate of psychiatric illness. They concluded, "The loss of a spouse through death, divorce, or separation is especially predictive of ill health."[18]

HOW THE FAMILY PROMOTES INDIVIDUAL ACHIEVEMENT

The greatest wealth a society has is its citizens. To thrive, a country needs the productivity of motivated citizens who press forward to achieve. The mystique of the "American work ethic" springs out of the innumerable individual accomplishments and innovations that have, collectively, made the United States the great nation that it is.

This, too, begins in the family. One very accomplished professional woman remembers hearing her father, Mr. Stone, say to his family on many occasions, "We Stones are hard workers!" The repetition left an indelible impression. It is this unique ability of the family to encourage the development of character in its members that makes it invaluable in preparing young people to be positive contributors to society. Values that are essential in the work force—discipline, respect for authority, perseverance—most often are forged in the family crucible.

Family also provides motivation. The historian and family expert, Alan Carlson, has said, "The family contains within its bounds the necessary positive incentives which make human beings behave in economically useful ways."[19] In echoes of Gilder, the point is that both men and women will strive harder to achieve and

---

15. William Doherty and Richard Needle, "Psychological Adjustment and Substance Use among Adolescents before and after a Parental Divorce," *Child Development* 61 (April 1991), 332–35; as cited in *Families First*.
16. Deborah Dawson, "Family Structure and Children's Health: United States, 1988," U.S. Department of Health and Human Services, Public Health Service, Centers for Disease Control, National Center for Health Statistics (June 1991), 7.
17. Dawson, "Family Structure," 3.

18. David Williams, David Takeuchi, and Russell Adair, "Marital Status and Psychiatric Disorders among Blacks and Whites," *Journal of Health and Social Behavior* 33 (1992), 140–57; as cited in *The Family in America* (Rockford, Ill.: The Rockford Institute, October 1992).
19. Carlson, *Family Questions*, xvi.

1. PARENTING AND FAMILY ISSUES

provide for those they love than they will for themselves.

And, as any teacher in any school across our nation will attest, the family is an irreplaceable foundation for education. The data underscore a family's integral role in preparing children for learning. More specifically, just as crime, health, and poverty are affected by family structure, the research leaves no doubt that children in two-parent families have a significant advantage in formal education. According to a U.S. Department of Education study of twenty-five thousand students, after controlling for socioeconomic status, race, and sex:

> [Students] from single-parent families were still more likely to fail to perform at the basic proficiency levels. They were about one-quarter to one-third more likely to perform below the basic reading and math levels and were more than two and a half times as likely to drop out of school as were students from two-parent families.[20]

(Additionally, Deborah Dawson found that children in single-parent homes were 40–75 percent more likely to repeat a grade and 70 percent more likely to be expelled from school.)

Although there are several factors contributing to the worsening state of American education, we cannot afford to turn a blind eye to the effect single parenthood is having on the readiness to learn, and the ability to learn, of millions of young children. If our country is to stay competitive, it cannot afford to have poorly educated citizens and workers. In particular, as we move into an increasingly sophisticated, highly technical economy, we will need equally sophisticated workers.

THE FAMILY'S PART IN FOSTERING LOYALTY

My brother and I as kids used to belt out the *Battle Hymn of the Republic* with childish enthusiasm . . . "Mine eyes have seen the glory of the coming of the Lord," building to a crescendo on the "Glory, Glory Hallelujahs." We loved that part, drawing out the emphasis on the high note with questionable musical effect. The song is indelibly linked in my mind with the laughter of my grandfather, a proud Marine and World War II veteran, who took great joy in teaching us patriotic songs. Now, many years after his death, the song tends to evoke for me, not laughter, but a wistfulness mixed with a deep and abiding pride in the heritage of courage and patriotism he gave us.

Like many veterans, my grandfather did not talk much about his war experiences. He did not have to. The mere fact of his service gave testimony to his devotion to his country. Semper Fi! No school civics lesson could compare with the example of patriotism set by my grandfather.

Years later, the last notes of Taps sounded over Arlington National Cemetery as the chaplain stepped over to hand my mother-in-law the folded flag. My father-in-law had died after a full life, surviving service in two wars. Beyond our circle of friends and family, up the hill as far as the eye could see, were row after row of rounded white tombstones, standing in mute testimony to the men and women who were dedicated to the defense of our country.

One last, essential need a country has of its citizens is devotion. Our Constitution says our Founding Fathers had come together to "provide for the common defense." The gravestones in Arlington Cemetery represent devotion to country. Where does that kind of sacrifice, that bravery, come from? What made so many willing to die?

Why were they willing? For love of country, surely. But that love and devotion is most often predicated upon the fact that the country hosts what we hold most dear: our families.

'LOVE THE ONE YOU'RE WITH . . .'

What kind of social construct produces a good citizen? Many view the sweeping changes occurring in families as a benign social progression. In the last few decades, more and more people

---
20. "Characteristics of At-Risk Students in NELS: 88," U.S. Department of Education, National Center for Education Statistics, 13–14; as cited in *Families First*.

3. Fount of Virtue, Spring of Wealth

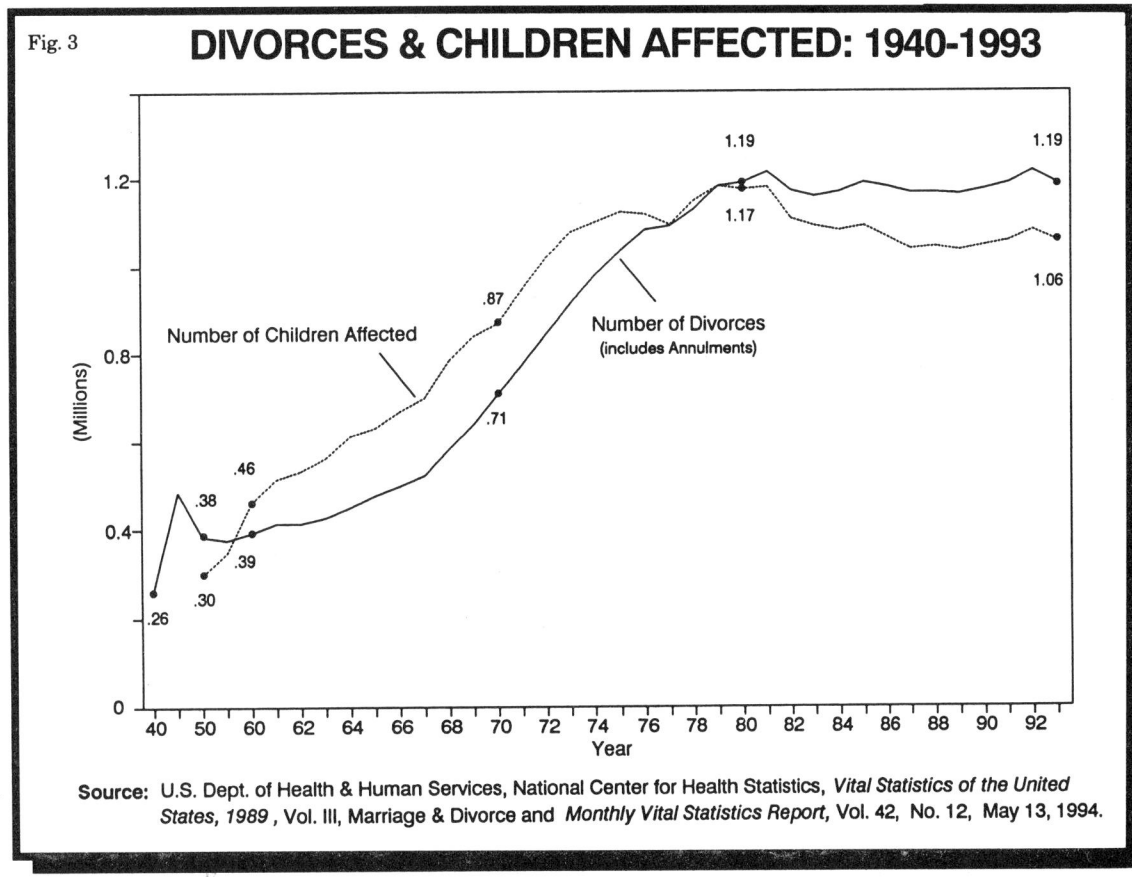

Fig. 3 **DIVORCES & CHILDREN AFFECTED: 1940-1993**

Source: U.S. Dept. of Health & Human Services, National Center for Health Statistics, *Vital Statistics of the United States, 1989*, Vol. III, Marriage & Divorce and *Monthly Vital Statistics Report,* Vol. 42, No. 12, May 13, 1994.

have joined the chorus of voices saying, "All you need is love." According to this viewpoint, family structure simply does not matter very much. Gloria Steinem, for instance, has stated that "family is content, not form." This rationalization accompanied the sixties' and seventies' changes in individual behavior—changes that gave rise to the skyrocketing rates of divorce and out-of-wedlock births. Partly causal factor and partly retrospective rationale, the cries of "family is what you make it" provide both a blazing path for those seeking individual self-actualization and an explanation for those seeking a context for their own bewildering circumstances. Unfortunately, the theory does not match up too closely with reality.

One man, now a successful professional in his midtwenties with a beautiful wife and a well-adjusted life, still vividly remembers the effect the loss of his father in a car crash had on him as a young teenager. He hears the public debate over family structure with strident voices claiming that single-parent families are "just as good" as two-parent homes and compares those claims to his own experience.

"Do they mean," he asks, "that my dad's death just didn't matter? How can they imply that growing up without my father was irrelevant?"

Another woman struggled through her parents' divorce as a teenager. As an adult, she built her own successful life—until she began contemplating marriage herself. Then, her parents' experience became a barrier, as it does for many children of divorce. Today, even with an adoring, devoted fiancé, she says she struggles with feelings of insecurity. "How do I know," she wonders, "that he won't leave me like my father left my mom?"

Even though the personal anecdotes are piling up and are supported by an ever-increasing body of social science research, many authorities in our society deny the intrinsic importance of the family. *Time* magazine, in a special issue

1. PARENTING AND FAMILY ISSUES

devoted to "Beyond the Year 2000," claimed that "the very term nuclear family gives off a musty smell." Leslie Wolfe, executive director of the Center for Women Policy Studies, said in that issue in an article entitled: "The Nuclear Family Goes Boom": "The isolated nuclear family of the 1950s was a small blip in the radar. We've been looking at it as normal, but in fact it was a fascinating anomaly."

The real surprise in Wolfe's statement is that she did not come right out and use the overworked sobriquet of "Ozzie and Harriet" as a target in her sneer at the 1950's. Typically, those attacking the nuclear family exercise no such self-restraint. In a 1989 report, Daniel Seligman of *Fortune* reported that:

> we sidled up to Nexis the other day and nonchalantly asked how many news stories in 1989 included the phrase "Ozzie and Harriet." Startling answer: 88 stories. Usual context . . . A politician was onstage reciting the news that the traditional nuclear family—the kind symbolized by the Nelsons during their marathon stint on black-and-white TV—was dead or dying.[21]

Almost exactly one hundred years after Friedrich Nietzsche declared that "God is dead," his intellectual descendants, in perhaps a natural succession of thought, are crying, "The family is dead."

Stephanie Coontz, who has written a book titled *The Way We Never Were: American Families and the Nostalgia Trap*, is one such pessimist. Perhaps her entire thesis—and that of the ideological movement she represents—can be summed up by this statement:

> Although there are many things to draw on in our past, there is no one family form that has ever protected people from poverty or social disruption, and no traditional arrangement that provides a workable model for how we might organize family relations in the modern world.[22]

Predictably, she joins the chorus attacking sitcom straw men: "1950's family strategies and values offer no solution to the discontents that underlie contemporary romanticization of the 'good old days.' Contrary to popular opinion, 'Leave it to Beaver' was not a documentary."[23]

The depth of the philosophical disagreement on the family can be seen in a further look at Senator Moynihan's statements on family. Despite his early and courageous recognition of the consequences of family breakdown, he, too, denies the existence of a specific "form" that is recognizable as *family*:

> It would be enough for a national family policy to declare that the American government sought to promote the stability and well-being of the American family; . . . and that the President, or some person designated by him, would report to the Congress on the condition of the American family in all its many facets—not of THE American family, for there is as yet no such thing, but rather of the great range of American family modes in terms of regions, national origins, and economic status.[24]

This debate over the value of the family and its place in society is nothing new. Nevertheless, at times, we retread ground that has proven sterile in the past, unwisely failing to learn from history.

Sociologists tell us that an almost inescapable component of the growth of modernity is a corollary rise in individualism. This has certainly been true in our own society. And individualism has had a marked effect on family life. Problems arise when "the sacredness of the family" is replaced by "the sacredness of the individual."[25] Harvard sociologist Carle Zimmerman has identified cycles of two or three generations where individualism has replaced an emphasis on family,[26] including eight major periods in Western history in which the family was viewed as old-fashioned. In each of these times, soci-

---

21. Daniel Seligman, *Fortune*, July 17, 1989; as cited in "Catching the Reruns: Ozzie, Harriet, and the Media," *Family Affairs* 2 (New York: Institute for American Values, Summer/Fall 1989), 13.
22. Stephanie Coontz, *The Way We Never Were* (New York: Basic Books, 1992), 5.

23. Coontz, *The Way We Never Were*, 29.
24. Moynihan, *Family and Nation*, 11.
25. Don McNally, "The Family in History," *Vanguard* (September/October 1980), 13.
26. McNally, "The Family in History," 13–14; citing Carle Zimmerman, *The Family of Tomorrow* (1949) and Christopher Dawson, *The Dynamics of World History*.

ety began valuing the individual over the concerns of the family as a unit. The results in each instance led to societal decay.

Examples of the failure of individualism include the Greek and Roman societies in the third century B.C. and the fourth century A.D., respectively. Greek society in the third century B.C. engendered the collapse of the patriarchal family, leading to an emphasis on the individual and devotion to public life that resulted in late marriages and small families. The final cultural breakdown led to Roman conquest. The Roman Empire in the fourth century A.D. made the same mistake, resulting in its fall.[27]

GOVERNMENT'S INFLUENCE ON THE FAMILY

As we approach the twenty-first century, the family and the government are in an uneasy alliance in America. Forward movement, however, has become a wobble rather than a smooth progression as the balance between family and government has been shaken askew. A battle rages—sometimes quietly, sometimes with guns blazing—as those who would increase the power of government jockey for position with those who resist in favor of family autonomy. David Blankenhorn, in looking at these combatants, gives this analysis:

> There is a particularly sterile argument about the role of government. The traditional argument between left and right has been over the size of government. The conservatives want less, the liberals want more, and that's the perennial argument. The debate regarding families is not really, should government be smaller or larger. What matters is the relationship of public policy to family well-being. What is the distribution of costs and benefits to families and what is the message of public policy about what we value and what we devalue about the importance of family in this society.[28]

The central question remains: Do we value families? While their role in strengthening society should be undisputed, it is not. The family is indispensable in the formation and continuity of a strong and stable society. Even in the face of those who view the family as irrelevant, the family stands on its own, with its own intrinsic strength. So much so, in fact, that some recognize the superiority of family over government in accomplishing societal goals.

The real battle is with those who see the family as competition and threat. Witness Judith Stacey, author of *Brave New Families*: "The 'family' is not here to stay. Nor should we wish it were. On the contrary, I believe that all democratic people, whatever their kinship preferences, should work to hasten its demise."[29]

Although Stacey's sentiments are based on the overarching philosophy of individualism, her conclusions fall in line with those who have, throughout history, seen the interests of family as diametrically opposed to those of society—and come down on the side of an omniscient, paternalistic government. Plato, Marx, Hitler, Stalin—all in their own way recognized the strength of the family. But rather than viewing that strength as a basis upon which to build a thriving, vital society, they sought to control and dampen its vibrancy.

History has proven the bankruptcy of opposition to the family. The way forward for individuals and for the nation is in rebuilding and reemphasizing the family. The society that does so will, in the doing, become stronger.

---

27. McNally, "The Family in History," 13–14; citing Carle Zimmerman, *The Family of Tomorrow* (1949) and Christopher Dawson, *The Dynamics of World History*.

28. David Blankenhorn, "The Relationship of Public Policy to Family Well-Being," *American Family* (August 1988), 3.
29. As cited by David Popenoe, "The Controversial Truth: Two-Parent Families Are Better," *New York Times*, 16 Dec. 1992, 21.

Things that go bump in the home

JOHN LEO

A Page 1 headline in the *Los Angeles Times* announces "A New Side to Domestic Violence." This "new side," apparently quite puzzling to the reporter, is that under mandatory arrest laws, a large number of women are now being arrested after domestic battles. In Los Angeles, arrests of women in such cases have almost quadrupled in eight years. In Wisconsin, the number of abusive men referred by the courts for counseling has doubled since 1989, while the number of abusive women referred for counseling increased 12-fold.

You could sense that the reporter was grappling with a baffling question: How is it that laws intended to protect women are producing so many arrests of women themselves? Luckily, he was able to come up with three explanations: a backlash against women, spiteful action by police officers who resent mandatory arrest laws, and outright male trickery.

Under the heading of trickery came tales of a man who smashed a brick on his head and blamed his wife, a man who bloodied himself by scratching his ear and a man, born with on odd bump on his head, who repeatedly showed the bump to police and got his wife arrested three times.

There is another explanation, one that has nothing to do with lucky head bumps or rogue cops. The explanation is this: If mandatory arrest laws are fairly applied, we will eventually see roughly equal numbers of men and women arrested, because the amount of domestic violence initiated and conducted by men and women is roughly equal. In fact, women may well be ahead.

Newsroom taboo. No, you haven't read this in your local newspaper, and certainly not in elite papers like the *Los Angeles Times*. The obvious reason is that publishing this news would create a severe political problem in the newsroom. To their credit, feminists made domestic violence a political issue. But they shaped this issue around a theory: This violence is an expression of patriarchy as a social force and marriage as a patriarchal institution. It is something men do to women because of the way society is organized.

An enormous amount of evidence now shows this paradigm is wrongheaded. But feminists are unwilling to adapt it to reality, and since the modern newsroom is supportive of feminism, news stories on domestic violence are carefully crafted, consistently unreliable and often just wrong.

Follow the work of the National Family Violence Survey. The original 1975 survey showed rather high rates of female-on-male domestic violence, but these were fitted to the paradigm and explained as understandable reactions to male violence. But the second survey in 1985 clearly showed equality in turning to violence: In both low-level assaults and severe assaults, only the wife was violent in a quarter of the cases, only the husband in another quarter, both in half of the cases. These findings came from self-reports.

This signaled a split in research: Feminist researchers keep churning out work that fits feminist theory, while independent researchers keep finding equality in the use of violence. Men are more dangerous—they kill partners twice as often—and more likely to inflict serious damage. But women are just as inclined to be violent with their partners as men are. (The rather high rates of violence in lesbian homes echoes this finding.) The equality findings undercut the feminist theory of partner violence as patriarchy in action, with its dark view of men and marriage. Instead, they support the common-sense view that violence between partners has more to do with problems of individuals in a difficult culture rather than with any vast ideological scheme.

The feminist insistence on using theory to mug facts has unfortunate results. One is that a generalized view of men as uniquely violent and dangerous to women ("men batter because they can," "the most dangerous place for a woman to be is in the home") has leached deep into popular culture. In a recent TV ad for girls' athletics, a young girl says if she plays sports, she will be more likely to leave an abusive relationship. A recent national list of what children want actually included the wish that daddies would stop hitting mommies.

In fact, children are now more likely to see mommy hit daddy. The rate of severe assaults by men on women in the home fell by almost 50 percent between the first National Family Violence Survey (1975) and the most recent update of data in 1992. It dropped from 39 percent per 1,000 couples per year to 20. Give the feminists credit for this. They did it mostly by themselves. But the rate of dangerous female assaults on males in the home stayed essentially static over that period—45 per 1,000 couples—and is now twice as high as the male rate. Give feminists responsibility for this, too. By defining partner violence as a male problem they missed the chance to bring about the same decline in violence among women.

Feminist studies of partner violence rarely ask about assaults by women, and when they do, they ask only about self-defense. Journalists, in turn, stick quite close to the feminist-approved studies for fear of being considered "soft" on male violence. The result is badly skewed reporting of domestic violence as purely a gender issue. It isn't.

Growing Up against the Odds

Despite billions of federal dollars, America's children face an uncertain and bleak future.

ROBERT ROYAL

Robert Royal is vice president of the Ethics and Public Policy Center, Washington, D.C.

Childhood used to be a time of relative security and innocence in America. Most children passed their lives in stable families that, despite their economic ups and downs, provided their young members with the kind of affection and discipline needed to become mature and responsible adults.

Neighbors, churches, and schools reinforced family lessons. Sex, drugs, and violence were only a distant threat. And children were given a chance to develop physically, emotionally, intellectually, and spiritually before entering the hard, adult world that all of us eventually must face.

It certainly is no longer the case for far too many young people, whose future is America's future. The kind of childhood that many people now living can still remember did not just happen naturally. It was a civilizational achievement.

Prior to relatively modern times, children were basically regarded as young adults. Though they were not often exposed to the kinds of social pathologies children commonly encounter today, they were set to work as soon as they were able and received little formal schooling except at the higher social levels. For mostly agrarian, premodern societies, this system worked well enough. It promoted stability and social integration for the young, but it also entailed almost complete social immobility.

Modern societies have provided unprecedented social and economic opportunities for ever-wider groups of people. But they imposed some new requirements on families and children. Childhood in early bourgeois democracies became easier in many ways. Children went to school instead of to the fields. Childhood development—with its suddenly proliferating special dimensions—took on new importance. Early life experiences became far more crucial to the new open societies since they could no longer rely on automatic training of children for work or their moral formation as responsible citizens. Yet the risks were amply rewarded.

What several historians have called the "invention of childhood"—the cultural creation of a period specially set aside from practical worries—meant a liberation for millions of children in the middle and lower classes to grow into the productive and prosperous adults their talents and ambitions suited them to be.

Some scholars have tried to dispute the sometimes overidyllic picture that has been painted of Victorian and early twentieth-century family life. To a certain extent, they are right to do so. No doubt, families during those times suffered from the full range of human sins and weaknesses. But as families operating under modern conditions, and as special enclaves for children being prepared for full participation in adult life—with its high demands for good work habits, vocational skills, and general knowledge—they were generally far healthier and effective socializing institutions than what we often are forced to call families today.

It used to be fashionable among radicals to criticize families for their narrow bourgeois values and patriarchal oppression. Now that we have had the full experience of what the absence of the limited but necessary bourgeois values and disciplines can mean for children's lives in modern societies, bourgeois respectability does not seem so bad, particularly given the frightening alternatives that have emerged.

Typically, what we now have are children growing up not only outside bourgeois exhortations to productivity, regular work habits, and religiosity, but children forced to grow up before they have had any real formation at all.

There is a difference between giving a young child the responsibilities of tending sheep (rigorous, but no moral danger) and exposing a young child to the now common dangers of drugs, sex, and violence (for many, the premature death of childhood). Premodern children, even if they were treated harshly by modern standards, still existed within the horizon of the adult world of responsibility and realism. But postmodern children often confront adult problems without benefit of adult supervision or sufficient emotional and moral maturity to make good decisions on their own.

Decline, by the numbers

What has all this meant? Many different things. But there are some concrete measures of what has happened to children in recent decades that show us to be, if not in a full social crisis, at least in the grips of something quite alarming.

In a now-famous story, *U.S. News & World Report* a few years ago went back to teacher surveys in the 1940s and discovered that the most serious public school disciplinary problems then in their order of importance were: talking out of turn, gum chewing, making noise, running in the halls,

Major Problems

* By the year 2000, 40 percent of all American births are expected to occur out of wedlock.

* Since 1960, the rate at which teenagers take their own lives has more than tripled.

* About 20 percent of teenage girls will have at least one abortion by age 20.

* Of all age-groups, children are the most likely to be poor.

cutting into lines, improper dress, and littering.

In 1990, a similar teacher survey revealed the most serious problems to be: drugs, alcohol, pregnancy, suicide, rape, robbery, and assault—this during one of the greatest expansions of expenditures on education at all levels in the history of America and perhaps the world. Whatever had happened for the good in those 50 years, some unprecedented evil had happened to the environment of schoolchildren.

That societal shift had effects outside of school as well. Take suicide. For most of us, childhood and adolescence were times of happiness and exuberance, but also inevitably times of stress, of finding out who we are and where we fit into the world. These stresses have led to suicide attempts at any age, either because of personal or social problems or medical conditions.

The medical conditions, we may assume, are less of a problem than ever before. Yet teenage suicide rates have gone from 3.6 per 100,000 in 1960 to 11.3 in 1990. Young males are far more likely—five times more likely—to commit suicide than young females.

But young males do not only direct lethal violence toward themselves. Rates of arrests of juveniles for violent crimes (mostly males) have closely paralleled the suicide rate, more than tripling between 1965 and 1990. Murders by young blacks doubled over the same period.

A factor that may explain a good portion of these phenomena is that 70 percent of juveniles in long-term correctional facilities grew up in homes without fathers. Compared with the murder, mayhem, and disorder of "alternative" families, the rigors of the old "patriarchal" family were mere kid stuff.

But absent fathers are far from being the only alarming parental problem. Somewhere between six and seven million schoolchildren come home every day to a house with no adult present at all—many in far from ideal, inner-city neighborhoods. Various problems naturally stem from this lack of supervision. One of the worst is that many teenage pregnancies, according to good survey data, get started in those after-school hours.

Like the other juvenile pathologies, rates of teen pregnancy and abortion have been soaring. In 1972, already a high year relative to previous decades, almost 5 out of every 100 teenage girls got pregnant in America. By 1990, the rate had zoomed to almost 10 out of 100. Put in con-

1. PARENTING AND FAMILY ISSUES

crete figures, probably more than a million unmarried American teenagers get pregnant every year, and of those about 40 percent have abortions.

When this yearly toll is added up and the consequences tallied for the whole of American society at any given moment, it is easy to see why illegitimacy, school dropout rates, young women in poverty as heads of households, violent crimes, and drug use by undisciplined youths—all mutually reinforcing and self-perpetuating pathologies—have become the most nationally pressing social problems.

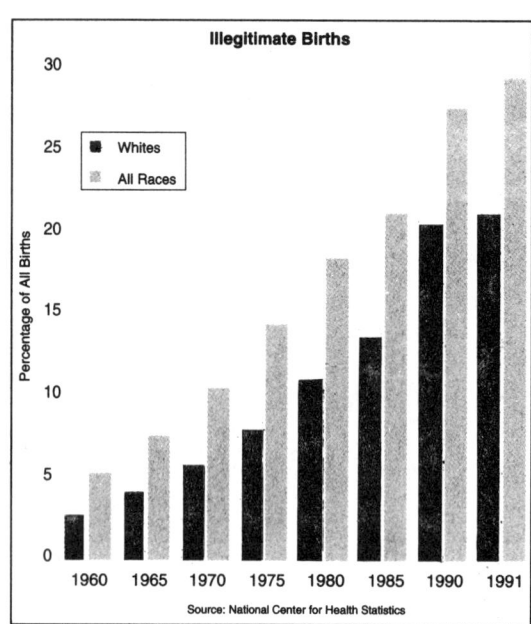

Causes and effects

The causes of all this are complex—but not so complex that we cannot discern and try to do something about them. Family breakdown is probably the area most Americans now agree poses the greatest threat to public life.

With a divorce rate of nearly 50 percent and illegitimacy rates of about 22 percent for whites and 68 for blacks (for a national average of around 30 percent), most children now come into the world with a handicap unprecedented in any culture on earth at any period of human history: no stable family to speak of. All the attempts to redefine and expand the notion of family in order to avoid stigmatizing people cannot get around this plain fact.

Radical feminists and advocates of alternative life-styles have tried, for example, to make it appear that the rise of the euphemistically designated "female-headed household" is a perfectly viable alternative to old patriarchal, male-headed, two-parent families. This is, at best, dangerously misleading.

Certainly many single women are heroic and successful parents and should be commended for their achievements. But they should not be recommended as models for others, not yet parents, to imitate. Nor can the risky business of single parenting be simply equated with two-parent or extended families. The documented results just do not support the contention.

Every pathological tendency in the culture has been visited on our children at earlier and earlier ages in recent decades. And we can see from the data that all social classes have been affected by it to a greater or lesser extent. The general breakdown in public morals, for example, has led to cheating in school at massively higher levels than ever before.

Paradoxically, cheating seems to be more widespread among middle- and upper-class children than among their poorer counterparts. Aside from that exception, the poorer and more vulnerable sectors of the society have been hit the hardest by social pathologies—as is always the case. Wealthier people living in more stable communities just have more layers of correction and insulation built into the social fabric.

By the time we get down to an inner-city woman who has to bring up one or more children—sometimes adolescent boys who need the presence of a father—in a rough neighborhood and add that she probably has to work long hours to support the family, all elements are present for a variety of mishaps.

This is not meant to be criticism of poor black women or women on welfare. It is a mere description of the challenges and potential failures that anyone, white or black, Hispanic or Asian, faces when the supplementary systems of families have disappeared. Statistical surveys have shown that children from all races and ethnic groups placed in the same circumstances of family breakdown and poverty tend to exhibit the same problems.

Riding the Third Wave

This breakdown could not have come at a worse time in American history. When America was still a predominantly agricultural or industrial nation, there were many jobs for men and women with limited or minimal skills. Lack of formal training in school or even an ability to read were not serious disqualifications for a host of honorable ways to earn a living.

Today, the mainstream of the economy requires literacy and specialized skills. One sign of where we are headed: A report in April 1995 documented that only about one-third of high school graduates are proficient in reading. The next third are below proper levels, and the final third are functionally illiterate.

Without some effort to redress this problem through remedial programs or other means, two-thirds of the teenagers graduating from high school every year in America will be unsuited for the primary work force.

Alvin and Heidi Toffler have recently described the new information society in which we are now living as the Third Wave, replacing the Second Wave of the industrial society, which in turn had replaced the First Wave of agricultural societies. They and their most famous admirer, House Speaker Newt Gingrich, are optimistic that the Third Wave will provide new opportunities that will naturally lead the young into a new age.

The Tofflers also regard the old nuclear family as appropriate for Second Wave industrialization, but—far too blithely—as only one of many viable family forms for the Third Wave. They may be right that other kinds of "families" will be common in the brave new world aborning. All the evidence we have so far, however, indicates that the welfare state cannot substitute for the family.

"Alternative" families, even at best, are equally unlikely to do well by children except for some rare and very lucky individuals. No amount of Third Wave opportunities alone, especially absent nurturing families across the social spectrum, will offset the continuing damage being done to children at present.

Some social critics, such as the writer Mickey Kaus in his provocative book *The End of Equality,* have argued more plausibly that, without some major changes, we may be on our way to a two-tiered society:

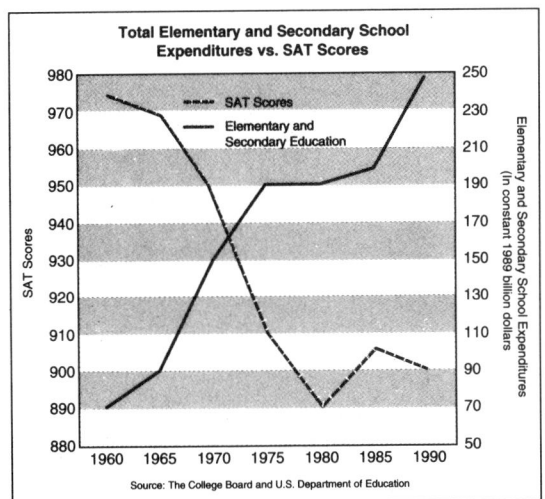

those from relatively intact families who are literate and participate in the social and economic mainstream, and a large mass of virtually illiterate proletarians who will be forced into low-paying, menial jobs largely without benefits like health-care insurance and retirement plans.

An America where two-thirds of the work force is a distant second class economically is not the America the Founding Fathers or our contemporary public culture would think tolerable. Nor is it likely to remain stable for long.

Restoring childhood and adulthood

Many of the people who fail in modern societies as adults do so because they never had the time and place to be children. Our kind of society makes high

> **What we now have are children growing up not only outside regular work habits, and religiosity, but children forced to grow up before they have had any real formation at all.**

demands and provides high rewards for adults.

But adulthood, except for retraining as society and the economy change, is usually not an optimum time for learning the things we usually learn better as children. Paradoxically, in the time of protection and innocence, we learn responsibility and common sense.

The habits of our first two decades have to carry us for the rest of a lifetime. Unless we find a way to make childhood possible again—either through a moral regeneration that restores the family or (far less likely) through alternative institutions—not only will we be destroying our children's present and future, but we may be concocting a dismal prospect for ourselves as well.

Why Leave Children With Bad Parents?

Family: Last year, 1,300 abused kids died—though authorities knew that almost half were in danger. Is it time to stop patching up dead-end families?

MICHELE INGRASSIA AND JOHN MCCORMICK

THE REPORT OF DRUG PEDdling was already stale, but the four Chicago police officers decided to follow up anyway. As they knocked on the door at 219 North Keystone Avenue near midnight on Feb. 1, it was snowing, and they held out little hope of finding the pusher they were after. They didn't. What they discovered, instead, were 19 children living in horrifying squalor. Overnight, the Dickensian images of life inside the apartment filled front pages and clogged network airwaves.

For the cops that night, it seemed like a scavenger hunt gone mad, each discovery yielding a new, more stunning, find. In the dining room, police said, a half-dozen children lay asleep on a bed, their tiny bodies intertwined like kittens. On the floor beside them, two toddlers tussled with a mutt over a bone they had grabbed from the dog's dish. In the living room, four others huddled on a hardwood floor, crowded beneath a single blanket. "We've got eight or nine kids here," Officer John Labiak announced. Officer Patricia Warner corrected him: "I count 12." The cops found the last of 19 asleep under a mound of dirty clothes; one 4-year-old, gnarled by cerebral palsy, bore welts and bruises.

As the police awaited reinforcements, they could take full measure of the filth that engulfed this brigade of 1- to 14-year-olds. Above, ceiling plaster crumbled. Beneath their feet, roaches scurried around clumps of rat droppings. But nothing was more emblematic than the kitchen. The stove was inoperable, its oven door yawning wide. The sink held fetid dishes that one cop said "were not from that day, not from that week, maybe not from this year." And though the six mothers living there collected a total of $4,500 a month in welfare and food stamps, there was barely any food in the house. Twice last year, a caseworker from the Illinois Department of Children and Family Services (DCFS) had come to the apartment to follow up reports of serious child neglect, but when no one would let her in, the worker left. Now, it took hours to sort through the mess. Finally, the police scooped up the children and set out for a state-run shelter. As they left, one little girl looked up at Warner and pleaded, "Will you be my mommy?"

Don't bet on it. Next month the children's mothers—Diane Melton, 31; Maxine Melton, 27; May Fay Melton, 25; Denise Melton, 24; Casandra Melton, 21, and Denise Turner, 20—will appear in Cook County juvenile court for a hearing to determine if temporary custody of the children should remain with the state or be returned to the parents. Yet, for all the public furor, confidential files show that the DCFS is privately viewing the 19 children in the same way it does most others—"Goal: Return Home."

Why won't we take kids from bad parents? For more than a decade, the idea that parents should lose neglected or abused kids has been blindsided by a national policy to keep families together at almost any cost. As a result, even in the worst cases, states regularly opt for reunification. Even in last year's budget-cutting frenzy, Congress earmarked nearly $1 billion for family-preservation programs over the next five years. Yet there is mounting evidence that such efforts make little difference—and may make things worse. "We've oversold the fact that all families can be saved," says Marcia Robinson Lowry, head of the Children's Rights Project of the American Civil Liberties Union. "All families *can't* be saved."

Last year there were 1 million confirmed

6. Why Leave Children with Bad Parents?

cases of abuse and neglect. And, according to the American Public Welfare Association, an estimated 462,000 children were in substitute care, nearly twice as many as a decade ago. The majority of families can be repaired if parents clean up their acts, but experts are troubled by what happens when they don't: 42 percent of the 1,300 kids who died as a result of abuse last year had previously been reported to child-protection agencies. "The child-welfare system stands over the bodies, shows you pictures of the caskets and still does things to keep kids at risk," says Richard Gelles, director of the University of Rhode Island's Family Violence Research Program.

Nowhere has the debate over when to break up families been more sharply focused than in Illinois, which, in the last two years, has had some of the most horrific cases in the nation. Of course, it's not alone. But unlike many states, Illinois hasn't been able to hide its failures behind the cloak of confidentiality laws, largely because of Patrick Murphy, Cook County's outspoken public guardian, who regularly butts heads with the state over its aggressive reunification plans. The cases have turned Illinois into a sounding board for what to do about troubled families.

The Chicago 19 lived in what most people would consider a troubled home. But to veterans of the city's juvenile courts, it's just another "dirty house" case. In fact, Martin Shapiro, the court-appointed attorney for Diane Melton, plans to say that conditions could have been worse. He can argue that Melton's children weren't malnourished, weren't physically or sexually abused and weren't left without adult supervision. He's blunt: "Returning children to a parent who used cocaine—as horrific as that might seem—isn't all that unusual in this building." If only all the cases were so benign.

What Went Wrong?

ON THE LAST NIGHT OF JOSEPH Wallace's life, no one could calm his mother's demons. Police say that Amanda Wallace was visiting relatives on April 18, 1993, with 3-year-old Joseph and his 1-year-old brother, Joshua, when she began raving that Joseph was nothing but trouble. "I'm gonna kill this bitch with a knife tonight," Bonnie Wallace later told police her daughter threatened. Bonnie offered to keep the boy overnight, but Amanda refused, so Bonnie drove them to their apartment on Chicago's impoverished West Side. It's unclear what forced Amanda's hand, but authorities tell a harrowing tale: at about 1:30 a.m., she stuffed a sock into Joseph's mouth and secured it with medical tape. Then she went to the kitchen, retrieved a brown extension cord and wrapped it around Joseph's neck several times. She carried her son to the living room, stood him on a chair, then looped the cord around the metal crank arm over the door. In the last act of his life, Joseph waved goodbye.

Amanda Wallace, 28, has pleaded not guilty to charges of first-degree murder. No one ever doubted that Amanda was deeply troubled. When Joseph was born, she was a resident at the Elgin Mental Health Center in suburban Chicago, and a psychiatrist there warned that Amanda "should never have custody of this or any other baby." Three times, the DCFS removed Joseph from his mother. Yet three times, judges returned him to Amanda's dark world. Six months after the murder—which led to the firing of three DCFS employees—a blue-ribbon report blasted the Illinois child-welfare system, concluding that it had "surely consigned Joseph to his death."

Even in the most egregious instances of abuse, children go back to their parents time and again. In Cook County, the public guardian now represents 31,000 children. Only 963 kids were freed for adoption last year. But William Maddux, the new supervising judge of the county's abuse and neglect section, believes the number should have been as high as 6,000. Nationwide, experts say, perhaps a quarter of the children in substitute care should be taken permanently from their parents.

But it's not simply social custom that keeps families together, it's the law. The Adoption Assistance and Child Welfare Act of 1980 is a federal law with a simple goal—to keep families intact. The leverage: parents who don't make a "reasonable effort" to get their lives on track within 18 months risk losing their kids forever. The law itself was a reaction to the excesses of the '60s and '70s, when children were often taken away simply because their parents were poor or black. But the act was also one of those rare measures that conservatives and liberals embraced with equal passion—conservatives because it was cheap, liberals because it took blame away from the poor.

By the mid-'80s, though, the system began to collapse. A system built for a simpler time couldn't handle an exploding underclass populated by crack addicts, the homeless and the chronically unemployed. At the same time, orphanages began shutting their doors and foster families began quitting in droves. The system begged to know where to put so many kids. It opted for what was then a radical solution: keeping them in their own homes while offering their parents intensive, short-term support—child rearing, housekeeping and budgeting. But as family-preservation programs took off, the threat of severing the rights of abusive parents all but disappeared. What emerged, Gelles argues, was the naive philosophy that a mother who'd hurt her child is not much different from one who can't keep house—and that with enough supervision, both can be turned into good parents.

In hindsight, everyone in Chicago agrees that Joseph Wallace's death was preventable, that he died because the system placed a parent's rights above a child's. Amanda could never have been a "normal" parent. She had been a ward of the state since the age of 8, the victim of physical and sexual abuse. Between 1976 and Joseph's birth in 1989, her psychiatrist told the DCFS, she swallowed broken glass and batteries; she disemboweled herself, and when she was pregnant with Joseph, she repeatedly stuck soda bottles into her vagina, denying the baby was hers. Yet 11 months after Joseph was born, a DCFS caseworker and an assistant public defender persuaded a Cook County juvenile-court judge to give him back to Amanda, returning him from the one of the six foster homes he would live in. The judge dispatched Amanda with a blessing: "Good luck to you, Mother."

Over the next two years, caseworkers twice removed Joseph after Amanda attempted suicide. But a DCFS report, dated Oct. 31, 1992, said she had gotten an apartment in Chicago, entered counseling and worked as a volunteer for a community organization. And though the report noted her turbulent history, it recommended she and Joseph be reunited. Joseph Wallace was sent home for the last time 62 days before his death, by a judge who had no measure of Amanda's past. "Would somebody simply summarize what this case is about for me and give me an idea why you're all agreeing?" the judge asked. Amanda's lawyer sidestepped her mental history. Nevertheless, the DCFS and the public guardian's office signed on. When Amanda thanked the judge, he said, "It sounds like you're doing OK. Good luck."

Murphy says that deciding when to sever parents' rights should be obvious: "You remove kids if they're in a dangerous situation. No one should be taken from a cold

---

JOSEPH WALLACE

"Mother has a history of impulsive behavior, inappropriate anger, difficulty in getting along with others and recurrent suicidal attempts," the caseworker wrote. Nonetheless, her written recommendation was: "Goal, return home." Police have since charged that Amanda, Joseph's mother, hanged her son with an electrical cord.

FROM CASEWORKER REPORT ON JOSEPH WALLACE

COURTESY FAYE AND MICHAEL CALLAHAN

1. PARENTING AND FAMILY ISSUES

house. But it's another thing when there are drugs to the ceiling and someone's screwing the kids." Ambiguous cases? "There haven't been gray cases in years."

No one knows that better than Faye and Michael Callahan, one of the foster families who cared for Joseph. When Joseph first came to them he was a happy, husky baby. When he returned after his first stretch with Amanda, "he had bald spots because he was pulling his hair out," Faye says. By the third time, she says, Joseph was "a zombie. He rocked for hours, groaning, 'Uh, uh, uh, uh'." The fact that he was repeatedly sent home still infuriates them. Says Michael: "I'd scream at those caseworkers, 'You're making a martyr of this little boy!'"

See No Evil, Hear No Evil

EARLY LAST THANKSGIVING, ARETHA McKinney brought her young son to the emergency room. Clifford Triplett was semiconscious, and his body was pocked with burns, bruises and other signs of abuse, police say. The severely malnourished boy weighed 17 pounds—15 percent less than the average 1-year-old. Except Clifford was 5.

This wasn't a secret. In a confidential DCFS file obtained by NEWSWEEK, a state caseworker who visited the family last June gave a graphic account of Clifford's life: "Child's room (porch) clothing piled in corner, slanted floor. Child appears isolated from family—every one else has a well furnished room. Child very small for age appears to be 2 years old. Many old scars on back and buttocks have many recent scratches." In April, another caseworker had confronted McKinney's live-in boyfriend, Eddie Robinson Sr., who claimed that Cliff was a "dwarf" and was suicidal—neither of which doctors later found to be true. Robinson added that Cliff got "whipped" because he got into mischief. "I told him that he shouldn't be beat on his back," the caseworker wrote. "Robinson promised to go easy on the discipline."

It's one thing to blame an anonymous "system" for ignoring abuse and neglect. But the real question is a human one: how can caseworkers walk into homes like Clifford's, document physical injury or psychological harm and still walk away? A Cook County juvenile-court judge ruled last month that both McKinney and Robinson had tortured Clifford (all but erasing the possibility that he'll ever be returned to his mother). But caseworkers are rarely so bold. In Clifford's case, the April worker concluded that abuse apparently had occurred, but nine days later another found the home "satisfactory." Says Gelles: "Caseworkers are programmed by everything around them to be deaf, dumb and blind because the system tells them, 'Your job is to work to reunification'."

Murphy charges that for the past two

SAONNIA BOLDEN

"The amount of stress and frustration has been reduced. Sadie appears to have a lot more patience with her children and she continues to improve her disciplinary techniques." The same day the worker wrote this, Sadie's daughter Saonnia died after boiling water was poured on her. An autopsy uncovered 62 injuries, many recent.

FROM CASEWORKER REPORT ON SAONNIA BOLDEN

years, Illinois has made it policy to keep new kids out of an already-clogged system. "The message went out that you don't aggressively investigate," he says. "Nobody said, 'Keep the ----ing cases out of the system'." But that, he says, is the net effect. "That's just not true," says Sterling Mac Ryder, who took over the DCFS late in 1992. But he doesn't dispute that the state and its caseworkers may have put too much emphasis on reunification—in part because of strong messages from Washington.

The problems may be even more basic. By all accounts, caseworkers and supervisors are less prepared today than they were 20 years ago, and only a fraction are actually social workers. Few on the front lines are willing, or able, to make tough calls or buck the party line. In the end, says Deborah Daro, research director of the National Committee to Prevent Child Abuse, "the worker may say, 'Yeah, it's bad, but what's the alternative? I'll let this one go and pray to God they don't kill him'."

In most cases, they don't. Nevertheless, children who grow up in violent homes beyond the age of 8 or 10 risk becoming so emotionally and psychologically damaged that they can never be repaired. "The danger," says Robert Halpern, a professor of child development at the Erikson Institute in Chicago, "is not just the enormous dam-

age to the kid himself, but producing the next generation of monsters."

Clifford Triplett is an all-too-pointed reminder of how severe the injuries can be. He has gained eight pounds, and his physical prognosis is good. But there are many other concerns. "When he came, he didn't know the difference between a car and a truck, the difference between pizza and a hot dog," says his hospital social worker, Kathleen Egan. "People were not introducing these things to him." Robinson and McKinney are awaiting trial on charges of aggravated battery and felony cruelty. McKinney's attorney blames Robinson for the alleged abuse; Robinson's attorney declined to comment. Clifford is waiting for a foster home. A few weeks ago he had his first conversation with his mother in months. His first words: "Are you sorry for whipping me?"

Band-Aids Don't Work

ACCORDING TO THE CASEWORKER'S report, 2½-year-old Saonnia Bolden's family was the model of success. Over 100 days, a homemaker from an Illinois family-preservation program called Family First worked with Sadie Williams and her boyfriend Clifford Baker. A second helper—a caseworker—shopped with Sadie for shoes and some furniture for her apartment; she evaluated Sadie's cooking, housekeeping and budgeting. She even took her to dinner to celebrate her progress. On March 17, 1992, the caseworker wrote a report recommending that Sadie's case be closed: "Due to the presence of homemaker, the amount of stress and frustration has been reduced. Sadie appears to have a lot more patience with her children and she continues to improve her disciplinary techniques."

What the Family First caseworker evidently didn't know was that, just hours before she filed her report, Saonnia had been beaten and scalded to death. Prosecutors claim that Williams, angered because her young daughter had wet herself, laid the child in the bathtub and poured scalding water over her genitals and her buttocks. Williams and Baker were charged with first-degree murder; lawyers for Baker and Williams blame each other's client. Regardless of who was responsible, this wasn't

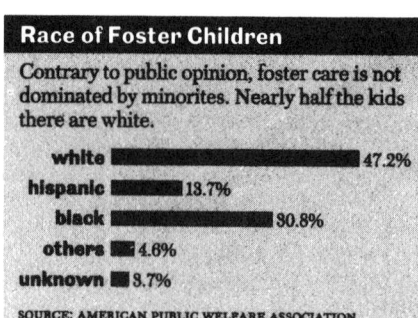

Race of Foster Children

Contrary to public opinion, foster care is not dominated by minorities. Nearly half the kids there are white.

- white: 47.2%
- hispanic: 13.7%
- black: 30.8%
- others: 4.6%
- unknown: 3.7%

SOURCE: AMERICAN PUBLIC WELFARE ASSOCIATION

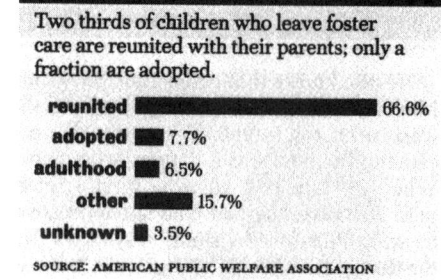

Where Do Children Go?

Two thirds of children who leave foster care are reunited with their parents; only a fraction are adopted.

- reunited: 66.6%
- adopted: 7.7%
- adulthood: 6.5%
- other: 15.7%
- unknown: 3.5%

SOURCE: AMERICAN PUBLIC WELFARE ASSOCIATION

6. Why Leave Children with Bad Parents?

A One-Man Children's Crusade

Twenty years ago, an angry young lawyer named Patrick Murphy wrote a book that exposed an injustice: state social workers too often seized children from parents whose worst crime was poverty. Today Murphy is the scourge of a child-welfare system that too often leaves kids with their abusive, drugged-out parents. He has not made the about-face quietly. In many cities, confidentiality laws protect caseworkers and judges from public outcries when their bad decisions lead to a parent's murder of a child. In Chicago, Murphy calls blistering press conferences to parcel out the blame. To those who say he picks on parents who are poor, black and victimized, he hotly retorts: "So are their kids."

Murphy is the Cook County (Ill.) public guardian, the court-appointed lawyer for 31,000 abused and neglected children. He's also a self-righteous crusader. Last year, campaigning to rein in one "family preservation" program, Murphy sent every Illinois legislator color autopsy photos of a little girl scalded and beaten to death after caseworkers taught her family new disciplinary skills. It's a loner's life, poring over murder files and railing at fellow liberals who think the poor can do no wrong. "A lot of people hate my guts," Murphy shrugs. "I can't blame them."

His views on family reunification changed because child abuse changed. Drugs now suffuse 80 percent of the caseload; sexual and physical assaults that once taxed the imagination are now common. Murphy believes that most families should be reunited—but the child-welfare agencies waste years trying to patch up dead-end families when they should be hurrying to free children for early adoption. Murphy, 55, blames such folly on bleeding hearts like himself, who once lobbied for generous social programs without working to curb welfare dependency and other ills. Now children of troubled families must pay the price—sometimes with their lives. "We inadvertently pushed a theory of irresponsibility," he says. "And we created a monster—kids having kids."

To Murphy's critics, that smacks of scorn for the less fortunate. "He's a classic bully," says Diane Redleaf of the Legal Assistance Foundation of Chicago, who represents parents trying to win back their kids. "Thousands of poor families are *not* torturing their children." Redleaf has drafted legislation that would force Murphy to get a judge's order each time he wants to speak about a case. That would protect children's privacy—and give the system a convenient hiding place. Murphy will fight to keep things as they are. His is the only job, he says, in which a lawyer knows that his clients are truly innocents.

J.M.

---

the first assault. The autopsy on Saonnia's visibly malnourished body found 62 cuts, bruises, burns, abrasions and wrist scars, among other injuries. Eleven were still healing—meaning they probably happened during the time the homemaker was working with the family.

Since Illinois's Family First program began in 1988, at least six children have died violently during or after their families received help. In many other instances, children were injured, or simply kept in questionable conditions. Such numbers may look small compared with the 17,000 children in Illinois who've been in the program. But to critics, the deaths and injuries underscore the danger of using reunification efforts for deeply troubled families. Gelles, once an ardent supporter of family preservation, is adamant about its failures. "We've learned in health psychology that you don't waste intervention on those with no intention of changing," he argues.

A University of Chicago report card issued last year gave the Illinois Family First program barely passing grades. Among the findings: Family First led to a slight *increase* in the overall number of children later placed outside their homes; it had no effect on subsequent reports of maltreatment; it had only mixed results in such areas as improving housing, economics and parenting, and it had no effect on getting families out of the DCFS system. John R. Schuerman, who helped write the report, says it's too simplistic to call Family First a failure. Still, he concedes that the assumption that large numbers of households can be saved with intensive services "just may not be the case."

Nevertheless, in the last decade, family-preservation programs have become so entrenched there's little chance they'll be junked. Health and Human Services Secretary Donna Shalala carefully sidesteps the question of whether it's possible to carry the reunification philosophy too far. Asked where she would draw the line in defining families beyond repair, she diplomatically suggests that the answers be left to child-welfare experts. "Nobody wants to leave children in dangerous situations," says Shalala. "The goal is to shrewdly pick cases in which the right efforts might help keep a family together." So far, not even the experts have come up with a sure way to do that.

Where Do We Go From Here?

POLICYMAKERS BELIEVE THAT IF THEY could just remove the stresses from a family, they wouldn't have to remove the child. But critics argue that the entire child-welfare network must approach the idea of severing parents' rights as aggressively as it now approaches family reunification. That means moving kids through the system and into permanent homes quickly—before they're so damaged that they won't fit in anywhere. In theory, the Adoption Assistance Act already requires that, but no state enforces that part of

---

CLIFFORD TRIPLETT

"I talked to him [Eddie Robinson, Cliff's mother's boyfriend] about Cliff and the old scars on his back. Robinson said . . . Cliff had a tendency to get into a lot of mischief," the caseworker noted. "This is why Cliff was whipped—however I told him that he shouldn't be beat on his back. Robinson promised to go easy on the discipline. (Said he wasn't doing the whipping.)"

FROM CASEWORKER REPORT ON CLIFFORD TRIPLETT

1. PARENTING AND FAMILY ISSUES

the law. Illinois is typical: even in the most straightforward cases, a petition to terminate parental rights is usually the start of a two-year judicial process—*after* the 18-month clean-up-your-act phase.

Why does it take so long? Once a child is in foster care, the system breathes a sigh of relief and effectively forgets about him. If the child is removed from an abusive home, the assumption is that he's safe. "There's always another reason to give the parent the benefit of the doubt," says Daro. "They lose their job, the house burns down, the aunt is murdered. Then they get another six-month extension, and it happens all over again. Meanwhile, you can't put a child in a Deepfreeze and suspend his life until the parent gets her life together."

In the most blatant abuse and neglect cases, parents' rights should be terminated immediately, reformers say. In less-severe cases, parents should be given no more than six to 12 months to shape up. "You have social workers saying, 'She doesn't visit her child because she has no money for carfare'," says Murphy. "But what parent wouldn't walk over mountains of glass to see their kids? You know it's a crock. You have to tell people we *demand* responsibility."

And if parents can't take care of them, where are all these children supposed to go? With just 100,000 foster parents in the system, finding even temporary homes is difficult. For starters, reformers suggest professionalizing foster care, paying parents decent salaries to stay home and care for several children at a time. Long range, many believe that society will have to confront its ambivalence toward interracial adoptions. Perhaps the most controversial alternative is the move to revive orphan-

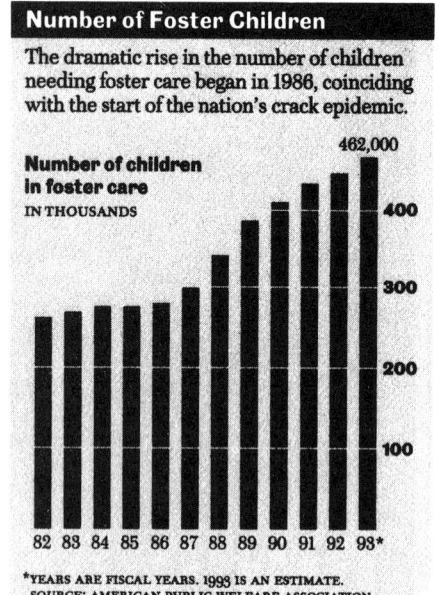

Number of Foster Children

The dramatic rise in the number of children needing foster care began in 1986, coinciding with the start of the nation's crack epidemic.

*YEARS ARE FISCAL YEARS. 1993 IS AN ESTIMATE.
SOURCE: AMERICAN PUBLIC WELFARE ASSOCIATION

ages, at least for teenagers, who are the least likely to be adopted. One of the fiercest supporters is Maddux, the new supervising judge of Cook County's abuse section. Maddux, 59, says that his own family was so desperately poor they once lived in a shanty with two rooms—one of which was an old car. When the family broke up, he and his younger brother went to live at Boys Town, Neb. He believes that many foster children today could benefit from the nurturing-yet-demanding atmosphere of group living. "I wasn't raised in a family after the age of 12," Maddux says. "I didn't miss it. Thousands of kids at Boys Town knew that being in a destitute, nonfunctioning family was a lot worse than not being in a family." In Illinois, some are taking the idea seriously—among the proposals is turning closed military bases into campuses for kids.

Ironically, Illinois could wind up with one of the best child-welfare systems in the nation. Pressed by public outrage over Joseph Wallace's death, state legislators last year passed a law that puts the best interest of children ahead of their parents'. Foster parents will be given a voice in abuse and neglect cases. And the DCFS is beefing up caseworker training, so that those in the field will learn how to spot dangerous situations more quickly.

Some of the toughest changes are already underway in Cook County. The much-criticized Family First program has been replaced with a smaller, more intensely scrutinized family-preservation project known as Homebuilders. And the county's juvenile-court system has been expanded so that there are now 14 judges, not eight, hearing abuse and neglect cases; that cuts each judge's caseload from about 3,500 to about 2,000 children per year. But reform doesn't come cheap. The DCFS budget has tripled since 1988, to $900 million, and it could top $1 billion in the next fiscal year.

Whether any of this can save lives, it's too soon to tell. In its report on Joseph Wallace's death, the blue-ribbon committee was pessimistic. "It would be comforting to believe that the facts of this case are so exceptional that such cases are not likely to happen again," the panel wrote with a dose of bitterness. "That hope is unfounded." The temptation, of course, is to blame some faceless system. But the fate of children really lies with everyone—caseworkers, supervisors, prosecutors, judges—doing their jobs.

Article 7

The three R's spell SUC·CESS'

(sək-ses') *n.* 1. The achievement of something desired, planned or attempted.

Family literacy programs help children with the developmental skills they need to succeed in school. Photo: National Center for Family Literacy.

SHARON DARLING

Millions of families across the United States seem trapped by a lack of education and ensuing poverty. An uneducated parent who lacks jobs skills cannot support a family. What began as an education problem becomes an economic problem for the whole family.

Children of undereducated parents are at grave risk of continuing the cycle. Fewer are in preschool programs, and more fail or drop out of school than do the children of more educated parents.

Family literacy programs address literacy across two generations, providing remediation for parents and prevention for children simultaneously. The primary goal of a family literacy program is to break the intergenerational cycle of a poor education and poverty. Family literacy does so by improving parents' basic skills, self-sufficiency and parenting skills, children's school readiness skills, and the quality of parent-child relationships. Four components of a family literacy program include parent literacy training, early childhood education, time together for parents and children, and parenting and life skills development.

Sharon Darling is president of the National Center for Family Literacy.

Education can help families break out of poverty when parents, as well as children, become literate.

A typical day in a family literacy program starts with parents and children leaving for school together. Parents attend adult basic education classes while the children participate in early education classes. Later, parents participate in programs to enhance their job and educational opportunities and promote self-sufficiency. They also learn positive parenting techniques. Then parents and children spend time together in the classroom, enabling them to develop better relationships.

Because family literacy preserves and strengthens families, creates self-sufficient families and expands work and training opportunities, it provides a model for a coordinated system of services that maximize scarce dollars for welfare reform. For families to escape the welfare cycle, literacy is key to long-term success. Statistics show a strong link between poverty and lack of education. Consider the following:

• 40 percent of female single parents have an eighth grade education or less,
• 75 percent of female heads of households with less than a high school education live in poverty.

Also consider that children's literacy levels are strongly linked to their parents' educational levels, especially their mothers', and there is an even stronger case for a family approach to education to break welfare dependency.

Literacy leads to success

Studies show that family literacy improves family life and improves the lives of families. Research shows that when families learn together in family literacy programs, they begin to read together, go to the library together and spend more quality time together. In a recent study, 79 percent of those who enrolled completed the program, and 83 percent of the adults in those families completed requirements for GED certification or a diploma.

The data also show that gains parents and children make in family literacy programs continue after they leave the programs. A follow-up study showed that one year after the program, 66 percent of the adults were

1. PARENTING AND FAMILY ISSUES

either employed, enrolled in another educational program, or had definite plans for continuing education.

In another follow-up study of different families two years after completing the program, 71 percent are either employed, enrolled in higher education or assuming a role as homemaker in a stable family. Among these adults, 38 percent more are employed than were before the program.

Family literacy parents take a much more active role in their children's education, and the children do much better in school than would have been otherwise expected. The parents volunteer in their children's schools, maintain contact with teachers and support their children's education activities at home. The children enter school ready to learn and progress in school.

All follow-up studies of family literacy children show a consistent pattern of performance. They are not being retained in grades nor being placed in special programs at near the rates otherwise expected for them. Their teachers consistently rate family-literacy children highly, with over 75 percent of these children at or above the average for their class on academic performance, motivation to learn, support from family, self-confidence and probable school success. They are rated even higher — generally over 85 percent at or above class average — on relations with other students, attendance and classroom behavior. Furthermore, more than half of the teachers initiate a response of "support from parents" when they are asked to list strengths of the children.

Although they come from a population of children classified as "at-risk" for school failure, a study of children who participated with their parents in the Kenan Trust Family Literacy Project found that label no longer applied. The study showed that after one year in the program, more than 90 percent of the children were judged by their teacher as ready for entry into kindergarten with no expected academic or social difficulties. Once in school, the percentage of these children rated average or above in their class by their teacher was 75 percent on overall academic performance, 84 percent on motivation to learn, 78 percent on support from parents, 90 percent in relations with other students, 87 percent in attendance, 89 percent in classroom behavior, 75 percent in self-confidence and 85 percent in probable school success.

Road to welfare reform

While family literacy may have a profound impact on individual families, its impact extends beyond the family.

Family Literacy programs foster teamwork between parent and child that carries over to school and home. Photo: National Center for Family Literacy.

It touches the communities in which these families live, the states in which the communities are located and the nation.

It might be useful to look at some indicators of the widespread lack of education:

• In 1993, the National Adult Literacy Survey made headlines with its revelation that 90 million American adults have literacy skills in the lowest two levels, making it difficult for them to fully function in society.

• In 1991, 4.4 million families received Aid to Families with Dependent Children, with an average payment of $389 per family per month. That added up to $20.5 billion in total assistance.

• And according to the Census Bureau, workers who lack a high school diploma have a mean monthly income of $452, compared to the $1,829 earned monthly by those with a bachelor's degree.

• The average annual cost per pupil in public school for 1990 was almost $5,000 — and that is only part of the cost incurred every time a child must repeat a grade.

Consider the societal impact of families getting off the welfare rolls and parents entering the work force, of adults who have increased earning potential because they have a higher level of education, and of children who are prepared for school success and don't have to repeat grades.

Addressing welfare and education reform through family literacy is an opportunity to address the following four tenets:

• Cease financial dependency on a government system

The National Adult Literacy Survey concludes that adults with proficiency within the two lowest levels of literacy are far less likely than their more literate peers to be employed full time, to earn high wages and to vote. Moreover, they are more likely to depend on food stamps, to live in poverty and to rely on nonprint sources for information about current events, public affairs and government. Individuals with poor skills do not have much to bargain with, they are condemned to low earnings and limited choices. Poor skills translate to welfare dependency.

• Provide work and training opportunities

Family literacy programs instruct parents in basic skills based upon their

7. Three R's Spell Success

Family Literacy programs give parents the opportunity to increase their skills in adult education classes. Photo: National Center for Family Literacy.

needs and goals for self-sufficiency. The instruction is presented in context with the literacy skills the parent needs to function as parent, consumer, employee and citizen. Parents are expected to pursue high-school diplomas, vocational opportunities and further education and training.

• **Preserve and strengthen the family unit**

Family literacy preserves the family unit by recognizing that the parent is the first and prime educator of a child. It helps parents get the skills they need for self-sufficiency and to help their children succeed.

Early results from evaluations of family literacy programs indicate that families gain from combining the four components of family literacy programs. Parents learn how to teach their children through play. They communicate with their children and often develop more positive, supportive relationships. Parents who bring these new skills into the home replace the legacy of failure with success.

• **Create programs that focus on outcomes for communities through coordination and collaboration of existing services**

A welfare recipient is more likely to transition to self-support when there is synergy in the community and coordinated services. Being able to manage one's affairs, being part of a community setting, and having access to both mental and physical health services can assist a welfare recipient in the quest for self-sufficiency. By coordinating systems, family literacy programs maximize resources offered to families.

Family literacy programs seldom rely on the resources of a single agency. They often do not require new dollars but a reallocation of resources from programs that often don't work. Family literacy builds on existing programs such as adult education, Head Start, family support centers, job training programs and early childhood programs. This holistic approach helps existing programs be more effective.

Families have a variety of needs that can be best met through a collaboration of services and resources. Family literacy programs identify family needs and provide the vehicle to implement community coordination. Family literacy programs deal with families. They are more than adult education or child-development programs. A family literacy program capitalizes on elements of both and facilitates a multifaceted approach that pulls similar programs together.

With Congress' resolve to balance the federal budget and return responsibilities for many welfare programs to states through block grants, states will be faced with redesigning services within fixed budgets. There are many public funding sources which may support family literacy services. Some can serve as primary funding sources by covering the cost of core services like adult education and preschool instruction. Others can serve as supplementary funding sources, for example, providing staff and in-service training or transportation for families.

Block grants could simplify delivery systems to better fit the needs of families and children. Family literacy is an intergenerational program that seeks to solve the problems of parents and children. It helps young children get the best possible start in life and at the same time helps their parents become economically self-sufficient.

Resources

For more, contact:
Linda Likins
Director of Policy Development
National Center for Family Literacy
325 West Main Street
Suite 200
Louisville, KY 40202-4251
Phone: (502) 584-1133
Fax: (502) 584-0172
E-mail: LKLIKINS@aol.com

Crime, Terrorism, and Violence

The probability is becoming greater that every American will at some time in his or her life be subjected to a criminal act that will involve some degree of violence. With the ever-increasing crime rate, especially in major cities, confidence in law enforcement is declining, and citizens feel forced to arm themselves and turn their homes into minifortresses. What must be done to make our streets safe to walk, our highways safe to drive, and our homes safe from unwanted intrusions? These are the concerns addressed by this unit.

The 1995 bombing of the Federal Building in Oklahoma City, the derailment of an Amtrak passenger train in the Arizona desert, the 1988 bombing of Pan Am flight 103, and the awareness that such things could happen at any time and any place, have heightened the concerns and fears of many Americans. Terrorism has traditionally been something that only happens somewhere else, not in America! This unit has been expanded to include articles that focus on what happens when extremists strike close to home. A shrinking economy and a tax base riddled with loopholes and inequities leave city, state, and federal governments struggling to raise cash to fight gangs, especially those that are well-organized, financed, and armed.

What should be done, what can be done, and how to determine the difference are hotly debated in "Terrorism in America." Orrin Hatch, senior senator from Utah, argues that we must take decisive action and take it now. Then, Doug Bandow, a senior fellow at the Cato Institute, cautions that the anger and anguish associated with the bombing in Oklahoma City must not result in a draconian overresponse.

In "Enemies of the State," Jill Smolowe looks at the rise of paramilitary groups in the United States. These secretive, paranoid, and obsessed groups, while not new to America, raise significant concerns for local, state, and federal regulatory agencies. They are highly trained, heavily armed, and have strong grievances that, if not addressed, are likely to cause them to become actively militant.

In "How Nation's Largest Gang Runs Its Drug Enterprise," the inner workings of the largest and most successful gang in the history of the United States are explored.

Jean Callahan, in her report "Forgiving the Unforgivable," argues that until victims can let go of their anger, frustration, anguish, and status as a victim, they remain captives of their abuser/criminal. To move on with their lives, they must cut their bonds and free themselves, and the only way this can occur is through forgiveness.

Nuclear weapons are a fact of life. It seems that they are with us to stay, so we must learn how to live with them. The fact that nuclear weapons exist may be the single greatest deterrence against their use. Michael May, in "Fearsome Security: The Role of Nuclear Weapons," asserts that the right time to maintain deterrence and to formulate ways to deal with nuclear weapons is during peacetime.

UNIT 2

Looking Ahead: Challenge Questions

What evidence is there to support the belief that the severity of punishment deters crime?

What effect does being a victim have on the individual?

What are the most productive ways of dealing with being a victim?

How might it be possible to eliminate violence?

How should the federal government respond to terrorism?

When did domestic terrorism first begin in America?

What problems face city, state, and federal governments in their efforts to eliminate drug abuse?

Why do gangs not only survive, but thrive, in major U.S. cities?

What seems to be the basis for the emergence of paramilitary extremist groups in the United States?

Would the disarming of America solve the crime problem? Defend your answer.

Explain why the United States should or should not unilaterally eliminate all its nuclear weapons. Explain why you agree or disagree that "the existence of nuclear weapons is the greatest deterrence against their use."

How would a functionalist's approach to the problem of crime, terrorism, and violence differ from the approach of a conflict theorist?

Terrorism

Let's Take Decisive Action

Orrin Hatch

Republican Sen. Orrin Hatch is the senior senator from Utah and chairman of the Senate Judiciary Committee.

The Dole-Hatch Comprehensive Terrorism Prevention Act of 1995 represents a landmark, bipartisan effort to address an issue of grave national importance—the prevention and punishment of acts of domestic and international terrorism. This legislation adds important tools to the government's fight against terrorism, and does so in a temperate manner that is protective of civil liberties. I believe this bill is the most comprehensive antiterrorism bill ever considered in the Senate.

This legislation increases the penalties for acts of foreign and domestic terrorism, including the use of weapons of mass destruction, attacks on officials and employees of the United States, and conspiracy to commit terrorist acts.

It gives the president enhanced tools to use his foreign policy powers to combat terrorism overseas, and it gives those of our citizens harmed by the terrorist acts of outlaw states the right to sue their attackers in our courts.

Our bill provides a constitutional mechanism to the government to deport aliens suspected of engaging in terrorist activity without divulging our national security secrets.

It also includes a provision that constitutionally limits the ability of foreign terrorist organizations to raise funds in the United States.

Our bill also provides measured enhancements to the authority of federal law enforcement to investigate terrorist threats and acts. In addition to giving law enforcement the legal tools they need to do the job, our bill also authorizes increased resources for law enforcement to carry out its mission. The bill provides $1.6 billion over five years for an enhanced antiterrorism effort at the federal and state levels.

The bill also implements the convention on the marking of plastic explosives. It requires that the makers of plastic explosives make the explosives detectable.

Finally, the bill appropriately reforms habeas corpus. Habeas corpus allows those convicted of brutal crimes, including terrorism, to delay the imposition of just punishment for years.

Several points, however, should be addressed. I have long opposed the unchecked expansion of federal authority, and will continue to do so. Still, the federal government has a legitimate role to play in our national life and in law enforcement. In particular, the federal government has an obligation to protect all of our citizens from serious criminal threats emanating from abroad or that involve a national interest.

We must nevertheless remember that our response to terrorism carries with it the grave risk of impinging on the rights of free speech, assembly, petition for the redress of grievances, and the right to keep and bear arms. We cannot allow this to happen. It would be cruel irony if, in response to the acts of evil and misguided men hostile to our government, we stifled true debate on the proper role of that government.

The legislation enhances our safety without sacrificing the liberty of American citizens. Each of the provisions of this bill

(continued on page 52)

Excerpted from the speech given by Senator Hatch as he submitted the Dole-Hatch Counterterrorism bill on May 29, 1995.

Terrorism in America

Let's Not Overreact

Doug Bandow

Doug Bandow is a senior fellow at the Cato Institute and a former special assistant to President Reagan. He is the author of The Politics of Envy: Statism as Theology (Transaction).

The reactions inside and outside of Washington to the Oklahoma City bombing were sadly predictable. Around the country was anger, desire for understanding, and hope for healing. In the halls of the White House and Congress was shock, followed by a race for political advantage and demand for more power. In short, everyone did what came most naturally—citizens worried about their country while politicians worried about their influence.

This reaction was evident in attempts to brand critics of government as contributing to a "climate of hate" in which violence might occur. Needless to say, it is in the interest of presidents, legislators, and bureaucrats alike to discourage criticism. And many were quick to use the tragedy in Oklahoma City in an attempt to place themselves beyond reproach.

Second, politicians of both parties began posturing with proposals for new "counterterrorism" legislation. These bills would vest the federal government with vast new powers to wiretap, investigate, deport, use the military, and rely on secret evidence. If people don't already have reason to fear government, they certainly will if these measures become law.

Before Congress acts precipitously, legislators should answer four questions. Is terrorism so serious a threat that it requires an immediate, draconian response? Has government policy contributed to violence, like the bombing of the Alfred P. Murrah Building in Oklahoma City? Do federal agencies require more power to combat terrorism? Does the law enforcement interest outweigh the rights and liberties of citizens that would be sacrificed?

LET'S CONSIDER FOUR QUESTIONS

1. Is terrorism so serious a threat that it requires an immediate, draconian response? The Oklahoma City bombing was a hideous act, but, thankfully, it represents the exception rather than the rule. There were no terrorist incidents in 1994, either actual or prevented. Of 11 incidents in 1993, 9 were committed by animal rights activists in one night. Over the past 11 years there has been only one incident of international origin, the World Trade Center bombing. The State Department reports that international terrorist attacks are at their lowest level in nearly a quarter century.

Of course, even one attack is too many. But the current level of terrorist activity provides no cause for Congress to act without due deliberation. Legislators need to recognize that law enforcement agencies today often abuse their power, thereby stoking violent passions. In any case, the police already possess expansive authority to combat terrorism; Congress should honestly assess whether increasing these powers would do anything to combat real crime. Finally, legislators need to remember that it is a free society that they are attempting to protect. Marginal gains in a campaign against minimal threats are not worth the sacrifice of fundamental liberties.

2. Has government policy contributed to violence, like the bombing of the Alfred P. Murrah Building in Oklahoma City? Nothing can justify terrorism. Nevertheless, public officials must recognize that distrust of government is not limited to

(continued on page 54)

2. CRIME, TERRORISM, AND VIOLENCE

(*continued from page 50*)

strikes a careful balance between necessary vigilance against the terrorist threat and preserving our cherished freedom. Several of the provisions deserve special mention.

WHAT ABOUT UNLAWFUL ALIENS?

First, I would like to discuss the Alien Terrorist Removal Act. I firmly believe that it is time to give our law enforcement and courts the tools they need to quickly remove alien terrorists from our midst without jeopardizing national security or the lives of law enforcement personnel.

This provision provides the Justice Department with a mechanism to do this. It allows for a special deportation hearing and *in camera, ex parte* review by a special panel of federal judges when the disclosure in open court of government evidence would pose a threat to national security.

Sound policy dictates that we take steps to ensure that we deport alien terrorists without disclosing to them and their partners our national security secrets. The success of our counterterrorist efforts depends on the effective use of classified information used to infiltrate foreign terrorist groups. We cannot afford to turn over these secrets in open court, jeopardizing both the future success of these programs and the lives of those who carry them out.

Some raise heart-felt concerns about the precedence of this provision. I believe their opposition is sincere, and I respect their views. Yet, these special proceedings are not criminal proceedings for which the alien will be incarcerated. Rather, the result will simply be the removal of these aliens from U.S. soil—that is all.

■
Congress has a responsibility to minimize the prospect that something like the Oklahoma City bombing can happen again.
■

Americans are a fair people. Our nation has always emphasized that its procedures be just and fair. And the procedures in this bill are in keeping with that tradition. The Special Court would have to determine that:

1. the alien in question was an alien terrorist;
2. an ordinary deportation hearing would pose a security risk; and
3. the threat by the alien's physical presence is grave and immediate.

The alien would be provided with counsel, given all information which would not pose a risk if disclosed, would be provided with a summary of the evidence, and would have the right of appeal. Still, in our effort to be fair, we must not provide to terrorists and to their supporters abroad the informational means to wreak more havoc on our society. This provision is an appropriate means to ensure that we do not.

Second, this bill includes provisions making it a crime to knowingly provide material support to the terrorist functions of foreign groups designated by a presidential finding to be engaged in terrorist activities.

Nothing in the Dole-Hatch version of this provision prohibits the free exercise of religion or speech, or impinges on the freedom of association. Moreover, nothing in the Constitution provides the right to engage in violence against fellow citizens. Aiding and financing terrorist bombings is not constitutionally protected activity. Additionally, I have to believe that honest donors to any organization would want to know if their contributions were being used for such scurrilous purposes.

PROTECTING HABEAS CORPUS

Finally, I would like to address an issue which has inappropriately overshadowed all of the other fine provisions of this legislation—the inclusion of the Specter-Hatch habeas corpus reform in this bill. Some have stated that the inclusion of habeas reform in this bill is political opportunism. Nothing could be further from the truth. The plain truth is, habeas corpus reform is entirely germane to this legislation. The president has asked for this reform. And the American people are demanding it.

Although most capital cases are state cases (and the state of Oklahoma could still prosecute this case), the habeas reform proposal in this bill would apply to federal death penalty cases as well. It would directly affect the government's prosecution of the Oklahoma bombing case:

1. It would place a one-year limit for the filing of a habeas petition on all death row inmates—state and federal inmates.
2. It would limit condemned killers convicted in state and federal court to one habeas corpus petition. In contrast, under current law, there is currently no limit to the number of petitions he may file.
3. It requires the federal courts, once a petition is filed, to complete judicial action within a specified time period.

8. Terrorism in America

Antiterrorism Bill Passes

- The Senate responded to the Oklahoma City bombing by passing major antiterrorism legislation June 7 that would expand law enforcement's powers and limit appeals by death-row inmates.

- The $2 billion measure, passed by a 91–8 vote, includes provisions sought by President Clinton to enlarge federal law enforcement agencies and the government's wiretapping authority and allow use of the military in emergencies involving chemical or biological weapons.

- A House version of the bill passed the Judiciary Committee on June 14 and is expected to reach the House floor after July 4.

- At the White House, Clinton praised the Senate's bipartisan vote—52 Republicans and 39 Democrats in favor—and expressed hope that the House could quickly pass the same bill so that any attack similar to the one that killed 168 people in Oklahoma City could be forestalled or prevented.

—*The Editor*

Therefore, if the federal government prosecutes this case and the death penalty is sought and imposed, the execution of sentence could take as little as one year if our proposal passes. This stands in stark contrast to the 8 to 10 years of delay we are so used to under the current system.

President Clinton vowed that justice in the wake of the Oklahoma tragedy would be "swift, certain, and severe." We must help him keep this promise to the families of those who were murdered in Oklahoma City by passing comprehensive habeas corpus reform.

The Comprehensive Terrorism Prevention Act of 1995 provides for numerous other needed improvements in the law to fight the scourge of terrorism, including the authorization of additional appropriations—nearly $1.6 billion—to law enforcement to beef up counterterrorism efforts and increasing the maximum rewards permitted for information concerning international terrorism.

CERTAIN, SWIFT, AND UNIFIED RESPONSE

The people of the United States and around the world must know that terrorism is an issue that transcends politics and political parties. Our resolve in this matter must be clear: Our response to the terrorist threat, and to acts of terrorism, will be certain, swift, and unified.

Ours is a free society. Our liberties, the openness of our institutions, and our freedom of movement are what make America a nation we are willing to defend. These freedoms are cherished by virtually every American.

We must now redouble our efforts to combat terrorism and to protect our citizens. A worthy first step is the enactment of these sound provisions to provide law enforcement with the tools to fight terrorism.

In closing, what is shocking to so many of us is the apparent fact that those responsible for the Oklahoma atrocity are U.S. citizens. To think that Americans could do this to one another! Yet, these killers are not true Americans—not in my book. Americans are the men, women, and children who died under a sea of concrete and steel. Americans are the rescue workers, the volunteers, the law enforcement officials, and investigators who are cleaning up the chaos in Oklahoma City. The genuine Americans are the overwhelming majority of us who will forever reel at the senselessness and horror of April 19, 1995.

It falls on all of us, as Americans in heart and spirit, to condemn this sort of political extremism and to take responsible steps to limit the prospect for its recurrence. Can Congress pass legislation that will guarantee an end to domestic and international terrorism? We cannot.

Nevertheless, Congress has a responsibility to minimize the prospect that something like this can happen again. We must resolve that anarchistic radicalism—be it from the left or the right—will not prevail in our freedom-loving democracy. The rule of law and popular government will prevail.

(Article continues.)

2. CRIME, TERRORISM, AND VIOLENCE

(continued from page 51)

fringe groups. A recent Gallup poll found that an astounding 52 percent of people believed "the federal government has become so large and powerful that it poses a threat to the rights and freedoms of ordinary citizens." Four out of 10 thought the danger was "immediate."

There is much to fear. Government misbehavior in Waco, Texas, against the Branch Davidians and in Ruby Ridge, Idaho, against Randy Weaver and his family was well-publicized and deadly. Yet rather than holding law enforcement officials accountable, the Clinton administration promoted one FBI agent, reprimanded for his role in both affairs, to deputy director. Congress has yet to hold hearings, despite abundant evidence of government agencies abusing their vast authority. [Waco hearings were slated to begin July 14.] These cases, along with numerous brutal and erroneous DEA, ATF, and local police raids, suggest that government power itself is a serious problem.

Placing even greater authority in agencies that have abused their trust would only exacerbate peoples' fears of Washington. Therefore, legislators should first concentrate on reforming the present system. Unnecessary powers need to be terminated; abuses need to be curbed; accountability needs to be reestablished. Only then, when people's liberties would be less at risk, should Congress consider expanding the authority of law enforcement.

3. Do federal agencies require more power to combat terrorism? Although Oklahoma City has become the justification for the pending antiterrorism bills, the alleged perpetrators of that bombing were quickly apprehended. So, too, were the bombers of the World Trade Center. Moreover, since 1989, law enforcement officials have prevented nearly as many terrorist attacks as have been committed, 23 compared to 31. There is no evidence that federal agencies need more power to respond to terrorist threats.

Federal law already bars financial support for foreign terrorist groups. Proposals to expand this prohibition, give the president unreviewable authority to designate groups as terrorist, and investigate people where no evidence of a legal violation exists are unjustified. Similarly, terrorist acts, like the Oklahoma City bombing, are already against the law. There is no need to expand the definition to include literally every crime—"any unlawful destruction of property"—for example, which may be best handled by local authorities (such as animal rights activists).

THE GOVERNMENT HAS ENOUGH AUTHORITY

The federal government already has wiretap authority for such crimes as arson and homicide; proposals to expand that power to almost any crime (including misdemeanors, in the Clinton administration legislation) have nothing to do with combating terrorism. After all, of 7,554 requests for wiretap authority submitted by the FBI since 1978, only one has been rejected by the special seven-member court that oversees the process. The many new powers being proposed by President Clinton, Majority Leader Robert Dole, and others are no more necessary to the prevention of terrorism.

4. Does the law enforcement interest outweigh the rights and liberties of citizens that would be sacrificed? Even if increased power might marginally improve gov-

---

A Hasty Response to Terrorism

- The Senate is congratulating itself for passing the Comprehensive Terrorism Protection Act of 1995 just seven weeks after the Oklahoma City bombing. But the Senate's hasty and ill-considered action has come at a price. Steps designed to protect Americans' physical safety will in many cases erode their liberties.

- The legislation, a grab-bag of bills proposed by the Clinton Administration and members of Congress, contains some sound provisions. It provides, for example, more FBI personnel and resources to combat terrorism, increased penalties for dealing in explosives used to commit crimes and measures to make bombs easier to trace.

- But the temptation was too strong, both in the White House and the Senate, to load the bill with tougher-sounding, more crowd-pleasing provisions. These include wide-ranging surveillance power for law enforcement agencies, crackdowns on suspected aliens and more blurring of the line between military forces and domestic peacekeeping police.

—Editorial, excerpted from *New York Times* June 9, 1995

ernment's ability to respond to terrorism, Congress must still weigh the benefits against the costs. For example, the FBI investigative guidelines were created for a reason: the agency's Counter-Intelligence Program (COINTELPRO) resulted in spying on literally millions of law-abiding Americans from the 1940s through the 1970s. Yet this orgy of surveillance did not make America more secure. Nearly 700 FBI operations yielded a grand total of four convictions.

Similarly, the wholesale federalization of crimes would make Americans less free without making them more secure. State law already covers violent crime; existing federal law reaches special offenses, such as threats against the president. Proposals to expand federal jurisdiction combined with a broadening of the much-abused RICO statute, enhanced restrictions on money laundering, loosening of restraints on wiretapping, and use of the military to enforce domestic law would provide numerous opportunities for government to abuse citizens' rights.

Expansion of wiretaps to almost any felony would also cost the American people more in lost liberty than any security they might gain. Today, the government is empowered to seek wiretaps in cases involving arson and homicide, typical ingredients of terrorist acts. Yet federal wiretaps rarely involve these issues. In fact, wiretapping is focused on, of all things, gambling, along with racketeering and drugs.

Other assaults on individual liberty that have been tied to terrorism include proposals to restrict habeas corpus, which requires the government to justify holding a citizen, and limit encryption software, which ensures the privacy of computer communication. Neither proposal is designed to combat terror-

■
Citizens worried about their country while politicians worried about their influence.
■

ism. After all, habeas corpus, such a jealously guarded right that the Constitution permits only Congress to suspend its application, and to do so only during "rebellion or invasion," applies to those already in government custody. And computers have played no role in any recent terrorist plot.

More closely tied to international terrorism is the proposal to allow special courts to use secret evidence, withheld from the defendant, in deportation proceedings of legal residents of the United States. Yet the right of "confrontation" is a critical procedural safeguard. Once Congress embarks upon the slippery slope of allowing the government to present arguments without giving the defendant a chance to directly respond, legislators could apply this principle against citizens in any case involving a serious crime: murder, arson, and the like. Either the courts would void such laws as unconstitutional, as they have done in similar cases in the past, or American citizens could end up appearing before a tribunal akin to that of Great Britain's hated "star chamber."

DON'T SUBVERT POSSE COMITATUS

The president and Senator Dole have proposed increasing the role of the military in terror-

8. Terrorism in America

ism cases—essentially, repealing the Posse Comitatus Act whenever the attorney general desires military assistance. At the same time, both proposals would eliminate jurisdictional restrictions on such agencies as the ATF, allowing them to act however they pleased against anything termed "terrorism."

But there are very good reasons for retaining a bright line between the military and domestic law enforcement. The Defense Department should not be diverted from its most important job of defending America from international foes. Soldiers are not trained in the niceties of civil liberties; involving the Pentagon will simultaneously militarize and centralize law enforcement, poor practices in a republic. In fact, abuses have been evident in the ongoing use of the National Guard in drug interdiction campaigns. Similarly, reducing restraints on specialized law enforcement agencies will encourage further malfeasance by bureaucracies that already exceed their rightful authority, without any concomitant improvement in domestic security.

Terrorism obviously poses a serious threat to a free society like our own. But legislators should tailor their response to meet the threat, not garner votes. Despite the hideous Oklahoma City bombing, America remains largely free of terrorism. Congress' first task, then, should be to investigate how renegade government agencies are abusing their power and creating grievances that some misguided people believe are properly addressed through violence. Only then should legislators consider expanding federal law enforcement authority, and then only if they can do so without undermining the basic freedoms that make this nation unique—and worth living in.

Enemies of The State

America's "patriots" have a tough list of demands: keep your hands off my land, my wallet—and my guns

Jill Smolowe

Annamarie Miller is a dedicated schoolteacher with an obvious love of history and ideas, who dresses fastidiously in neatly pressed shirts and slacks and is inclined to exclaim "Gosh!" when she gets excited. Though hardly a menacing presence, Miller, 27, is a determined renegade who refuses to take any authority figure's word at face value. It all began, she says, during her student days at California State University at Chico. "I became disillusioned by the revisionism of history," she says. "A lot of stuff they were teaching me twisted the truth." Inspired by campaign literature, she began to question the "truths" of authorities far more powerful than her college professors. The Federal Reserve Board, for instance. Why had it never been audited? Had it perhaps already bankrupted the country? Or the Social Security Administration. Was it going to collapse before Miller was old enough to collect? Through such questions, Miller gradually arrived at a hard "truth" of her own. The constitutional rights of all Americans, she believes, are threatened by an overgrasping, irresponsible government.

She doesn't keep it to herself. Each Tuesday at 7 p.m., Miller broadcasts that unflinching message via public-access TV to an audience of 50,000 viewers in Northern California. Along with her husband Scott, a retail clerk, and his brother Randall, a chef, she uses their half-hour show, *The Informed Citizen,* to warn of threats to the American way of life. Among them: a conspiratorial U.S. government that is surrendering its sovereignty to the U.N.; efforts by police and gun-control advocates to disarm citizens; and a tax burden that is robbing Americans of their hard-earned income. Her aim, she insists, is simply to inform and motivate. "A lot of people," she says, "are willing to give up their rights and freedom out of fear."

Before the bombing in Oklahoma City, few Americans would have thought that either Miller or her show posed a serious threat to the civic order. Unlike many other citizens who identify themselves as "patriots"—an amorphous, far-right populist movement of both armed militias and unarmed groups that harbor a deep distrust of government—Miller does not spend her weekends running around in camouflage, shooting at imagined enemies. Nor does she buy into every wild conspiracy theory that crackles along the patriot grapevine, like last week's alert that the Oklahoma catastrophe—which "patriots" suspect involved three bombs, not one—was a government plot to enable President Clinton to proclaim martial law and divert public attention from forthcoming Whitewater hearings. Indeed, Miller's attitude toward the Oklahoma City culprits—"I say hang 'em"—sounded much like the President's.

But as Americans try to understand the social currents that could wash a homegrown terrorist up to the front doors of the Alfred P. Murrah Federal Building, they are taking a second look at people like Miller. It's not because Miller shows any signs of violent tendencies herself, but because she is one of the disseminators of a virulent antigovernment philosophy that may have helped plant thoughts of insurrection in someone else's head. Miller's own first thought upon hearing of the bombing was, "Oh, my gosh, I hope some idiot calling himself a patriot didn't do this." She admits that her own unarmed group, the Sons of Liberty, had attracted a "loose cannon," a young man who tried to join last summer. "He was saying things like, 'We ought to blow up the federal building,'" she recalls. The Sons of Liberty promptly tossed him out.

But what ideas do these fringe characters decamp with? Where do they go? And

Would you describe the members of militia groups as:	Describes	Doesn't describe
DANGEROUS	80%	11%
A THREAT TO OUR WAY OF LIFE	63%	26%
CRAZY	55%	33%
WELL INTENTIONED	30%	58%
PATRIOTS	21%	55%

Do you think the Federal Government should spy on the militias in order to monitor their activities?	
YES	68%
NO	26%

From a telephone poll of 600 adult Americans taken for TIME/CNN on April 27 by Yankelovich Partners Inc. Sampling error is ± 27%. "Not sures" omitted.

9. Enemies of the State

what, if anything, are they up to now? What is most perplexing is the vague intersection between those so-called patriots who merely spread the word Paul Revere-style, those who are arming themselves in anticipation of a fight to defend their rights, and those who are already taking aim at perceived enemies. In recent months there have been several incidents between armed, angry citizens and government agents that have prompted officials to take these groups more seriously.

Last September, for instance, three men driving a Saturn were stopped in Fowlerville, Michigan, by a police officer after their vehicle crossed the center line. According to the town's police chief, Gary Krause, an officer found the car packed with weapons, including a .357-cal. revolver, three assault rifles, three 9-mm semiautomatic pistols and 700 rounds of ammunition. The three men identified themselves as bodyguards of Mark Koernke, the self-promoting militia propagandist. "They said they had just completed maneuvers," Krause recalls. While those three men didn't show up for their arraignment six days later, dozens of militia members did, turning out in camouflage fatigues and taunting police officers.

How many people have reached that breaking point with their Federal Government—and are they acting alone or together? If you count just the people who are arming themselves against the day when U.N. tanks roll through the heartland to establish the one-world order, estimates range only as high as 100,000. But if you include all the people in as many as 40 states who respond to the patriot rhetoric about a sinister, out-of-control federal bureaucracy—all the ranchers fed up with land- and water-use policies, all the loggers who feel besieged by environmentalists, all the underemployed who blame their plight on NAFTA and GATT—then the count soars upwards of 12 million. "People are drawn in under this soft umbrella of anger at the government and soon taken into the more violent part of the movement if they continue to express interest," says Mary Ann Mauney of the Atlanta-based Center for Democratic Renewal, which monitors hate groups.

Unfortunately, newcomers to the movement will find few guideposts that signal, This way the true believers, that way the dangerous zealots. The ranks of the antifederalist insurgency include plenty of the former: tax protesters, home schoolers, Christian fundamentalists and well-versed Constitutionalists. But the groups also contain an insidious sprinkling of the latter, including neo-Nazis and white supremacists. What binds these diverse elements is a fervent paranoia. The most fearful patriots believe that Soviet fighter jets are on standby in Biloxi, Mississippi, that frequent flyovers by "black helicopters" signal an imminent occupation by the armies of a one-world government, and that stickers on some interstate highways are coded to direct the invading armies. They also regard such federal agencies as the FBI, the Bureau of Alcohol, Tobacco and Firearms and particularly the Federal Emergency Management Agency as the shock troops for an all-out war on personal liberties.

WHILE FEAR IS A COMMON DEnominator, not everyone worries about the same things. Tom Metzger, who founded White Aryan Resistance in 1980 after breaking with the Ku Klux Klan, ridicules talk of a military invasion. "Ninety percent of that stuff is nonsense," he says. "We've got 10 million Mexicans flooding into this country, and the militias are worried about repainted helicopters." For their part, many militia groups aggressively weed out racists, and a few even have minority members. According to its leader, Fitzhugh MacCrae, New Hampshire's Hillsborough County Dragoons includes blacks, Latinos and Asians, and favors good works like shoveling snow for the elderly. "I'm pro-choice and I donate money to PBS," he says. "How subversive is that? But I also support the Second Amendment. It is the only amendment that empowers the rest of them."

Indeed, the right to bear arms seems to be the one altar where moderate Constitutionalists and armed zealots can worship comfortably side by side. "There's a real fear that once the Second Amendment is abridged, the First will be the next to go," says Scott Wheeler, a writer for the U.S. Patriot Network. Despite the reverence for guns, however, "the vast majority of people in the militias are not violent or dangerous," says James Aho, a sociologist at Idaho State University who has interviewed 368 members of the radical right.

They do, however, have an unusually vigorous commitment to self-defense. "Within two years, I expect to see the Constitution suspended. We will be prepared to defend it," says Norman Olson, an independent Baptist minister who together with real estate salesman Ray Southwell founded the Michigan Militia. Toward that end, Olson has led army-style maneuvers on an 80-acre tract of scrub pine and meadow dotted with obstacle courses and bunkers. Most of those who come for the training sessions are middle-aged, white, family men who must struggle to support their loved ones and struggle even harder to catch their breath during Olson's exercises, which require them to traverse rugged terrain shouldering semiautomatic rifles.

Despite the popularity of these exercises, the militia stripped Olson of his command last Friday after he sent inflammatory faxes to the news media blaming the Oklahoma explosion on the Japanese government. When a TIME reporter knocked on Olson's door later that day, Olson appeared in a blue bathrobe. "Why are you bothering me?" he asked. "Can't you see I'm trying to stop World War III?" The next day, both Olson and Southwell, who had helped prepare the fax, resigned under pressure from the militia.

AS BEFITS THESE GO-IT-ALONERS, militia members favor decentralization in their own ranks. The movement has "no national structure, no central command and no central leadership, either recognized from within the movement or without," says Jonathan Mozzochi, executive director of the Coalition for Human Dignity in Oregon. Partly, he believes, this is because it is a "grass-roots upsurge," but the lack of clear structure is intentional as well.

In 1987, Robert Miles, a former Ku Klux Klan Grand Dragon who was convicted in 1971 of burning school buses in Pontiac, Michigan, articulated the idea of "leaderless cells," an organizational structure of small autonomous groups that effectively thwarts infiltration and defuses culpability. "Miles compared his new concept to a spider web," says Richard Lobenthal of the Anti-Defamation League of B'nai B'rith. "You can put your hand in it and it gives, and when you remove the hand, it is still there."

That web is further fortified by the information revolution, which enables people to disseminate their ideas widely, cheaply and often under the safe cloak of anonymity. Bomb recipes have been transmitted across computer bulletin boards. CB and shortwave radios enable militia members not only to communicate between themselves but also to monitor the communications of law-enforcement officials. In the gray split-level house that serves as the nerve center of the Michigan Militia there are 15 phone lines, four computer, multiple fax machines, a professional printing press and a full television-production facility.

Less well-funded patriots mount inexpensive programs on public-access or satellite TV. Or they can make and market their own videotapes—a propaganda tactic that insulated the patriot evangelists from any direct blame for the antigovernment acts they may inspire. The handful of celebrities on the patriot circuit—people like Koernke, attorney Linda Thompson and Militia of Montana founder John Trochmann—all have tapes in circulation that promote their theories about the plot to take over the world. In a two-hour video called *America in Peril: A Call to Arms*,

2. CRIME, TERRORISM, AND VIOLENCE

AMERICA'S MILITIAS: A PRIMER

CATALYSTS Of The MOVEMENT

Oct. 24, 1945: United Nations founded, beginnings of "one-world government"

1958: John Birch Society founded; advocates anticommunism, minimal Federal Government, abandonment of Federal Reserve System, U.S. withdrawal from the United Nations

June 3, 1983: Tax resister and Posse Comitatus member **Gordon Kahl**, wanted in the killings of two U.S. Marshals in North Dakota, is "martyred" in shoot-out with a county sheriff and federal agents in northwest Arkansas

Sept. 11, 1990: In an address to a joint session of Congress after Iraq invades Kuwait, President George Bush proclaims a new world order based on multinational action under U. N. aegis

Aug. 31, 1992: Idaho white separatist **Randy Weaver** surrenders after a stand-off with federal agents in which his wife, his 14-year-old son and a U.S. Marshal are shot and killed. A jury later acquits Weaver and co-defendant Kevin Harris on charges of killing the Marshal. No one is indicted in the deaths of Weaver's wife and son. He now lives in Grand Junction, Iowa.

April 19, 1993: FBI launches an attack on the Branch Davidian compound in Waco, Texas, and more than 70 Davidians are killed as the settlement burns to the ground

Nov. 30, 1993: President Clinton signs the Brady bill into law; it requires a five-day waiting period for all handgun purchases

FAVORITE CONSPIRACY THEORIES

According to many militia members, the U. N. plans to conquer the U.S. using the National Guard and L.A. gangs to disarm the public.

Recent chemical spills are practice runs for a much larger series of disasters, faked by the government, to draw people out of their homes and enable U.N. forces to enter homes and seize guns.

Before the U.N. takes over, the Federal Emergency Management Agency (FEMA) will head up an interim government.

The Amtrak repair yards in Indianapolis will be used as a huge crematorium to dispose of political dissidents.

Black helicopters have been buzzing Western states on missions of surveillance for the invading U.N. troops.

Salt mines beneath Detroit hold a division of Russian troops waiting for the order to rise and take over the U.S.

Small colored bar-code stickers found on the back of road signs will help direct the invading troops.

The government has installed electronic devices in car ignitions to stall autos on the day the new world order takes over.

Paper currency has bar codes on it so government agents can drive by each house with secret scanners and count how much money each family has.

GEOGRAPHY OF ZEALOTRY

United States Militia Association Blackfoot, Idaho
Founder Samuel Sherwood identifies closely with the Weaver siege. He has said that "Civil War could be coming, and with it the need to shoot Idaho legislators."

Almost Heaven Kamiah, Idaho
This armed community created by **Bo Gritz** will be made up of about 30 families. It hopes to be self-sufficient and obey all laws "unless they go against the laws of God and common sense." Gritz has written, "The tyrants who ordered the assault on the Weavers and Waco should be tried and executed as traitors."

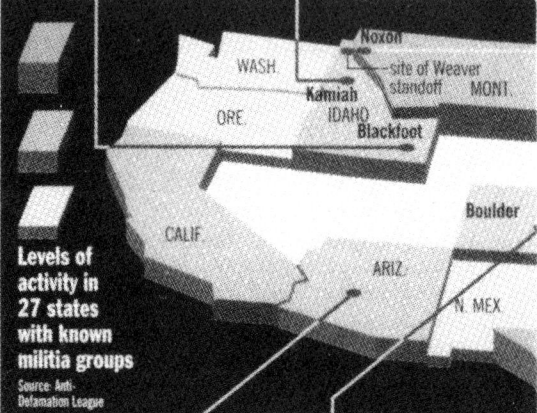

Levels of activity in 27 states with known militia groups
Source: Anti-Defamation League

Police Against the New World Order Arizona
Leader Jack McLamb, a former Phoenix police officer, puts out the monthly *Aid and Abet Police News-letter*, which discusses "constitutional issues for lawmen," and he broadcasts from his home outside Phoenix an hour-long call-in radio program.

Guardians of American Liberties Boulder, Colo.
Leader Stewart Webb, has a history of anti-Semitism and has appeared on right-wing radio shows discussing his conspiracy theories. The group describes itself as a network of American citizens, formed to ensure the government is free of corruption and to safeguard the U.S. Constitution.

ENEMIES LIST

Council on Foreign Relations:
New York City-based foreign policy group whose board includes the heads of Xerox and AT&T and former Federal Reserve chairman Paul Volcker; patriot activists see the council as an advocate of destroying U.S. sovereignty

United Nations:
The Trojan horse of one-world government

Bureau of Alcohol, Tobacco and Firearms:
Accused by militia members of oppressive enforcement of federal gun laws

Attorney General Janet Reno:
Ordered the attack on the Branch Davidian's Waco, Texas, compound

Trilateral Commission:
Long the bête noire of Pat Robertson and the John Birch Society, the commission is a conclave of American, European and Japanese business leaders and politicians that was established and partly funded by banker David Rockefeller in 1973. Prominent Trilateralists: Jimmy Carter, George Bush, Bill Clinton, Henry Kissinger

Federal Emergency Management Agency:
Established during the Carter Administration to deal with hurricanes, earthquakes and other disasters but actually "the most powerful organization in the United States," according to anonymous patriot Internet postings; seen by patriots as a tool of the Trilateral Commission "to seize control of the reins of government through emergency fiat"

9. Enemies of the State

**Militia of Montana
Noxon, Mont.**
MOM is one of the most visible and extreme militias, created by **John Trochmann,** his brother David and nephew Randy to protest the Weaver siege. They feel citizens must form unorganized militias to protect themselves. They also advertise and distribute books, tapes and videos.

**Michigan Militia Corps
Harbor Springs, Mich.**
Started in April 1994 in response to gun-control legislation. **Norman Olson,** a Baptist minister and gun shop owner, was its commander until his ouster last week. It believes the U.N. plans to lead the U.S. into a socialist world government. The corps claims a membership of 12,000.

**Mark Koernke
"Mark from Michigan," Dexter, Mich.**
A leading propagandist of the militia movement. Until last week he had an hour-long shortwave radio show, *The Intelligence Report,* five nights a week over World Wide Christian Radio WWCR. He has ties to the Militia of Montana and John Trochmann.

**American Justice Federation
Indianapolis, Ind.**
Chaired by attorney **Linda Thompson** after the federal assault on the Branch Davidians. It is "dedicated to stopping the New World Order and getting the truth out to the American public." She called for an armed march on Washington and treason trials for congressional traitors.

**Constitution Defense Militia
N.H.**
A small, well-organized group led by Edward Brown. New Hampshire law provides for an unorganized militia. Brown is opposed to gun control, the U.N. and the Federal Government.

TELLING THEM APART

Aryan Nations: White-supremacist group based near Hayden Lake, Idaho, headed by the Rev. Richard Butler

Survivalists: People who prepare for imminent breakdown of the economy and government by stockpiling food, water, guns and ammunition and moving to wilderness hideouts in Idaho, Northern California and Montana

Patriots: Network of antifederal activists; they stress the 10th Amendment to the Constitution, which says powers not delegated to the Federal Government are reserved to the states or to the people themselves

Militias: Groups that arm themselves to defend the Constitution, which they believe is in peril

Christian Identity: Movement that believes Northern Europeans are the chosen people of the Old Testament and Jews are the offspring of Satan; calls the Federal Government the "Zionist Occupational Government"

Posse Comitatus: Tax-resistance movement that included many Midwesterners who were hard pressed by farm crisis of the early- and mid-1980s; believed that IRS was a tool of Zionist international bankers

Wise Use: Movement based in the West and financed by mining and timber companies, seeks to lift restrictions on grazing, logging and mining on federal lands

Texas Constitutional Militia
Set up by Jon Roland, it began in early 1994 and held its first "muster" on April 19, 1994, Waco's first anniversary. Its coded E-mail network serves as an information highway for the underground. Roland claims to have penetrated the government's electronic intelligence.

**Florida State Militia
Stuart, Fla.**
Founded by Robert Pummer. Its handbook says, "We have had enough—enough drugs and crime, violence and bloodshed, enough Waco . . . and government attacks on Christian Americans." It warns, "Buy Ammo now! You will not be able to get it later!"

**Citizens for the Reinstatement of Constitutional Government
Monroe, N.C.**
Albert Esposito, its leader, urges his group to amass the four B's: Bibles, bullets, beans and bandages. The group aims to "make the Holy Bible and the U.S. Constitution the law of the land."

**Blue Ridge Hunt Club
Virginia**
Founder James Roy Mullins and other members were arrested in July and charged with possession and sale of a short-barreled rifle and unregistered silencers. Officials said Mullins formed the club to arm its members in preparation for war with the government.

Koernke, an Ann Arbor janitor who goes by the handle "Mark from Michigan," ominously reviews the "evidence" of one-world conspiracy. At FEMA, he asserts, fewer than 64 employees are engaged in disaster work; the other 3,600 are "there to manage the system after they take over." The incursion is inevitable, he argues, and the only choice is "to lock and load."

Koernke unequivocally denies any involvement in the Oklahoma blast, but he has capitalized on the atrocity to rally his followers around the idea of government complicity in the explosion. On his shortwave radio broadcast the day of the bombing, he exhorted true believers in Oklahoma to grab their video cameras and shoot footage of the site. "Document *what* agencies were coming in and out. As a matter of fact, [my wife] Nancy and the kids, watching the initial footage of this, saw what appeared to be United Nations observers' badges." The next day, patriot computer bulletin-board systems were rife with messages like this one on the Citizenship BBS in California's San Fernando Valley: "This was orchestrated by the shadow government (i.e., Trilateralists, ATF, FEMA, etc.) to whip the public into such a frenzy that Americans will BEG to surrender their privacy for some government-provided protection from terrorism." At week's end, the shortwave station carrying Koernke's broadcast dropped his show.

Other patriots are worried that their own agendas will be confused with the as-yet-unexplained agenda of the Oklahoma bomber. "If you get one crazy out there who doesn't have a brain, then everybody gets lumped in," complains Dean Compton, 33, who heads an armed militia in California's Sierra Nevada foothills. In dread of just such an event, he announced the formation of a group in March called the National Alliance of Christian Militia. Compton, who claims 85% of the militia movement is Christian, says the new alliance is an attempt to distinguish their efforts from "the hate groups and the Klan."

2. CRIME, TERRORISM, AND VIOLENCE

SUCH DISTINCTIONS, HOWEVER, ARE not always apparent. Two days after the bombing, 550 patriot Christians gathered in Branson, Missouri, for the International Coalition of Covenant Congregations Conference. "I mingled with a lot of people there, and there was not a shred of sympathy for what happened in Oklahoma," an attendee told TIME. According to this source, some participants felt the carnage was understandable retaliation for the 1993 deaths of children during the government raid in Waco, Texas.

California's Compton asserts that "mainstream" militias are working round the clock to assist in the investigation of the bombing. "We want worse than anybody to make sure these guys come to justice." But, he warns, "if the government gets heavy-handed" in its search, "we'll have some problems." Easygoing and articulate, Compton, the father of three, discusses his apocalyptic convictions with patience and occasional humor. Yet he is girding for guerrilla-style warfare against his own government. He's got a 9-mm semiautomatic pistol strapped to his hip, a wad of emergency cash and enough ammunition to fight a small battle. In the back of his battered Chevy Silverado, he packs a green .223-cal. Sporter assault rifle, a $200 Kevlar helmet, a CB radio, walkie-talkies, camouflage uniforms and 15 days of provisions.

Compton, who quit his job as a real estate agent in the lumber community of Shingletown to run armed-militia training camps, says all these preparations are for the sake of his children. "I decided I'd do everything in my power to make sure that when my son grows up, he'll have the same freedoms I had," he explains. His children, who range in age from 3 to 11, are schooled at home by Compton and his wife. Compton fled the San Francisco Bay Area 10 years ago to settle on a 120-acre mountain ranch, and his rage against the government seems to have grown in this region where more jobs than trees have been felled in recent years. "Three local lumber mills have been closed in the past year because of spotted owls," he claims. Compton also complains that he can't dig up the manzanita bushes on his own land because of local ordinances. "But," he adds cynically, "you sure better pay taxes on it, or they'll take it all away from you."

That part of Compton's rant is familiar territory for those caught up in the modern Sagebrush Rebellion, a land-rights movement that is spreading rapidly in Western states. Over the past few years, offices of the Bureau of Land Management and the Forest Service have come under increasing attack by ranchers, farmers and loggers fed up with federal rules about land use, water rights and endangered habitats. In Nevada, where more than 80% of the state is public land, federal employees have been refused service in restaurants, taunted at public gatherings and harassed with vulgar gestures. In March a bomb exploded in the forest rangers' district headquarters in Carson City, shattering windows and damaging the office of the chief ranger.

Do you agree citizens have the right to arm themselves in order to oppose the power of the Federal Government?	Yes
WHITE MALES	34%
ALL OTHERS	22%

Do you agree Americans have the constitutional right to buy and store large amounts of weapons?	Yes
WHITE MALES	34%
ALL OTHERS	18%

Increasingly these local-control activists are finding common cause with the militias. "Both have an antifederal outlook," says Dan Barry of the Environmental Working Group in Washington. "They run into each other because they have similar priorities." Indeed, their complaints are often indistinguishable. "We've been pushed so far by rules and regulations, the feds are in our pockets so deep, people are outraged," says Ronny Rardin, a commissioner in New Mexico's Otero County. The land-reform rebels have also been developing an appetite for militia-style conspiracy theories. "The New World Order will be running our lives through the United Nations!" warns a fund-raising letter for the National Federal Lands Conference, one of the leading groups.

Meanwhile, the conspiracy theories grow wilder by the day. In one, Queen Elizabeth is working through British conglomerates to regain control of the colonies—witness the purchase of Burger King and Holiday Inn by British companies. There is also rumored to be a global conspiracy to implant newborn babies with microchips. Robert Brown, editor and publisher of *Soldier of Fortune*, a must-read for many gun advocates and survivalists, has tracked some of these wacky conspiracies and discounted them, including one much vaunted black helicopter sighting. But that hasn't quieted the patriot grapevine. "It's all bull. But they don't want reasonable explanations," he says, "because they don't fit their preconceived notions."

The movement's communications network is often more sophisticated than its judgment, which means that any mysterious incident gets blown into a conspiracy epic. Take, for instance, the "invasion" of Okanogan County in northwest Washington State. It began last Labor Day when a local cattle rancher stumbled across a backwoods military camp teeming with men in fatigues. Word quickly spread that the invasion of U.N. troops had finally begun. When concerned citizens showed up at Sheriff Jim Weed's office, Weed grabbed the telephone and soon learned that the men in cammies were actually border-patrol officials conducting a joint operation with Canadian authorities. By then, though, panic had spread throughout the state, prompting phone calls from state senators and representatives. To this day, there are some patriots who still don't believe Weed's explanation. "I was accused by one person of being seen getting off a U.N. helicopter at an airport wearing a blue helmet," Weed says.

In the wake of Oklahoma City, many U.S. agencies are stepping up their watch of militia activities, but they may only feed the patriots' paranoia about government. If investigators start knocking on the doors of militia members, warns Ron Cole, who is a lecturer on the patriot circuit, "it could conceivably turn into an armed struggle against the government."

Chip Berlet, who tracks right-wing populism for Political Research Associates, based in Cambridge, Massachusetts, is not alone in drawing parallels between America's patriot movement and Germany's Weimar Republic. "You see the rise of a large group of disaffected middle-class and working-class people with a strong sense of grievance," he says. "None of the major parties speak for them." If their grievances aren't resolved, he warns, they are likely to become more militant. The message from the militias is largely the same: whether it takes a whisper or a shout, we will be heard.

—Reported by Sam Allis/Boston, Edward Barnes/Petoskey, Michigan, Patrick Dawson/Billings, David S. Jackson/San Francisco, Scott Norvell/Atlanta and Richard Woodbury/Denver

Article 10

How Nation's Largest Gang Runs Its Drug Enterprise

Ann Scott Tyson

Staff writer of The Christian Science Monitor

CHICAGO

At 13, Greg obediently stood "security" watch over Gangster Disciple drug turf on Chicago's North Side for $50 a day. At 14, he went to jail for shooting a rival Vice Lord in the back to protect his gang.

Today, with a beeper on his waistband and a revolver in his hand, the 16-year-old shows up on schedule to sell drugs for the GD at one of its hundreds of Chicago venues. Being late will bring a fine or "violation," a brutal beating by gang enforcers.

> **The GD is 'one of if not *the* largest and most successful gang in the history of the United States.'**
> —*James Morgan, DEA*

"Are you searching, brother?" Greg asks a passerby. "I got it. I got the rock [crack cocaine]. I got the boy [heroin]."

Customers, nick-named "geezers" and "hypes," saunter through the ripped out doorway of the high-rise tenement where Greg works. Inside, other gang members search the buyers before sending them up dark stairways to be "served."

After a few hours on the job, Greg has sold drugs worth $500. The gang takes $350 and gives Greg $150, based on a standard 70-30 split. Gang narcotics revenues total $5,000 on an average day at the building, one of many in the GD's city wide domain.

The lucrative business operating out of this Chicago tenement provides a look at the illicit inner-working of perhaps the nation's largest gang-run narcotics network.

Mafia-like in structure, military-like in discipline, it is a $100-million-a-year enterprise that stretches into 35 states.

The business is propelled by profit-hungry drug dealers and craving addicts, ready supply and insatiable appetite.

Yet the backbone of the GD's retail operation—and the key to its status as perhaps America's most powerful super-

'Gangster Disciples' Chicago Territories

This map was developed from a detailed computer list of Gangster Disciple leaders, their territory in metro Chicago, and their "count" (number of gang members under them). Allegedly ordered by GD chief Larry Hoover in 1993, the document was discovered by federal agents in 1995 in a folder marked "L. Sr." at a rap-concert firm run by Mr. Hoover's wife. It reveals the size and scope of the GD organization.

- Governor: 'Rob' Count: 341
- Governor: 'Terry' Count: 317
- Governor: 'Boo-G' Count: 609
- Governor: 'J.D.' Count: 920
- Governor: 'Bible' Count: 573
- Governor: 'Charlie' Count: 360
- Governor: 'Sherman' Count: 1205
- Governor: 'King' Count: 411
- Governor: 'Kadafhi' Count: 761
- Governor: 'Fool' Count: 1193

Source: TrialGraphix, Chicago

2. CRIME, TERRORISM, AND VIOLENCE

gang—is the top-down organization that commands drug turf and dealers like Greg with an iron fist.

"The Gangster Disciples are one of if not *the* largest and most successful gang in the history of the United States," says James Morgan, special agent in charge of the US Drug Enforcement Administration in Chicago.

"They are incredibly well-disciplined and trained," he says of the 30,000-strong gang. In contrast, he says, other big gangs such as the California-based Crips and Bloods are "like posses, like roaming bands of drug dealers with no leaders."

The GD has "a very sophisticated battle plan and a very sophisticated organization," Chicago-based US Attorney James Burns told President Clinton during a May briefing on the gang. "When you look at the structure... it's absolutely incredible."

Dismantling this huge, militaristic hierarchy is the goal of a joint US and local law-enforcement investigation aimed at bringing down the GD. Last year, federal authorities indicted 39-high ranking gangsters and collaborators on drug-conspiracy charges. Ten were convicted in March, three are fugitives, and 26 others, including GD chief Larry (King) Hoover, await trial in October.

The federal crackdown on the Gangster Disciples' leadership is fueling internal strife within the gang. A rash of more than a dozen murders and scores of shootings have rocked the gang since January as top-level members struggle for power, falling-outs occur with gang branches in other cities, and rank-and-file disputes go unchecked.

Meanwhile, the gang is attempting to ruthlessly close ranks. For instance, GD board member Darryl (Pops) Johnson and four other gang members were indicted June 20 for allegedly murdering two fellow GDs suspected of breaking a code of secrecy and aiding US investigators. One of those murdered, Charles "Jello" Banks, was a cooperating federal witness when he was killed.

Police in GD territory also report a spate of people shot in the leg or abdomen, a typical punishment for insubordination.

Despite evidence of discord at the top, the GD's drug trade so far remains remarkably efficient and intact, with 24-hour sales ongoing in much of Chicago, police say.

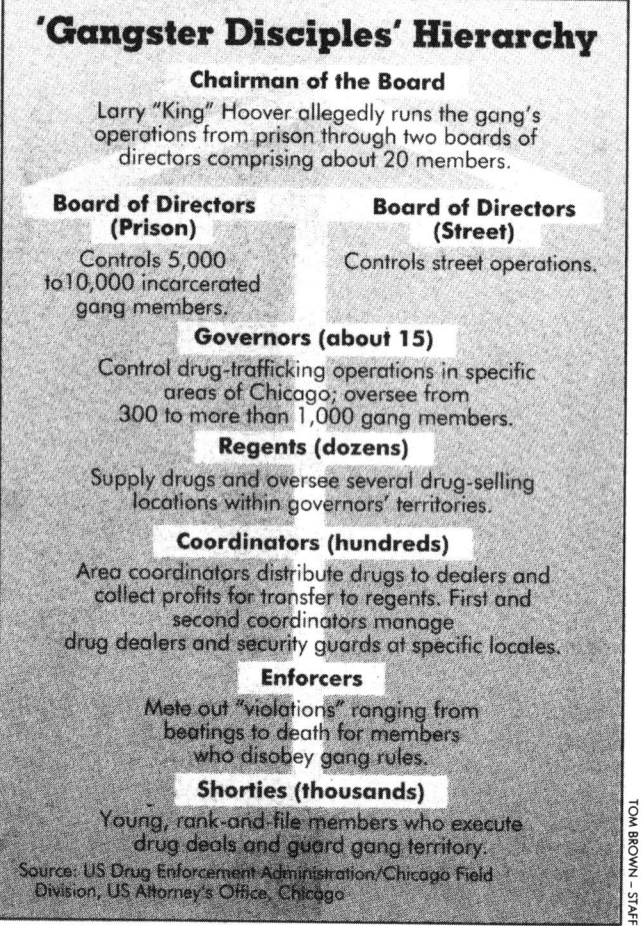

'Gangster Disciples' Hierarchy

Chairman of the Board
Larry "King" Hoover allegedly runs the gang's operations from prison through two boards of directors comprising about 20 members.

Board of Directors (Prison)
Controls 5,000 to 10,000 incarcerated gang members.

Board of Directors (Street)
Controls street operations.

Governors (about 15)
Control drug-trafficking operations in specific areas of Chicago; oversee from 300 to more than 1,000 gang members.

Regents (dozens)
Supply drugs and oversee several drug-selling locations within governors' territories.

Coordinators (hundreds)
Area coordinators distribute drugs to dealers and collect profits for transfer to regents. First and second coordinators manage drug dealers and security guards at specific locales.

Enforcers
Mete out "violations" ranging from beatings to death for members who disobey gang rules.

Shorties (thousands)
Young, rank-and-file members who execute drug deals and guard gang territory.

Source: US Drug Enforcement Administration/Chicago Field Division, US Attorney's Office, Chicago

TOM BROWN – STAFF

Whether US law-enforcement agencies can permanently put the wealthy and well-armed Gangster Disciples out of business is a crucial test for the government as gangs proliferate across the country.

"Gang violence has spread to every corner of America," US Attorney General Janet Reno said last month. Ms Reno announced the results of the first national survey of gang activity, which showed an estimated 650,000 gang members and 25,000 gangs nationwide. Gang problems are worsening in 48 percent of the communities surveyed and improving in only 10 percent, she said.

The magnitude of the government's challenge is clear from an inside look at the GD and its rise from a renegade prison gang of a few dozen men in the 1970s to a mature criminal enterprise that dominates large areas of Chicago's inner city and suburbs.

"Over the last quarter century, one gang in particular has... evolved better than the rest," says US Attorney Burns, "and that's the Gangster Disciples."

From the beginning, discipline has been central to the GD's mystique and culture.

Tommy, a veteran GD with a crossed pitchfork tattoo on his right arm, recalls joining the gang soon after Mr. Hoover formed it in prison in 1974 from a splinter group of the original Black Disciples. (Hoover is currently serving a 150- to 200-year state prison term for a gang-related murder.)

"It was very strict. You had to have total respect," says Tommy. He was "blessed" into the gang by a committee of six members after memorizing the 16-rule gang code written by Hoover. "The main man [Hoover] made the laws. You gotta know the laws," he says. (Tommy's real name and the names of the other gang members interviewed have been withheld.)

The canons prohibit members from using addictive drugs, gambling on credit, engaging in homosexual rape, being a bad sport, and stealing from or showing disrespect to other members. Personal cleanliness and exercise are also required.

Every morning at 5 a.m., Tommy remembers, he gathered in the field of his public-housing project with a group of young black GD members to do jumping jacks and jog.

Hoover, a self-proclaimed student of Al Capone, modeled the GD organization to a degree after Chicago's Italian Mafia. At the top is the "chairman" (Hoover) and two boards of directors, one operating in prison and the other on the street. Below them, governors control up to 1,500 members each in specific territories, which are further subdivided among regents and coordinators.

In the 1970s and early 1980s, the GD used its manpower mainly to guard and expand its turf against rival gangs. Tommy rose to become a coordinator. Bloody gang wars were common, and the main payback for Tommy's loyalty was a promise of status and protection.

But when crack cocaine began flooding Chicago in the late 1980s, the lucrative drug trade emerged as the gang's priority. The GD refined its organization to run a citywide network of dope-dealing franchises. The gang now buys drugs from Colombian cartels in 100- to 200-kilo shipments and distributes them down the ranks, while profits and dues called "street taxes" flow back up.

Today, the gang lures young recruits, many of them poor and jobless, with the promise of ready cash. These youths shoulder much of the risk for the gang's business, but profit far less than the top executives, who prosecutors say generally don't handle drugs or weapons.

Jeff, 14, and Mike, 13, are two recent recruits, or "shorties." Both grew up in Chicago's Cabrini-Green public-housing project, which is dominated by the GD and known here as "the killing fields." Gang members say about 40 percent of young men at Cabrini belong to the GD.

Childhood memories for these teens include annual barbecues put on by the GD at Cabrini and treats of popcorn or potato chips handed out on the litter-strewn playground—all paid for with drug money, gang members say.

Now Mike and Jeff have bigger responsibilities. Every day, for $75, they work three-hour "security" shifts around Cabrini-Green. First they pick up guns—powerful handguns like TEC 9s, .357 Magnum semiautomatic revolvers, and Uzi machine guns—from a car that cruises the area. Standing on corners or outside buildings, they look out for police or rival gang members.

"It's a job. You're protecting your community," says Mike, sitting astride his 10-speed bike at Cabrini one morning, "I'm not gonna let them [rivals] kick down my door and kill my family. I'm gonna shoot them," he says, pointing to a building a block away controlled by the rival Cobra Stones.

Despite their bravado, both Mike and Jeff admit they are scared of going to jail. "I don't want to get beat up and raped," says Jeff, a soft-spoken seventh grader wearing the GD colors of black and blue. Jeff joined the gang only a few months ago after failing to find another job. The money, he says comes in handy for buying clothes and food.

Greg, the 16-year-old drug dealer, is one step higher up the chain of command. He reports to the GD coordinator for his building. The coordinator distributes drugs, collects the "street tax" all dealers must pay, sets the schedule, and runs the security force. Many dealers, like Greg, work to save enough money to start selling drugs for themselves in addition to selling for the gang.

"I see the people comin' up. I feel I can do it too, gettin' money and stuff," Greg says.

Hardened by jail and struggling in school, Greg's ambitions and sense of security now seem inseparable from the GD. "I never go anywhere alone. I plan on staying in the gang forever," says the second-generation gang member. "I want to be a board member, like the second man to the 'King' [Hoover]," he smiles, touching a gold earring worn in GD fashion in his right ear.

Still, Greg's cocky optimism is tempered by the disillusionment of more experienced GD members like Sammy. Sammy remains loyal to Hoover, but he says drug money and greed have corrupted the GD and compromised its written ideals of unity and self-determination as "Brothers of the Struggle."

Jailed himself for drug dealing, the father of three is now trying to gain distance from the gang, but at a high cost. A few months ago, after Sammy refused to stand security, gang enforcers beat him so viciously that he was hospitalized with a head injury.

"I feel boxed in," Sammy says one summer evening outside his apartment. "There are a lot of guys with potential, but we tend to let it slip away by being on the street."

While incarcerated in a state-run prison, Hoover continues to run his narcotics ring. But federal prosecutors hope that if Hoover and his top lieutenants are convicted and sent to a higher-security federal prison, they would be effectively cut off from the street organization. They also expect that GD middle-managers and foot soldiers like Sammy to defect.

"I see an escalating fleeing from the front lines," says Assistant US Attorney Ron Safer, the lead prosecutor in the GD trials.

Sammy agrees, but his vision of the future is more apocalyptic. "If Larry Hoover went to [federal] jail, I think the organization would crumble. There would be constant shooting. Everyone would want to control the city," he says. "It would be a dangerous sight to see."

Forgiving the Unforgivable

Survivors of crime and abuse are learning an unlikely method of freeing themselves from their anguish: forgiveness. It's a skill, say therapists, that can benefit us all.

Jean Callahan

Former NEW AGE JOURNAL *articles editor Jean Callahan is a Boston-based freelance writer. Her article "Spiritual Adventures in the Borscht Belt" appeared in the May/June issue.*

Four years ago, Phyllis Hotchkiss woke up in the middle of the night with a terrible ache in her chest. Her nineteen-year-old son, Brian, who had gone out to a carnival earlier in the evening, still wasn't home, and she was worried. A few hours later, two policemen rang the doorbell and told her that Brian was dead. His head had been bashed in, probably with a tire iron, although the murder weapon was never located. For months the police had no suspect, either, but Hotchkiss knew—the way a mother knows—that his killer was the neighbor with whom Brian had spent his last hours. Eventually that neighbor would be arrested, tried, and convicted, but in the months after Brian's death, his freedom tortured Hotchkiss. With alarming cruelty, he would station himself across the street from her home, staring and laughing at family members as they came and went. When the truth about Brian's death finally came out, it made no more sense than did his murderer's persecution of the family: Brian had simply balked at pitching in money for gas; the two young men had argued, then fought; some time later, Brian was killed.

Today, taped to the wall above the computer in the Hotchkisses' suburban Massachusetts home office is a photograph of Brian's murderer. Push pins are stuck into his eyes, his cheeks, his neck. "On bad days," says this small, soft-spoken woman in her mid-forties, "I move them around, add more, or push them in further."

Phyllis Hotchkiss has lived through a parent's worst nightmare, and a day doesn't go by when she doesn't feel hatred for the remorseless young man who took her son's life. "I will never forgive him for Brian's death," she says. It was rage, in fact, that gave her the energy to form a local chapter of Parents of Murdered Children, a group that has supported dozens of other people who have faced similar tragedies.

But several months ago, something happened that seems to be softening Hotchkiss's anger. As part of her work with the parents' group, she initiated a series of weekly meetings with prisoners—mostly convicted murderers—at the Bay State Correctional Institute in nearby Norfolk, Massachusetts. She had proposed the meetings to the prison chaplain, she explains, so that she and other women whose children have been murdered could "tell our stories and make sure these men realize how much they've hurt people." She expected to see the prisoners as "human garbage," the way she sees Brian's murderer, but instead she was touched by the courage the men showed in facing the grieving mothers. She was especially moved by one man who broke down sobbing when he described himself as a victim as well as a criminal because his own son had been murdered.

As each series of meetings ends, Hotchkiss and the other women hold a simple ceremony in which the men are given a remembrance—a tiny gold angel lapel pin. "It's never too late to change a life," she says. "If I can touch one prisoner's heart, maybe he won't hurt anyone else."

Talking to Hotchkiss now, it's clear that a subtle shift has occurred as she has opened her own heart to these strangers. She is surprised herself at how the meetings with these prisoners have affected her. She seems to be on the verge of redefining herself, not only as the mother of a murder victim but also as someone who can integrate that terrible experience, who can remember Brian without constantly feeling the pain of his loss. Phyllis Hotchkiss is not ready to forgive her son's murderer yet, but she has nonetheless begun to experience what a number of psychotherapists describe as the healing power of forgiveness.

ONFRONTED WITH such a story, many of us will shudder—perhaps as much in response to the word *forgiveness* as to the callousness of the crime. In such a case, forgiveness, at least as the term is widely understood, would seem entirely inappropriate—even dangerous. Shouldn't people who commit crimes be condemned and held accountable for their wrongdoing? Wouldn't "forgiving" a brutal killer merely send a message that society will tolerate those who murder, steal, rape, or in some other way slash the social fabric? Isn't Phyllis Hotchkiss's sense of righteous wrath a natural human—and wholly justifiable—response to being so cruelly victimized?

11. Forgiving the Unforgivable

Some things, we might argue, are just unforgivable. Drawing upon the same logic, we might similarly justify our own long-term anger and resentment about far less catastrophic tragedies—the boss who fired us precipitously, the alcoholic father who was never there for us emotionally, the girlfriend who unceremoniously dumped us just when we needed her most. Even those who are at peace with their personal histories can, in a heartbeat, label any number of social offenses absolutely "unforgivable": Rape. Torture. Child abuse. Drunk driving. Infidelity.

There may be compelling reasons to cling to our grudges and animosities, and significant satisfactions in doing so. But we hold on, say those who advocate forgiveness, at a very high price—emotionally, physically, even spiritually. "There are people who wake up in a rage every morning over something that happened thirty or forty years ago," notes Robin Casarjian, psychotherapist, author of the recently published *Forgiveness: A Bold Choice for a Peaceful Heart*, and a leading voice in the growing forgiveness movement. "If anger is running your life, who is it hurting?"

For more than a decade, Casarjian has been helping people let go of old anger and resentments through the practice of forgiveness. She's presented the idea to recovering addicts and alcoholics, to prisoners, to mothers living in homeless shelters with their children, to public school teachers, and to groups in the corporate, health-care, and military communities. At the beginning of each forgiveness workshop—no matter what type of audience she's addressing—Casarjian must confront her listeners' inevitable resistance to the concept.

"So often when people think about forgiveness they think about what it's going to do for someone else," she says. "They say, 'I'm not going to forgive them, after what they did,' as if forgiving them would be doing the other person a favor. What they don't realize is that forgiveness is really an act of self-interest. We're doing ourselves a favor, because we become free to have a more peaceful life—we free ourselves from being emotional victims of others."

Robert Enright, a University of Wisconsin professor who has documented connections between forgiving and mental health in groups of elderly women, college students, and middle-aged adults, would agree. Enright has found that people who forgive long-standing grievances decrease their levels of anxiety and depression and raise levels of self-esteem. Their blood pressure tends to drop and the tension in their faces eases. The greatest emotional transformation occurs, Enright says, among those for whom the stakes are highest: someone, say, who has been deeply wounded in a very close relationship.

Mind-body research also suggests that holding onto anger and resentment can make you more than depressed and anxious; it can also make you physically sick. For years, psychoneuroimmunologists have shown how emotional stress changes the way the immune system functions. People under stress routinely develop more colds, flus, and infections, and stress may help trigger autoimmune diseases such as lupus and rheumatoid arthritis. Describing the cancer-prone personality in their book *Getting Well Again*, O. Carl Simonton, M.D., Stephanie Matthews-Simonton, and James Creighton cite "a tendency to hold resentment and a marked inability to forgive" as part of the process through which some people develop serious illnesses. And in his books *The Trusting Heart* and *Anger Kills*, Duke University medical school professor Redford Williams, M.D., makes a convincing case that of all the Type A traits, clinging to anger is the one most likely to trigger the development of heart disease. He also says that people who relax and learn to let go of their anger are likely to live longer even when they do get sick.

OK. So forgiveness may have its benefits. But what if you just can't bring yourself to do it? Not to worry, says Casarjian. "When people tell me that, I ask them not to say 'I *can't* do it,' but to say 'I haven't been able to.'" Forgiveness, Casarjian explains, seldom occurs spontaneously, like some mystical bolt of lightning that strikes if you're lucky. Rather, it's a skill that can be practiced and learned. But before it can be learned, it's important to understand what forgiveness does and does not mean.

OST OF US LEARNED the word *forgiveness* in childhood religious training. In the Old Testament, God forgave the Israelites for worshiping false idols, but not until he'd sent them reeling through the desert as punishment. In the New Testament, Jesus forgives those who crucify him, a superhumanly hard act to follow. Our early efforts at forgiving are usually false starts, no deeper or more sincere than a five-year-old's mumbled "That's OK." But when advocates such as Casarjian use the term *forgiveness*, they are not talking about patriarchal noblesse oblige, turning the other cheek, condoning offensive behavior, letting the bastards off the hook, pretending everything's fine when really it isn't, making nice, denying anger, or any of the other popular definitions. "Forgiveness does not in any way justify or condone harmful actions," explains noted Vipassana meditation teacher Jack Kornfield in his new book, *A Path with Heart*. "While you forgive, you may also say, 'Never again will I knowingly allow this to happen.'" Casarjian concurs: "You can even forgive someone and at the same time work very hard at getting them convicted for the crime they've committed."

Nor does forgiveness mean that you're inviting the person to commit the act all over again. "People often have an underlying fear that they're setting themselves up to be hurt again, because they aren't clear about their own boundaries," says Casarjian. "They don't realize they can forgive a person and at the same time confront them and really challenge them to be accountable for growing up." Nor does it mean you have to become best friends with the person who has harmed you. "Just because you forgive someone," says Casarjian, "it doesn't mean you have to have dinner with them or see them or even talk to them again."

In fact, forgiving someone doesn't require that you see or talk to the person at all, since forgiveness is basically an inner process. Writes Kornfield: "Forgiveness is simply an act of the heart, a movement to let go of the pain, the resentment, the outrage that you have carried as a burden for so long. It is an easing of your own heart." Gerald Jampolsky, M.D., a psychiatrist who works with children facing catastrophic illnesses at the Center for Attitudinal Healing in Tiburon, California, describes forgiveness as a matter of perception—"seeing the light instead of the lamp shade"—seeing through the dumb, thoughtless, crazy things people do to the soul shining within each person. Casarjian likes to amend Jampolsky's definition by saying that forgiveness really means seeing the light as well as the lamp shade. You can acknowledge somebody's faults and still keep your heart open to that other human being.

For Casarjian, this choice to see in a new way—to "see with spiritual eyes rather than physical eyes"—is the crux of forgiveness. "I don't just look at a person and say they're a jerk. I look at them and, yes, I see that, but I also see their behavior in

65

the context of a much greater reality about them—I see a fundamental goodness that may be buried under the way they're acting." The side benefit of this practice, she says, is that when you start seeing the goodness in others, it reflects back on yourself. You begin to experience your own goodness as well.

FORGIVENESS, IT SHOULD be noted, remains a controversial subject, especially among survivors of child abuse and many therapists who work with them. Because the victims of childhood brutality find it so hard to believe that someone they love would hurt them, because they need parents' love and approval no matter how badly the parents have treated them, and because they want to stop hurting and be part of a family again, they are all too willing to forgive, these therapists say.

Having witnessed the emotional and physical costs of premature forgiveness, many therapists who work with survivors of child abuse and incest actually caution against forgiving. Judith Herman, M.D., author of *Trauma and Recovery*, maintains that in cases of childhood incest, forgiveness is not possible until the perpetrator has made amends. Alice Miller, the renowned Swiss psychotherapist, actually calls forgiveness a form of collaboration with the perpetrator in cases of incest and other child abuse. Speaking from her own experience in *Breaking Down the Wall of Silence*, Miller writes that "it was precisely the opposite of forgiveness—namely, rebellion against mistreatment suffered . . . that ultimately freed me from the past."

The problem with Herman's thinking, say forgiveness advocates, is the fact that perpetrators seldom make amends. And, although they agree with Miller that rebellion may be the most healthy response to mistreatment in the short run, it limits development when it becomes a general response to life. "People who have been deeply and profoundly abused, especially as children, may well need to work on rage and grief for a long time," says Casarjian, emphasizing that no one can ever tell anyone else that it's time to forgive. Says Kornfield: "When you have been deeply wounded, the work of forgiveness can take years. It will go through many stages —grief, rage, sorrow, fear, and confusion —and in the end, if you let yourself feel the pain you carry, it will come as a relief, as a release for your heart."

CASARJIAN SUGGESTS starting out the practice of forgiveness slow and easy. "When I first introduce the notion, I say, Don't even think about forgiving the people you have a real historical charge with—say, your mother or father." Instead, she suggests practicing on what she calls "neutral territory"—with people you feel you don't even know: "people in elevators, or strangers walking down the street." But, she warns, you'll soon find that there is little neutral territory: "What becomes particularly interesting is to observe that our 'ego' minds or personalities have judgments about so many people. I don't like the way that person looks or the way she dresses, or he looks like a nerd. We make character judgments about people based on their behavior or their dress."

One of the most enlightening, and certainly the most amusing, ways Casarjian has developed for helping people step out of this maze of judgment to experiment with the transformative powers of forgiveness is something she calls "the forgiveness walk." She asks you to simply walk through an ordinary half-hour of your life with an open heart, assuming that everyone you encounter is a fundamentally decent, good human being. As you meet each person or situation, assume the best and send out gentle thought-messages of love and respect. Try this exercise and you will probably be surprised to discover how quickly and how often the world gets the better of you. From ill-mannered drivers to incompetent sales clerks, from recalcitrant coworkers to your own dear children, the opportunities for irritation are infinite.

Who among us has never flown into a rage over some petty injustice? "Forgiving the big stuff is relatively easy," observes Marianne Williamson, author of the best sellers *Return to Love* and *A Woman's Worth* and a teacher of the forgiveness-heavy *Course in Miracles*. "What's really hard is forgiving the courier for not delivering the Federal Express package on time, or forgiving the telephone company because the line was supposed to be turned on at two o'clock and now it's five. In a crisis, something kicks in. You get a special grace. But on a daily basis, forgiving is harder."

With a little practice, says Casarjian, you'll soon begin to understand how remarkably the world changes if you just step back a bit and respond to minor aggravations intentionally and lovingly, keeping everyone's basic innocence in mind. Your mood will lighten, you'll have more fun, and you'll make much more profound connections with the people you meet.

Another first step toward forgiveness, says Casarjian, is to sit down and write a list of all the benefits you get from holding onto anger. Is it a way to prove you're "right"? Is it an effort to control a situation? Is it a way to be heard? Is it a way to avoid the risks of opening up to other feelings? Bodywork can help, too, she says. Practices such as therapeutic massage, bioenergetics, Alexander Technique, or Feldenkrais Method exercises can free up emotions held in the joints and muscles. Once you've dredged up all those hostile feelings, what do you do with them? Expressing anger can be a crucial prerequisite to forgiving. Communicate your hostile feelings in a positive way, whenever possible. When you can't communicate directly, write angry letters you have no intention of mailing. Devise a ritual to dispel the bad vibrations. Scream at the top of your lungs while you drive down the interstate with the windows rolled up and the radio blaring. Run a marathon. No matter how you choose to do it, the point is to work through your rage instead of holding onto it.

Once anger has been explored and expressed, says Carsarjian, you can more readily move beyond it to forgiveness by adopting the following viewpoint: Almost all human behavior is driven by a primal need for love and respect. Although at first it may seem absurd to imagine this about, say, the creep who steals your car, just try to imagine that each mean or seemingly stupid thing you see somebody do is an expression of fear and a very limited, very warped way of looking for love. If you can succeed even occasionally at this reframing, Casarjian says, if—just once in a while —you can weigh vulnerabilities, weaknesses, new stresses, old traumas, you will begin to make fewer harsh judgments.

There are times when expressing your anger directly by talking it through with the other person isn't desirable—or even possible. Say, for example, you harbor old resentments toward an ex-lover you haven't seen for years or a family member who is now dead. In such situations, Casarjian recommends a meditation she calls the Love and Forgiveness visualization.

After getting in a quiet, meditative state, you picture in your imagination a safe and comfortable place, one in which you feel peaceful and strong. Then you mentally invite someone you'd like to forgive to join you in this place. Take some time to talk in your imagination to this person,

11. Forgiving the Unforgivable

sharing the truth of your experience, including how that person has hurt and disappointed you. Then listen to the truth of that person's experience, trying to let go of blame and judgment.

When someone is openhearted and willing to heal a rift, says Casarjian, such a visualization can lead to tremendous change in a relationship, without the other person taking any part in the process. Jack Kornfield discovered this several years ago through the remarkable experience of a child abuse survivor who had attended one of his meditation retreats. "She'd been angry, depressed, and grieving for many years," he writes in *A Path with Heart*. "She had worked in therapy and meditation through a long process to heal these wounds. Finally in this retreat she came to a place of forgiveness for the person who had abused her. She wept with deep forgiveness, not for the act, which can never be condoned, but because she no longer wished to carry the bitterness and hatred in her heart.

"She left the retreat and returned home and found a letter waiting in her mailbox. It had been written by the man who had abused her, with whom she had no contact for fifteen years. While in many cases abusers will deny their actions to the last, in spite of forgiveness, something had changed this man's mind. He wrote, 'For some reason I felt compelled to write to you. I've been thinking about you so much this week. I know I caused you great harm and suffering and brought great suffering on myself as well. But I simply want to ask your forgiveness. I don't know what else I can say.' Then she looked at the date at the top of the letter. It was written the same day she completed her inner work of forgiveness."

Y FAR, THE MOST POWerful and mysterious act of forgiving is self-forgiveness. Without even knowing it, Robin Casarjian says, most of us severely limit the joy we experience because we hold ourselves accountable for "sins" committed in the past. By far the most extreme examples of this can be found among men in prison, with whom Casarjian has been working for the past several years. For many people, she acknowledges, the idea of convicted rapists and murderers forgiving themselves is "as unacceptable as the actual committing of the crime." But if the objection is that the guilty should suffer emotionally for their sins, in fact the hard work of self-forgiveness is far more wrenching for these men than is simply continuing a life of denial, Casarjian notes.

One of the techniques she asks the men to try is called mirror-work, an exercise that involves looking yourself in the eye in a mirror and asking, Who am I? "I sometimes do not even look at myself when I am shaving," one prisoner wrote when he first looked at the man in the mirror. "There is something in my own eyes I want to avoid. What I feel is confusion and what I see are the bones of a lifetime of regret."

When Casarjian asks prisoners to write letters to themselves about forgiveness, litanies of pain and anger pour out. One begins like this: "Dear Ron, I forgive you for hurting yourself both physically and mentally. I forgive you for fucking up two marriages, for being bulimic, and for taking so many drugs. I forgive you for being so mean to your stepchildren and to your own kids and for not letting them love you." Another ends this way: "I know you've felt guilty about your past and present situations, but it's OK now. It's OK to feel guilt but it shouldn't be 'held onto' and 'lingered in' throughout the rest of your life. You can forgive yourself for those times. You didn't forget them or repeat them. You've held onto them long enough to LEARN from them. Growth is the most important aspect of making a mistake. You've done that."

Far from a superficial act of pardoning themselves, the process of self-forgiveness causes these men to suffer greatly, Casarjian says. In examining their lives and their actions, often for the first time, they are forced to acknowledge deep guilt and remorse.

IT IS NOT SURPRISING THAT ISSUES OF forgiveness often arise in the middle years. The process of self-evaluation and reflection that characterizes this period of life usually helps people develop more realistic self-images. And as we get in touch with our own weaknesses, something paradoxical happens. Other people's weaknesses are easier to understand: We develop empathy. And, when we become more aware of others' pain, our own pain seems easier to accept. "Everyone in this life encounters pain," says Enright. "A crucial part of carrying on is accepting it." Those who accept life's pain instead of blaming it on the actions of others are relieved of a greater pain—the pain of bitterness and hostility. And they also avoid a further risk, notes Enright: that the pain will be passed on. Miserable married couples toss their anger back and forth to each other and at the children. A major family feud that is never resolved is handed down from generation to generation, a secret, unintentional heirloom.

In the end, forgiveness proceeds a step at a time. No matter how earnestly you try, waves of anger will sweep you away from time to time. "It's not a matter of black and white," says Casarjian. "At first, there may be just a few moments in the day when you are more understanding, when you can live closer to love than to fear. Then maybe if you embrace that state of being, a year from now there may be an hour of greater compassion in your day. In another year, more. The work gradually gets integrated." Be gentle with yourself, she says, and, as you continue to work on forgiveness, you will become more peaceful, more insightful. When you resent someone you're bound to them. When you forgive, you're free.

# Fearsome Security

THE ROLE OF NUCLEAR WEAPONS

MICHAEL M. MAY

Michael M. May is professor of engineering-economic systems and co-director of the Center for International Security and Arms Control at Stanford University. He is director emeritus of Lawrence Livermore National Laboratory and has published widely on nuclear weapons and arms control, most recently in American Scientist, *November-December 1994.*

A Hiroshima-size nuclear bomb would be about a thousand times as destructive as the Oklahoma City truckload of explosives that destroyed a large modern building and killed nearly 200 people. It could destroy about one square mile of the city, an area roughly 20 by 20 blocks, and kill perhaps 200,000 people, the destruction depending on such things as building construction and weather. A typical thermonuclear weapon, of which there are many in the weapon states' arsenals today, could destroy several square miles of a large city and kill as many as a million people. Either kind could destroy a large air base and everything on it, a large tank attack, or an aircraft carrier.

What is the role of these things? What does the United States need them for? What does anybody need them for?

The official U.S. answer, given last fall by Clinton administration spokesmen, is somewhat informally translated, "We're not sure, but there are a lot of nuclear weapons in Russian and other hands, there is a lot of unrest in the world, so we'll reduce the numbers gradually, in tandem with the Russians, and try to contain proliferation, but still keep several thousand around just in case." This position has been criticized as lacking vision. Much more fashionable in intellectual and academic circles has been a debate between abolitionists, who want to do away with nuclear weapons as soon as possible, and "marginalizers," who want to keep a few around but at the edge of polite policy discussion, unseen and unheard from, except in connection with the occasional rogue and pariah state.

However out of fashion, the U.S. position is right, and the abolitionists and marginalizers are wrong. But defending the U.S. position requires more unpalatable arguments than administration spokesmen could use, especially a few months before the recently concluded nuclear Non-Proliferation Treaty extension conference took place this spring.

Peace—The Essential Security Interest

The most basic argument is that nuclear weapons (and only they) transform peace among the most powerful states in the world from something that is nice to have but secondary to essential security interests as seen by governments, into the essential security interest of governments and governed alike. It does this at some risk of potential catastrophe. If we were in a world of stable states with no rival territorial and other interests that could not be dealt with by empowered and respected international institutions, this risk would be the main matter, and marginalization of nuclear weapons would indeed be in order. But we are not.

Contrary to what many wish, the states of the world are not becoming law-abiding citizens of one world, at least not in essential security matters. In such matters, the United States, as do other states, relies ultimately on its own forces, for the good and sufficient reason that there is nothing else reliable around. As a result, traditional security-oriented behavior abounds today. What is going on in Central Europe, in East Asia, in the Middle East would be familiar to Metternich and Bismarck. This is not to deny that our world is in many respects different from theirs, in particular that it is interactive, reactive, and interdependent as never before. But it is insecure now in much the ways it was then. And now, as then, having the most powerful weapons and deterrent plays an essential role in attaining the number one security policy objective of the United States, which is to preserve our central interests abroad without involving us in war.

Nuclear weapons are not all that is needed to make war obsolete, but they have no real substitute. Modern weapons such as those used in the Gulf War promise victory to the side that has them and leave the other side eager to build or buy them. Students of international politics have long noted this security dilemma: in the measure that a states's search for security through conventional armaments and alliances is successful, it makes its neighbors and rivals insecure and leads to wars that may be disastrous for everyone. To make things worse, unscrupulous politicians maintain themselves in power by preying on and exacerbating these longstanding fears among neighbors, as we now see tragically around the globe. Gulf War weaponry offers nothing new in this regard.

Nuclear weapons do something different. They cheaply and predictably destroy whatever both sides are fighting for. It is just that ability to destroy the battlefield as well as the enemy, to leave war without winners, that makes them essential. Abolitionists and marginalizers have nothing realistic to offer to replace them in that role.

War Is Not Obsolete

But surely war is over among advanced industrialized nations, whose prosperity and very survival rests on the maintenance of peace among themselves? Unfortunately, there is no sign that the forces that led to the most disastrous wars in the past are spent.

The world balance of military power (a consequence, among other things, of the balance of economic power), is changing and will continue to change. Reasonably or not, that makes other powers insecure. World War I and World War II were the way the major and most civilized powers in the world handled the changing balance of power occasioned by the growth of Germany and Japan in relation to the others. It need not have happened that way. It shouldn't have happened that way. As early as 1913, thoughtful people pointed out that Europe was so interdependent economically that war would destroy it and that war was therefore obsolete. They were right about the first observation and wrong about the second. Why?

Most popular analyses of these wars, especially now, on the 50th anniversary of our victory in World War II, focus on who was right and who was wrong, on the fanatical or evil groups who were the more obvious sources of the catastrophes. But that is like focusing on the match and not on the tinder. It is good drama but poor analysis. Why did a few such leaders have the power to send tens of millions to their death? Why was the system, staffed for the most part, then as now, with ordinary, not particularly incompetent or evil men, unable to handle their challenge?

The answers are the stuff of international relations libraries, but at the very heart is the security dilemma noted above: so long as security depends ultimately on rival goods, goods that have to be competed for, such as territory, alliances, access to scarce resources, then more security for one state will mean ultimately less for the other. Maintaining a peaceful balance would be to everyone's advantage, at least for advanced nations no longer bent on acquiring virgin lands for plunder. But that makes peace a public good. Like all public goods, it requires an agreed-on authority able to coerce rebels and free-riders into supporting it if it is to be provided. Reason alone has never been enough: to be the last state to defect is just too dangerous.

There was no such authority earlier this century. Is there one today? The United Nations Security Council? The International Court of Justice? Maybe someday, but not this century or perhaps the next. When have we or any other major nation submitted to these institutions when our central, or sometimes even our peripheral, interests were involved?

What we are seeing instead is the major powers of the world, the United States, Russia, China, Japan, the European Union, mixing universalist global or regional initiatives with the carving out of

2. CRIME, TERRORISM, AND VIOLENCE

> Winston Churchill warned at the beginning of the atomic age that safety could be the sturdy child of terror, and that we should not give up atomic weapons until we were sure and doubly sure that we had something better to take the place of terror in that respect.

spheres of influence in much the same way their predecessors on the world stage did. Spheres of influence no longer involve gunboats and marching armies most of the time—though sometimes they still do. Usually they involve such things as supplier-client relationships in arms and nuclear fuel, alliances and stationing of troops abroad, rule-writing in international agreements, and support in negotiations with rival powers.

The major security questions of the day are all handled with considerable attention to this traditional concern: the expansion of NATO; the tip-toeing around Serb aggression; Russia's policy in its "near-abroad," a balance of power term if ever there was one; China's policy in the South China Sea; continuing U.S. willingness to pay with blood and money to keep its suzerainty over Latin America and the Gulf States; China's blocking of UN sanctions in North Korea, sanctions that would have increased U.S. influence on the peninsula and decreased the value of Chinese security guarantees; Russia's and China's offer of nuclear help to Iran.

Even nuclear nonproliferation policies, which contain an element of universalist values, reflect this attempt to acquire or augment spheres of influence. How explain otherwise China's relative unconcern about a nuclear North Korea or Russia's about a nuclear Iran? Nuclear weapons in the hands of these countries would mainly limit U.S. power projections in the area, which necessarily depend on such concentrated and therefore vulnerable military assets as air bases, ports, and aircraft carriers. They would do relatively little to limit the influence of the big near-neighbors.

A Perilous Balancing Act

None of this makes the states involved evil. All of it is understandable. But it is dangerous. It is fraught with risks of escalation, whether owing to the accession to power of leaders who stake their political future on extremist or irredentist positions, or owing to perfectly well-meaning and ordinary leaders finding themselves in a corner none of them predicted or can deal with.

It also has the consequence that a number of states are less secure now than they were during the Cold War. The Cold War carried with it oppression and suffering, but it provided a stable security framework for many states located in traditionally troubled areas—traditionally troubled because in between, or of interest to, multiple larger states. These states are now having to look to new arrangements for their security. Some will find it in reliable alliances with larger powers, some will not. The latter are good candidates to look to nuclear weapons for their security, especially if, as is the case for Iran and North Korea, they are isolated from much of the world community.

We are told by abolitionists and marginalizers that the main threat to the United States are these isolated states, that nuclear weapons are not otherwise relevant to our security situation, and that therefore now is a good time to give them up or marginalize them. If marginalizing carries the somewhat trivial meaning of not talking about the nuclear balance constantly, of not measuring our security solely by that balance, that is a good idea and would have been a good idea during the Cold War. It might have allowed us to see something other than a monolithic threat in the Soviet Union. But if marginalizing means not paying careful attention to the influence of nuclear weapons on the security situation of various states, including us, it is a bad idea.

Nuclear weapons do not of themselves create the authority to provide and maintain peace, but they impose a penalty, obvious to all, not least to volatile democratic electorates, for overlooking this truth. Thereby, they have and will continue to have a profound influence on our security. They will do so in at least two ways: they will continue to put a considerable premium on caution when one nuclear state deals with another in matters either or both consider central, and, partly as a result, they will continue to tend to freeze lines of demarcations between spheres of influence where these lines matter to either or both of the contending major states.

Thus, in Central Europe, in East Asia, in the Middle East, not only do local powers have a strong incentive not to get caught in some euphemistically-called "buffer zone," but also the major powers are equally strongly impelled to make sure there is no misunderstanding about who gets to do what where. As a result, for the next several years, possibly decades, we are going to witness again a definition and consolidation of spheres of influence that will remain irreversible until some cataclysmic change, such as recently rocked the former Soviet Union, takes place in one of the major powers. Reversing it by force otherwise would be too dangerous to attempt.

The Czechs, Hungarians, and Poles are thus perfectly right to press for early incorporation into NATO and the EU. They have only a limited time window, while Russia cannot press its demands too forcibly. The United States may think it is pressuring the present Iranian regime to behave somewhat more acceptably. Instead, it may be giving up Iran to Russian influence for a long time to come. And the *New York Times* is wrong to press for recognition of Taiwanese independence on ideological grounds: China has said it would go to war to prevent Taiwanese independence, and we will not go to war to bring it about. Taiwan is on the Chinese side of the divide, as Ukraine is on the Russian side. These sides exist, however we may dislike it.

Millennial Hopes Deferred

All this makes unpleasant reading. It is not the brave new world of international commerce and cooperative endeavors for solving global problems. That new world is indeed here, and, on most days, occupies and will continue to occupy the attention of governments and political commentators and analysts. But that world will also continue to bring out conflicting interests, and these interests will continue to be pursued by independent states with the full political backing of their populations. The ultimate limit on this pursuit will be the fear of nuclear escalation, not some supranational law.

Nuclear deterrence is thus a fact of modern life. It is inherent in the technology. The only question is how it will be handled. My guess is that it will not be handled better if the United States backs out of the job of deterrence. And it will not be handled better if, by some unlikely chance, everyone were to back out of the job of deterrence, and the task of implementing it were left to some future government in an emergency situation scurrying to resurrect the caution-inducing deterrent it once had. Peacetime is the right time to maintain the deterrent and to formulate how we will deal with it.

Thus, not only is abolition impossible under present circumstances, it would be a mistake even if it were possible. And marginalization is meaningless: nuclear weapons do not and should not enter daily calculations but do and should enter calculations of ultimate security. Deterrence was far stronger and more stable in the last 50 years than most lay commentators thought. Perversely, perhaps because it worked so well, many are now willing to give it up. But people should not give up fire insurance because fires have been rare, although it is reasonable to look for lower premiums and to work on less risky means of fire containment.

We don't, however, have these less risky means. Winston Churchill is not particularly in favor these days. But Churchill, whatever his failings, understood what led to war. He warned at the beginning of the atomic age that safety could be the sturdy child of terror, and that we should not give up atomic weapons until we were sure and doubly sure that we had something better to take the place of terror in that respect. Look around: we have nothing better to take its place.

Aging, Health, and Health Care Issues

UNIT 3

According to the Clinton administration, access to quality health care should be the right of every American citizen, which is not the case at this time. Thirty-five million individuals have no health insurance, and many more are underinsured. Those trying to do something about it are confronted with the harsh realities of limited resources. The pressure to cut the deficit, balance the budget, clean up the environment, fight crime, stop the decline of our cities, and maintain the deteriorating infrastructure of America has placed severe constraints on the amount of federal monies available to fund an ever-expanding number of individuals needing care. This demand is further increased by the consequences of pandemics such as AIDS, a growing proportion of the aged with noncurable chronic disorders, and an ever-advancing technology that can save the lives of very premature infants and extend the lives of terminally ill individuals. This unit looks at some of the problems facing individuals and society as we agonize over the hard choices that must be made. One of the best summaries of the problems created by entitlements is contained in "Will America Grow Up before It Grows Old?" The "baby boomers" are beginning to age, and their sheer numbers will bankrupt the Social Security system and place the future of America in serious trouble unless major changes are made. These changes must include balancing the budget, reining in senior entitlements, raising the retirement age, boosting individual and pension savings, and altering tax schedules.

On the other end of the age continuum are severely premature infants. Sharman Stein's report, "The Cruelest Choice," looks at the consequences of medicine's ability to "save" extremely premature babies. The lives of babies as young as 23 weeks and weighing only 12 ounces have been saved, but at horrendous financial costs to their parents and society (over $5.6 billion a year). In addition, a significant proportion of severely premature infants suffer from brain abnormalities, are disabled, and exhibit a wide range of learning disabilities, which will cost society even more.

"A New Look at Health Care Reform" suggests that if meaningful changes in health care are to occur, we must seriously confront the attitude that medical care is "free" or at least very low in cost. If health care is free, then the attitude that "nothing is too good for me and mine" drives consumer attitudes and actions. Murray Weidenbaum suggests that competition needs to be introduced into medicine, that people need to pay for the medical coverage they desire, that advertising be permitted and encouraged, and that consumers be informed of available options.

The article "Guns, Money & Medicine" reports that by the year 2003, gunshots will have surpassed car accidents as the leading cause of injury deaths in the United States. Gunshots are extremely difficult and expensive to treat. It is estimated that they cost over $20 billion each year, and this does not include the costs associated with long-term disability. As most of those treated cannot pay the expenses associated with their gunshot wounds, the costs are absorbed either by the hospital or by society. These escalating costs are threatening to bankrupt hospitals and the health care system.

Thomas Szasz, author of "Mental Illness Is Still a Myth," is a most vocal critic of psychiatry and its treatment of those displaying unusual and dysfunctional behaviors. He cogently argues that psychiatry, and the activities of most psychiatrists, could be best understood if viewed as a branch of law or a type of secular religion, but not as a "science."

Looking Ahead: Challenge Questions

Why has Social Security become such a critical issue in the 1990s? What major changes must be made in Social Security entitlement programs?

Is adequate health care in the United States a right or a privilege? Which should it be? Explain your answer.

What reforms, if any, should the U.S. government be making in health care?

What criteria must we consider in attempting to save severely premature infants?

How have automatic weapons impacted on the health care system? How have injuries resulting from gunshots impacted on medicine?

What are the implications of classifying psychiatry as a branch of the law or a form of secular religion rather than as a branch of science?

In what significant ways would the approaches of individuals espousing each of the three major sociological theories differ in studying aging and various aspects of the health care issues?

What conflicts in values, rights, obligations, and harms underlie the issues and problems covered in this unit?

WILL AMERICA GROW UP BEFORE IT GROWS OLD?

The long gray wave of Baby Boomers retiring could lead to an all-engulfing economic crisis—unless we balance the budget, rein in senior entitlements, raise retirement ages, and boost individual and pension savings. Yet politicians of both parties say that most of the urgently necessary reforms are "off the table."

PETER G. PETERSON

Peter G. Peterson is the chairman of The Blackstone Group, a private investment bank. His book *Will America Grow Up before It Grows Old? How the Coming Social Security Crisis Threatens You, Your Family and Your Country,* is published by Random House (1996).

A NATION OF FLORIDAS

BEEN to Florida lately? You may not realize it, but you have seen the future—America's future, about two decades from now. The gray wave of senior citizens that fills the state's streets, beaches, parks, hotels, shopping malls, hospitals, Social Security offices, and senior centers is, of course, an anomaly created by our long tradition of retiring to Florida. Nearly one in five Floridians is over sixty-five. But early in the next century a figure like that won't be exceptional. By 2025 at the latest the proportion of all Americans who are elderly will be the same as the proportion in Florida today. America, in effect, will become a nation of Floridas—and then keep aging. By 2040 one in four Americans may be over sixty-five.

When we consider the great demographic shift that will shape our national future over the next fifty years, we are speaking not of a mere transition but of a genuine transformation. Just fifteen years from now the first batch of Baby Boomers will hit sixty-five, bringing changes—economic, political, social, cultural, and ethical—that will transform American society. This transformation will challenge the very core of our national psyche, which has always been predicated on fresh beginnings, childlike optimism, and aspiring new generations. How we cope with the cultural dimensions of this challenge I will leave to others—to sociologists, political scientists, historians, and philosophers. I am none of these. I am a businessman who has long participated in public debates over the political economy of rising living standards. What concerns me most about America's coming demographic transformation is simply this: on our present course we won't be able to afford it.

To provide for the largest generation of seniors in history while simultaneously investing in education and opportunity for the youth of the twenty-first century, we must reject the prevailing "entitlement ethic" and return to our former "endowment ethic," which generated America's high savings, high growth, and rising living standards in the past. Endowment implies "stewardship"—the acceptance of responsibility for the future of an institution. But given our current emphasis on individual self-

13. Will America Grow Up before It Grows Old?

fulfillment, we must, in addition to endowing the future of our nation and its institutions, endow our individual futures and those of our children, because no one else is going to do it for us. What I am talking about is *self-endowment*.

"Hope I die before I get old," The Who sang in their classic sixties anthem, "My Generation." That statement, like so many slogans of the Baby Boomers' youth culture, was wishful thinking. The generation that once warned "Don't trust anyone over thirty" is now passing fifty.

The real question is, Will America grow up before it grows old? Will we make the needed transformation early, intelligently, and humanely, or procrastinate until delay exacts a huge price from those least able to afford it—and confronts us with an economic and political crisis to which there is no longer a win-win solution?

DEMOGRAPHICS IS DESTINY

WITHIN the next fifteen years the huge generation of Baby Boomers, whose parents brought them into the world with such optimism, will begin to retire. As they do, they will expect the munificent array of "entitlements" that were guaranteed (again with such optimism) to every retiring American with no anticipation of the ever-growing length of retirement as life expectancy increases or the ever-rising expectations of independence, affluence, health, and comfort of life in retirement. But consider who is expected to pay for this late-in-life consumption: the relatively small "bust" generation in whose productive capacity we have failed to invest. Neither the founders of Social Security sixty years ago nor the founders of Medicare thirty years ago imagined the demographic shape of America that will unfold over the next several decades. Ponder the following:

• With 76 million members, the Baby Boom generation is more than half again as large as the previous generation. To get some idea of how much the number of seniors could grow by the time the youngest Baby Boomers turn seventy, think of the entire population of California and the New England states combined. Or think of it this way: the number of Social Security beneficiaries will at least double by the year 2040.

• In 1900 only one in twenty-five Americans was over sixty-five. The vast majority of these people were completely self-supporting or supported by their families. By 2040 one out of every four or five Americans will be over sixty-five, and the vast majority will be supported to some degree by government entitlements.

• In 1960 there were 5.1 taxpaying workers to support each Social Security beneficiary. Today there are 3.3. By 2040 there will be no more than 2.0—and perhaps as few as 1.6.

• The number of "young old" (sixty-five to sixty-nine) will roughly double over the next half century, but the number of "old old" (eighty-five and over) is expected to triple or quadruple—adding the equivalent of an entire New York City of over-eighty-five-year-olds to the population. Nearly three quarters of those over eighty-five will be single, divorced, or widowed—the groups most likely to need extensive government assistance.

• In 1970 children under five outnumbered Americans aged eighty-five and over by twelve to one. By 2040 the number of old old will equal the number of preschool children, according to some forecasts.

• The extraordinary growth of the old old population will add especially to federal health costs. This is because the average annual medical-care bill rises along a steep curve for older age groups. The ratio of Medicare and Medicaid spending on the old old to spending on the young old is about 2.5 to 1.

• In 2030 only about 15 percent of the over-sixty-five population will be nonwhite. But about 25 percent of younger

IN 1900 ONLY ONE IN TWENTY-FIVE AMERICANS WAS OVER SIXTY-FIVE. BY 2040 THE FIGURE WILL BE ONE IN FOUR OR FIVE.

Americans will be nonwhite. This will create a potentially explosive situation in which largely white senior Boomers will be increasingly reliant on overtaxed minority workers.

• In order to provide the same average number of years of retirement benefits in 2030 that were contemplated when Social Security was set up, in the 1930s, the retirement age would have to be raised from sixty-five to seventy-four by 2030. But this projection—daunting as it is—assumes that future gains in longevity will slow as average life expectancy approaches the supposed "natural limit" to the human life-span. Many experts now question whether such a limit really exists. Summing up research at the National Institute on Aging, the demographer James Vaupel goes so far as to suggest that we are now on the threshold of a "new paradigm of aging," in which the *average* life expectancy could reach 100 or more.

Of course, the United States is not the only country facing an "age wave." Indeed, the age waves in most industrial countries are approaching faster than ours, and—to judge by official projections—could have an even worse impact on their countries' economies and public budgets. But these

3. AGING, HEALTH, AND HEALTH CARE ISSUES

other countries enjoy long-term defenses that we lack. Unlike the United States, most can actually budget their public spending on health care, and so have much greater control over this potentially explosive dimension of senior dependency. Unlike the United States, most generally tax public benefits as they do any other income. And unlike the United States, most have fairly healthy household savings rates (generally well over 10 percent of disposable income, as compared with about five percent here), and so can absorb public-sector deficits much better than we can.

Most important, unlike the United States, these other countries are unencumbered by the illusion that their people have some sort of inalienable right to live the last third of their adult lives in subsidized leisure. In other countries what government gives can be taken back if doing so is deemed to be in the public's long-term interest. In 1986, when Japan enacted a major reduction in future pension benefits, the Ministry of Health and Welfare issued a concise justification that cited "equity between the generations." Few if any objections were heard. In a statement issued the day he assumed office, Japan's new Prime Minister, Ryutaro Hashimoto, referred to the "imminent arrival of our Aging Society" as a priority imperative. Citing much longer life-spans and a much reduced fertility rate, he told the Diet in his opening speech that Japan would have to "overhaul those social arrangements premised upon a life-span of twoscore and ten to suit our new expected life-span of fourscore." Do we recall any American President ever making such a statement at *any* point in his term, let alone in the equivalent of an inaugural address?

Australia has made employee pensions mandatory, increasing coverage from under 40 percent to nearly 90 percent of the work force. Iceland has means-tested its social-insurance system. Germany has enacted, and France, Sweden, Italy, and the United Kingdom are debating, increases in the retirement age. Some of these changes have provoked fierce controversy—or even widespread protest, as happened in France last winter. But the disagreement is almost always over how best to allocate public resources. No one questions that government has the right to reduce benefits.

Even many developing countries with populations still much younger than our own are preparing for their demographic future with astonishing resolution. In South Korea the household savings rate runs at about 35 percent; "Working to make a better life for the next generation" is a typical company motto. Account balances in Singapore's Central Provident Fund—the country's mandatory pension-savings system—now total nearly three quarters of GDP. In Chile the average worker owns $21,000 worth of assets in the fifteen-year-old national funded retirement system—a sum about four times the average annual Chilean wage. Argentina, Peru, and Colombia are following Chile's lead and setting up funded systems of their own. Here, nothing has been saved in any national retirement system for any worker to own.

UNSUSTAINABLE PROMISES

THE economist—and sometime humorist—Herbert Stein once said, "If something is unsustainable, it tends to stop." Or, as the old adage advises, "If your horse dies, we suggest you dismount."

We cannot sustain the unsustainable. Nor can we finance the unfinanceable. By 2013, when Baby Boomers will be retiring en masse, the annual surplus of Social Security tax revenues over outlays will turn negative. By 2030, when all the Boomers will have reached sixty-five, Social Security alone will be running an *annual* cash deficit of $766 billion. If Medicare Hospital Insurance is included, and if both programs continue according to current law, the combined cash deficit that year will be $1.7 trillion. The horse, in other words, will be quite dead. By 2040 the deficit will probably hit $3.2 trillion, and by 2050, $5.7 trillion. Even discounting inflation, the deficit that year for these two senior programs will come to approximately $700 *billion*—four times the size of the entire 1996 federal deficit. Long before that time we will have had no choice but to dismount.

Wall Street has yet to react to these obviously unfinanceable numbers. When will it? Since financial markets try to anticipate events, the reaction will surely come years before the first Boomers start retiring on Social Security, in 2008. How severe will the reaction be? Should the markets conclude that America has lost any chance to deal with this challenge in advance, we will almost certainly see a full-scale economic emergency as interest rates roar into outer space.

Apologists for the status quo dismiss these numbers as "mere projections." So let me emphasize that the numbers I have used for Social Security and Medicare are *official* projections, calculated by federal actuaries and economists working for the Social Security and Health Care Financing Administrations. The same experts also calculate an alternate and much worse "high-cost" projection, which has historically proved to be more accurate than the forecasts I have used here. Moreover, the retirement and medical-care needs of the Boomer generation are by no means hypothetical. The Social Security Administration's former chief actuary A. Haeworth Robertson points out that fully 96 percent of senior benefits payable over the next seventy-five years will go to people who are already alive (and therefore countable) today.

Well, say the skeptics, if we can't borrow trillions of dollars, maybe we can raise taxes a bit and muddle through. But this isn't a viable option either. Let's start with the political fact that both parties in Washington are currently hawking a tax *cut*, though they disagree about its size. A tax increase is unmentionable. Then consider the magnitude of the tax increases we would need. By 2040 the cost of Social Security

13. Will America Grow Up before It Grows Old?

NEARLY TWO FIFTHS OF ALL SOCIAL SECURITY BENEFITS NOW GO TO HOUSEHOLDS WITH INCOMES THAT ARE ABOVE THE U.S. MEDIAN.

as a share of worker payroll is expected to rise from today's 11.5 percent to 17 or 22 percent—depending on whether you accept the official or the high-cost projection. Add both parts of Medicare, which currently cost the equivalent of 5.3 percent of payroll but are growing so rapidly that they will eventually overtake and surpass Social Security in size, and we're talking about 35 to 55 percent of every worker's paycheck before we even start to pay for the rest of what government does.

Obviously, tax increases of this size would destroy the economy. More to the point, they would kill the taxpayers. There is also the interesting question of whether American taxpayers could be expected to comply with them. The experience of runaway pension systems in Latin America and Eastern Europe teaches us that when payroll taxes begin even to approach these levels, tax evasion becomes widespread and much of the economy moves into the tax-exempt "gray market." In other words, it may be impossible to fund the future cost of our current benefit promises no matter how willing we are to legislate higher tax rates.

The senior lobby asserts that whatever the economic consequences, future American workers are duty-bound to fulfill their side of an ill-defined "contract between generations." Yet one group's "earned right" to a benefit is another group's "unearned obligation" to pay a tax. It is to this second group that our children and grandchildren belong. Understandably, they are suspicious of a binding "contract" to which they never agreed. According to a 1994 poll, Americans under thirty-five are much more likely to believe in UFOs than to believe that they'll ever receive Social Security benefits.

There's an old adage about robbing Peter to pay Paul. In the entitlement shell game we're proposing to rob Peter Jr. to pay Peter Sr.—even when the Peter Sr. in question may not need the money. In fact, Peter Jr. is already paying plenty. Because so much of Social Security is tax-free (and because retirees no longer pay FICA taxes), a typical retired couple on Social Security in 1994 with $30,000 in total cash income paid, on average, only $790 in federal taxes. Meanwhile, their son and daughter-in-law, struggling to raise a child on the same income, had a total federal tax burden of $7,035, if you include both the FICA tax they paid and that paid by their employers. No other industrial nation tilts its tax system away from the elderly—or tilts its benefits system toward the elderly—as much as the United States does.

The present system's true believers dress up Social Security and Medicare in the reassuring rhetoric of "insurance" and "pensions" and claim that beneficiaries are only getting back what they paid in. They're wrong. The majority of today's beneficiaries are getting back *far more* than they ever paid in FICA contributions: given an average life expectancy, the average one-earner couple retiring today will get about $123,000 more out of Social Security than the average earner and his or her employers ever paid into it, plus interest. Omit the employer's contribution and calculate only the payback on the personal taxes paid by the employee, and the windfall rises to $173,000. With Medicare thrown in it rises to nearly $310,000, much of that tax-free. These are not "earned benefits" but unearned windfalls that our children will have to pay for and certainly will never enjoy themselves.

Moreover, since FICA contributions have never been saved by the federal government, the point is moot: regardless of what a worker paid in, the federal trust funds now possess on that worker's behalf nothing but claims on future taxpayers. The term "trust fund" may suggest a vault in which one's Social Security taxes are stacked up, to be paid out later. But the Social Security "trust fund" is the ultimate fiscal oxymoron. Its "assets," which we are told will keep the system "solvent" until 2030, consist of nothing more than Treasury IOUs—claims against future generations. When it comes time to redeem these claims and the interest they have accumulated, where will the Treasury find the cash? Either by borrowing from the public or by raising taxes. Either way, absent any policy change future taxpayers will have to pay again for today's Social Security "surplus."

If the Social Security and Medicare balance sheets were evaluated according to private-sector accounting standards, both would be declared disastrously insolvent. How disastrously? Consider that the federal government has already promised to today's adults $8.3 trillion in future Social Security benefits beyond the value of the taxes they have paid to date—a figure more than 250 times as great as the much-decried "unfunded liabilities" of *all* private-sector pension plans in America! If federal law required Congress to fund Social Security the way private pensions must be funded, the annual federal deficit would instantly rise by some $675 billion. Add in our lavish and unfunded federal-employee pensions and the deficit would rise by $800 billion. Add in Medicare and it would rise by more than $1 trillion. If private-sector executives ran their pension systems this way, they would be thrown in jail for wholesale violation of federal pension-plan regulations.

Meanwhile, Congress has attempted to ban what policy wonks call "unfunded mandates"—federal laws that impose costs on the states without providing funding for them. That's fine—but worrying about such mandates while ignoring Social Security and Medicare is like mis-

3. AGING, HEALTH, AND HEALTH CARE ISSUES

taking Woody Allen for Arnold Schwarzenegger. Social Security and Medicare are *the mother of all unfunded mandates.*

It's time to face up to the fact that trustfund accounting is a hoax, that Social Security is in fact a pay-as-you-go system. Payroll taxes go directly to today's beneficiaries; benefits come directly from today's workers. Since FICA is a tax, and tax revenues are fungible, any annual surplus of FICA taxes over benefits is used to cover other government spending. A trustfund ledger for such transfers is a waste of time. Does it really help anyone to know that Social Security is a bit richer and the Treasury is a bit poorer? Given the apparent congressional appetite for constitutional amendments, why not consider one banning government trust funds?

As Federal Reserve Chairman Alan Greenspan has summed it up, the only bottom line that really counts is government's total borrowing balance with the public—otherwise known as the annual consolidated budget deficit or budget surplus. Transferring IOUs from the right pocket to the left pocket does nothing to bridge Social Security's and Medicare's enormous funding shortfall.

Along with this melancholy list of fiscal unsustainables we should consider some troubling moral unsustainables. Social Security was established to protect the elderly from indigence late in life—to prevent a "poverty-ridden old age," in the words of Franklin D. Roosevelt. If we allow it to go bankrupt by paying benefits to middle-class and affluent Americans, many of whom can live well enough without these benefits, what will happen to those who really need them? Among Social Security recipients whose incomes are under $20,000, Social Security accounts for *more than half* of the total. In spite of this sobering dependence, many political leaders imply by their inaction that it's fine to wait until trillion-dollar deficits have devastated our economy, and then slash benefits at the last minute. By doing so we would then deprive Americans at all income levels of the chance to plan for their futures. Millions of lower-income beneficiaries would be stranded in what might be called a demographic Depression, as the safety net that Social Security was enacted to provide suddenly vanished. Future historians may record that Social Security's "defenders" were the ones who most wanted to exempt the program from a balanced-budget amendment and thus from gradual and timely reform.

Paul Tsongas likes to say, "It's not enough for our children to love us. We should want them to respect us." When our children look into the Social Security trust fund and find nothing there but IOUs with their own names listed as payers, they will surely wonder how we could have treated them so shabbily.

THE INESCAPABLE BOTTOM LINE

NOTWITHSTANDING its strengths, real and imagined, the U.S. economy since the early 1970s has failed at what matters most: *raising productivity.* Why should the average American care about such a seemingly abstract concept? Because working longer hours—or putting everyone's spouse (or child) to work—is not the way to raise living standards. A higher standard of living means producing more while working the same or a lesser number of hours—in other words, being more productive. Only thus can real (after inflation) hourly compensation and take-home pay rise. The astute economist Paul Krugman once summed it up this way: "Productivity isn't everything, but in the long run it is almost everything. A country's ability to improve its standard of living over time depends almost entirely on its ability to raise its output per worker."

Since the early 1970s real national income per full-time worker (as calculated by the Commerce Department) has grown by approximately 0.4 percent a year. Total worker compensation has grown at about the same meager pace. This rate of growth is so low that a debate rages among economists over whether—after accounting for inflation and the rising costs of employer-paid health care—the typical U.S. worker has seen any perceptible wage growth since 1973.

We can no longer ignore what economists from Adam Smith to Karl Marx to Alfred Marshall to John Maynard Keynes to Paul Samuelson have insisted is the bottom line: sustained productivity growth requires investment, and no country can sustain high rates of investment without saving. These economists all understood that productivity growth depends on many underlying conditions, such as technological innovation and efficient markets, but they all agreed that capital accumulation is essential to productivity growth—and is, moreover, the one condition over which society can exercise direct control. Few experts disagree, especially when "capital" is defined, as many economists define it, to include such intangible collective investments as infrastructure, research, education, and training. Yet we now face public budgets strained to the breaking point by the costs of demographic aging, which will crowd out all forms of capital accumulation—private and public, material and human. Without fundamental policy reform a graying America cannot be a saving America.

But thrift is precisely what we've forgotten. From an average of 8.1 percent of GDP in the 1960s, the net national savings rate dipped to 7.2 percent in the 1970s and then plunged to 3.9 percent in the 1980s and to 2.3 percent thus far in the 1990s. Net domestic investment has fallen in tandem, from 7.3 percent of GDP in the 1960s to 3.5 percent in the 1990s—a decline that would have been much steeper if we had not switched from investing abroad to borrowing abroad.

Our structural deficits drain our already shallow pool of private savings—and hence crowd out private investment. To the extent that we try to control these deficits by reducing "discretionary" federal spending (a category that includes most future-oriented programs), they also crowd out public investment. Out of every nondefense dollar the federal government now spends, only about five cents builds tangible things that remain after the fiscal year is over. Recently a General Accounting Office study suggested that we must invest $112 billion to bring the infrastructure of schools back to acceptable levels. But where can we find such a sum when entitlements and interest on old debts crowd out everything else?

Long before the Boomers reach retirement age, we're preparing to cut everything from Head Start and school lunches to rapid transit and space shuttles in order to pay the rising cost of senior entitlements. Despite the radical rhetoric in Washington, the recent budget plans I have seen don't reverse but accelerate our current fiscal trajectory. Each of them proposes to slash appropriated domestic spending in real dollars while only gently restraining the growth in senior entitlements. Even in Congress's plan senior benefits in 2002 would consume still another record share of the budget—nearly 50 percent of noninterest outlays, up from 40 percent today and just 17 percent in 1965. This is in a benign demographic period, when the relatively small Depression generation (born before VJ Day) is still retiring and the relatively large Boomer generation (born after 1945) is still working and paying taxes. And remember: this is the *Republican* plan, widely attacked as a "declaration of war" on America's seniors.

To break out of this slow-growth, low-investment cycle we must set a higher productivity goal and then dedicate the resources required to meet it. A sensible objective would be to increase the rate of growth in real per-worker national income by a percentage point, from the post-1973 average of 0.4 to about 1.5 percent a year. Even this substantial increase would not equal American growth rates of the 1950s and 1960s, or match Japan's record during the 1970s and 1980s. But it would come close to returning U.S. productivity growth to its average rate over the past century—and it would bring growth close to the rates of most of our European competitors. If we raise productivity to 1.5 percent, twenty years from now national income per worker would be nearly $10,000 higher in today's dollars, and federal revenues (at the same tax rates) would be nearly $400 billion higher, than will be the case if we continue on our current course.

Conventional economic theory suggests that this ambitious goal requires a shift of six to eight percent of GDP from consumption to savings, giving us a long-term savings and investment rate of about 10 percent of GDP. But where will these extra savings—an average of at least $4,500 per U.S. household annually—come from? About a third can be financed by balancing the federal budget and keeping it balanced. The rest will have to come from greater private saving.

13. Will America Grow Up before It Grows Old?

AMERICA'S SAVINGS GAP

THUS we come to what we Americans as individuals can and must do for ourselves and the nation—ichthyology from the standpoint of the fish. There are four main sources of income for those over the age of sixty-five: continued employment, government benefits, private pension income, and accumulated personal savings. As we shall see, the adequacy of each of these sources is uncertain.

When it comes to our retirement plans, we are a nation in denial. About nine out of ten Boomers say they want to retire at or before age sixty-five (about six out of ten before age sixty). More than two thirds say they will be able to live "where they want" and live "comfortably" throughout their retirement years. A stunning 71 percent expect to maintain in retirement a standard of living the same as or better than what they enjoyed during their working years.

Yet probe them more deeply about their retirement dreams, and most Boomers admit that they are terrified that neither they nor their government is saving enough. Some two thirds confess that they've never even calculated how much they need to save for their retirement, and an amazing 86 percent acknowledge that "future retirees will face a personal financial crisis 20 years from now." Yet at the same time, they do not expect or even want much from government. Nearly nine out of ten Boomers agree that "the government has made financial promises to [their] generation that it will not be able to keep." For every Boomer who says that government should shoulder the "main responsibility for providing retirement income," five say that individuals should. They will very likely get their wish. From all the numbers we have seen, it is obvious that government retirement benefits (mainly Social Security, Medicare, and Medicaid) are likely to be severely reduced by the time most Boomers retire.

What could take their place? Thirty years ago experts hoped that private pensions would become a universal supplement to Social Security. Such hopes never panned out. Today less than half of all U.S. private-sector workers are covered by pensions. Overall coverage has been flat since the early 1970s, and in recent years coverage has actually dropped sharply for younger men. This stems from long-term changes in the work force and in the nature of work—part-time work, working at home, multiple careers. Rates of pension coverage have always been highest for full-time career jobs, unionized jobs, and jobs in government and large corporations—in short, for jobs that are becoming increasingly scarce. As for Americans lucky enough to have pensions, they will be surprised, if not seriously disappointed, by how little their plans have set aside for them: the typical defined-benefit pension plan for average-earning workers with thirty years of service replaces just one

3. AGING, HEALTH, AND HEALTH CARE ISSUES

AMERICANS SEEM TO THINK THEY HAVE AN INALIENABLE RIGHT TO LIVE THE LAST THIRD OF THEIR ADULT LIVES IN SUBSIDIZED LEISURE.

third of pre-retirement earnings—an amount that is not indexed for inflation.

Clearly, retiring Boomers will have to rely heavily on the remaining source of retirement income: private savings apart from pensions. But this source may be the most uncertain of all, for it is questionable whether the average American is saving *anything* on his own: what one household saves in a bank account or a nonpension mutual fund scarcely offsets what another household borrows. Whenever the stock market or housing prices rise, many households may feel that they're saving enough. But our aggregate personal-savings rate, except for pensions, is now barely positive.

Many have argued that the current bust is attributable to the passage of so many Baby Boomers through the years of household formation, and that saving will turn up again as Boomers reach the traditionally high-saving middle years. But for this explanation to be valid, the personal-savings rate should have bottomed out by the mid-1980s—and climbed back again. Many Boomers have already entered the traditionally peak saving years. But the savings decline persists, contrary to predictions of a demographic reversal.

In 1992, according to the Federal Reserve Board, 43 percent of U.S. families spent more than their income; only 30 percent accumulated assets for long-term saving. In 1993, according to a Merrill Lynch analysis of Census Bureau data, half of all families had less than $1,000 in net financial assets—a figure that had not risen over the previous decade, even in nominal dollars. Among adults in their late fifties, the age at which workers are staring directly at retirement, median savings are still shy of $10,000. Even optimists admit that a bleak future awaits the approximately one third of all Boomers who are expected neither to accumulate financial assets nor to receive a private pension.

Ironically, the Baby Boom is the best-educated, most sophisticated, most well-traveled generation in our history. This irony provides still another illustration of the depth of our denial.

B. Douglas Bernheim, of Stanford University, concludes that Boomers on average must *triple* their current saving if they want to enjoy an undiminished living standard in retirement. And if one assumes a 35 percent reduction in Social Security benefits (which seems more than likely if not inevitable), then Boomers will have to *quintuple* their saving. A recent study by the Committee for Economic Development, *Who Will Pay for Your Retirement? The Looming Crisis*, comes to a similarly stark conclusion.

If it's true that the promise of late-in-life government benefits helped to suppress private savings in the past, maybe the growing expectation of cuts in government benefits will help to boost private savings in the future. Though economic theorists debate the point, people do take government subsidies into account when deciding how much to save. By thirteen to one, households say that they would save more if they knew that future Social Security benefits were going to be cut.

Finally there is the prospect of inheritance, that magic cure-all for any generation's retirement worries. In recent years Boomers have been cheered by a spate of upbeat stories about the "$10 trillion inheritance boom" that today's affluent seniors are expected to pass on. These Boomers may not have noticed the bumper stickers one sees in resort areas frequented by seniors: I'M SPENDING MY CHILDREN'S INHERITANCE. But even if the hoped-for hand-off takes place, there's a problem. Because this wealth is highly concentrated among relatively few families (what Donald Trump calls the "Lucky Sperm Club"), bequests may *average* as much as $90,000 per Boomer but will amount to only about $30,000 for the median Boomer. Muffy and Duffy will do fine, but for most of this generation the typical inheritance will just about cover the costs of settling Dad's estate and pay off a few lingering medical bills.

Dan Yankelovich, the dean of American opinion surveyors, has wisely said that our collective denial is not due to emotional or moral pathology. Rather, it is a case of "cognitive denial," by which he means a failure to make connections between how we prefer to see reality and what reality actually is. Clearly, this denial is manifest at the national level and at the personal level.

PRAYING FOR PRODUCTIVITY AND OTHER GOOD THINGS

ARE there any favorable trends under way that might moderate these bleak forecasts? Perhaps. But there is less to them than meets the eye. First, consider productivity growth, which determines real wage growth and hence tomorrow's tax base. Those who preach that high tech will bail us out and that we can avoid saving and investing our way back to economic growth tell us not to worry: we're in the midst of a productivity revolution. But we have good reason to worry. For one thing, after the Commerce Department recently updated its methodology, it became clear that the much-touted productivity gains of the 1990s are just

13. Will America Grow Up before It Grows Old?

about typical of earlier business-cycle recoveries over the past twenty-five years. For another, the Social Security Administration's best estimate of future deficits presupposes a permanent one-third improvement in productivity over our actual historical record since 1973. In other words, productivity growth will have to accelerate simply to ensure that the future isn't *worse* than the SSA's already unsustainable official projection. It would have to accelerate still more to ensure that things turn out better.

Well, if the productivity revolution—at least as it is now unfolding—won't save us, maybe the new baby "boomlet" will. It's true that current fertility rates, of about 2.0 to 2.1 lifetime births per woman, are a bright spot when compared with the low rates of 1.7 to 1.9 recorded during the "birth dearth" of the 1970s and 1980s. But even if these higher fertility rates turn out to be lasting, they won't have much effect on federal tax revenues until the mid-2020s—long after fiscal meltdown is scheduled to occur. Even then the positive impact will be small. To stabilize the ratio of retirees to workers, U.S. fertility would have to surge to 3.0 or higher—in other words, back to the Baby Boom levels of the 1950s and early 1960s, which no one expects. For one thing, the share of American women who say that a family of four or more children is "ideal" has plummeted from nearly 50 percent to about 10 percent since the 1950s. For another, the United States already has one of the highest fertility rates in the developed world. Average fertility in other major industrial countries is now 1.6; in Germany and Italy it is 1.3.

Well, then, if not babies, what about immigrants? Isn't importing more young workers a viable solution to America's aging? Again, not really. Immigrants, too, eventually grow old—and thus begin adding to Social Security and Medicare costs. To make a substantial dent in the costs of America's aging, huge and destabilizing waves of immigration would be required. In fact, to cancel out the projected growth in the Social Security payroll-tax rate over the next half century, today's level of net immigration would have to roughly quintuple, to about five million a year, beginning now. The reality, of course, is that America is in no mood to reopen Ellis Island.

Finally, consider health-care spending. Some point to the recent slowdown in medical-*price* inflation (as measured by the Consumer Price Index) and conclude that our problem is behind us. Not so. First, what matters is total *expenditures* on health care, and thus far in the 1990s real federal health-benefit outlays have not slowed at all. Second (and once again), the bleak official projections *already* assume a dramatic turnaround in recent trends. Over the past quarter century real Medicare spending per beneficiary has increased at the rate of five percent a year—several times as fast as real per capita income growth. Over no five-year period since 1970 has the growth in spending been less than three percent a year. Yet the Health Care Financing Administration's official projection assumes that the growth in real per-beneficiary Medicare spending will slow to about one percent a year by 2020. This projected cost-containment "triumph" is timed to occur just as aging Baby Boomers begin to increase the demand for every imaginable health-care service.

So let's hope—or pray—for productivity gains, higher fertility rates, and market-imposed discipline on health-care costs. But let's not forget the rosy scenarios of the 1980s that never came true and the problems we never grew our way out of. Public policy must be based on prudent expectations about the future—and prudence suggests that on our current trajectory the future may be worse than the bleak official forecasts.

No matter how clearly Social Security actuaries tell us that financial trouble looms ahead, politicians on both sides of the aisle are convinced that "middle-class" entitlement programs constitute the "third rail" of American politics: "Touch it and you're toast." So denial persists. It would be pleasant to blame this denial on Washington and say that the rest of us know better—that all we have to do is elect more-principled public servants who will dare to confront these issues. But the problem is interactive—the politicians and the people have all become gifted deniers.

Consider this irony: the public enthusiasm for budget balancing and cuts in "wasteful" programs is inversely proportional to the cost of those programs. Ninety-four percent of those polled in one recent survey favored slashing foreign aid, 77 percent wanted to cut public-housing funds, and 75 percent wanted to cut the space program. Yet these programs together make up only about three percent of the federal budget. Meanwhile, only 14 percent of respondents wanted to cut Social Security, and only 22 percent favored cutting Medicare. Yet these two programs together account for a staggering one third of the budget.

Or consider how we deny the truth about entitlement programs. In justifying every new benefit increase and every refusal to accept slower growth in expenditures for the elderly, the senior lobby talks as if "old" meant "poor." But elderly Americans now have the highest level of per capita household wealth of any age group—and, counting in-kind income such as health benefits, a lower poverty rate than younger adults. Although old-age benefits were originally intended to be a safety net for the truly needy, today's entitlement system more closely resembles a well-padded hammock for middle- and upper-class retirees. One third of Medicare benefits, nearly two fifths of Social Security benefits, and more than two thirds of federal pension benefits now go to households with incomes above the U.S. median. Back in the early 1960s the typical seventy-year-old consumed about 30 percent less (in dollars) than the typical thirty-year-old; today the typical seventy-year-old consumes nearly 20 percent more.

It is obvious that this senior affluence is not evenly distributed. Millions of seniors would be destitute without federal benefits. There is also no guarantee that this affluence

3. AGING, HEALTH, AND HEALTH CARE ISSUES

will continue for future generations of elders, which is why Boomers must prepare for their own retirement now. Households that are not saving enough must confront and act on their retirement-income needs. In a recent study Public Agenda found that only 20 percent of U.S. households are "planners" who deliberately save toward a quantitative goal. The rest—"strugglers," "impulsives," and "deniers"—leave their future more or less to fate.

Younger Americans need to understand how great a change in saving behavior is required, but that this change will hardly be unbearable *if they start now.* Thanks to compound interest, even small sacrifices count. A recent study published in *Fortune* magazine found that if a couple at age forty decide to go out to dinner and a movie only twice a month instead of four times, and put the savings into a 401K plan, they will net $169,500 for their retirement at sixty-five. Paying off credit-card bills when they come in instead of incurring finance charges will yield another $121,400.

But if Boomers don't start providing personally for their retirement, then their golden years will hold nothing like the life of leisure that most of them seem to expect. In *The Retirement Myth*, Craig Karpel warns that the generation we met in the 1980s as "yuppies" may reappear around 2020 as "dumpies"—destitute, unprepared mature people wandering the streets with signs reading WILL WORK FOR MEDICINE.

COMING TRANSFORMATIONS

MODERN Americans are inverse Victorians. The Victorians, of course, were famous for their prudishness about sex. But they were loquacious in planning for their old age and eventual death. A dignified death and a proud cemetery site represented important social values. Their detailed wills were a boon to Britain's legal profession. We are just the opposite: We will talk to almost anyone and say almost anything about our sexual experiences. Yet we deal with aging and mortality as reluctantly as the Victorians dealt with sex.

Because we have difficulty talking about our collective aging, the social, cultural, and economic transformations that will be caused by it will come as a shock and a surprise to many of us. "Shake the windows and rattle the walls" is what Bob Dylan wrote about Baby Boomers when they first came of age. My purpose in this essay is to suggest how aging Boomers might shake the windows and rattle the walls of our society one more time.

The Retirement Transformation

As recently as 1950 most men who were physically capable of doing so continued to work past the age of sixty-five; fully a third of those aged seventy and up were still in the labor force. Today just 16 percent of elderly men work. This trend toward early retirement is no longer affordable. Moreover, given the growing number of Americans who reach the late sixties and seventies in good health and with valuable skills, it is no longer enlightened social policy.

As Robert Butler, a former director of the National Institute on Aging, puts it, America must develop a new vision of "productive aging" in which "work expectancy" increases along with "life expectancy." We seek satisfying love and sex after sixty—why not satisfying work as well? The old idea of a rocking-chair retirement is dead, and it is time for the new idea of an active yet aimless and dependent retirement to die as well.

The open question is when and how this transformation will occur. Should we change the Social Security retirement age to sixty-eight? seventy? seventy-two? When will we tell those who will be affected, so that they can begin to adjust their life plans? And how will employers keep so many seniors on the payroll? What private-sector management and training programs will make senior employees more attractive? How are we going to change the perverse Social Security incentives that discourage seniors from remaining in (or re-entering) the work force? What jobs will best suit seniors who continue working, and how can we maximize their availability? How do we revamp traditional career patterns to allow for semi-retirement, phased retirement, and "un-retirement"?

The Health Transformation

On the eve of the New Deal all levels of government spent roughly $1.00 annually on health care for the typical older American. By 1965 the figure had risen to roughly $100, by 1975 to roughly $1,000, and by 1995 to roughly $7,000. Thirty years ago America spent more on national defense than it did on health care. Health care is expected to consume 18 percent of GDP by 2005—at least five times what we are likely to spend on defense. And that's before the special multipliers of the age wave—especially the huge growth in the old old, who are most likely to require extensive acute and chronic care—even begin to kick in.

Americans prefer to believe that high and rising health-care costs are primarily the result of waste, fraud, and abuse. If only we got rid of all the unnecessary tests and treatments, or slashed the excessive paperwork, or got tough on Medicaid cheats and profiteering drug companies, then presto, the

IF WE CAN'T BORROW TRILLIONS OF DOLLARS, COULD WE RAISE TAXES A BIT AND MUDDLE THROUGH? THIS IS NOT A VIABLE OPTION.

problem would be solved. But experts know that the real causes are far more intractable: fabulous (and fabulously expensive) new medical technologies, cost-blind benefit and insurance systems that exempt most Americans from having to make choices about treatment, and the American tendency to disdain limits, including the ultimate limit—death itself.

Heirs of Ponce de León, in search of the Fountain of Youth? Perhaps that's too harsh. But no other country switches on multimillion-dollar MRI scanners for routine complaints (we have eight times as many MRI units per capita as Canada), commits terminally ill patients to intensive-care units, or performs heart bypasses on septuagenarians at anywhere near the rates we do. Americans, a European once observed, like to think everything is an option—even death.

The problem is that it is almost impossible to pinpoint aspects of our lavish style of medicine that are "wasteful" in the sense that they are of absolutely no medical benefit. Little of what physicians do is based on certain knowledge of the outcome; most involves judgment calls about unknown probabilities. Henry Aaron, the director of the economic-studies program for the Brookings Institution, speaks for most thoughtful observers when he writes that "sustained reductions in the growth of health-care spending can be achieved only if some beneficial care is denied to some people."

In the end the long gray wave will leave us no choice but to rethink what we mean by health. Is it a consumer good that can be purchased on demand at the doctor's office, or is it a lifelong investment? Should that investment be a personal choice, or should it be regarded as a public duty? How much should government be responsible for health care and how much should individuals? Most important, what share of public resources do we wish to spend on health care for ourselves, and how much do we wish to dedicate to such economic and social goals as productivity-enhancing R&D and a better education for our children?

No other transformation presents such profound ethical questions. Who will decide what costly heart transplants and similar death-defying high-tech operations are appropriate for the growing elderly population, especially the burgeoning old old group? When, and how, will society determine that even if an eighty-five-year-old can enjoy another year of life through an expensive high-tech intervention, this may be the wrong value to pursue—especially when so many children lack even basic health-care coverage?

The Youth Transformation

In an aging America everything will depend on the skills, education, productivity, and civic good will of younger generations—for their labor must support the elderly. Yet nothing seems less obvious than their capacity to rise to the challenge we are passing on to them. They will be relatively few in number. They will inherit a huge national debt and a high and rising payroll-tax burden. To make matters worse, many more of these future adults than today's adults are growing up in families, neighborhoods, and schools plagued by economic hardship and social dysfunction.

Since 1973 the real median income of households headed by adults aged sixty-five and over has risen by more than 25 percent, while the real median income of households under age thirty-five has fallen more than 10 percent. Counting all sources of income, poverty in America is three times as likely to afflict the very young as the very old. The United States is the global leader in the life expectancy of eighty-five-year-olds but has fallen near the bottom of the industrial world's rankings in rates of infant mortality, marital breakup, child poverty, child suicide, hours of school-assigned homework, and functional illiteracy. Meanwhile, per capita federal spending on the elderly towers eleven to one over federal spending on children. The appropriate response to the outrageous is to be outraged, yet we seem oblivious of this devastating disproportion.

How can we remain an economic superpower when nearly a third of our children are born out of wedlock and few of their fathers are willing to assume legal, financial, and moral responsibility for them? How will America prosper in a competitive technological and information-based global economy when its children grow up to exhibit school-dropout rates and rates of functional illiteracy that are among the highest in the industrial world? How do we answer Senator Daniel Patrick Moynihan's haunting question: "Will we be the first species that forgets how to raise our young?" Or, to paraphrase Churchill, "Have we ever asked so much from so few, having done so little to prepare them for their burdens?"

We're talking not about physical capital but about human and social capital: the intact families, work habits, education, and high-tech skills upon which any hope of increasing productivity ultimately rests. If we are going to rely on just 1.6 to 2.0 workers to support every retiree, as the SSA forecasts suggest, we should want today's children to become the best educated, most skilled, and most productive citizens imaginable. How does that square with our current rush to cut discretionary spending and defund social programs, from Head Start to vocational schools, that have long provided education and training? How can we generate the funding and the political support to educate our young in today's overburdened economy? How can we make the twenty-first century the century for our children?

The Political Transformation

Today's seniors, represented by powerful lobbies and voting in disproportionate numbers compared with the young, are already a potent political force. Will the rapid growth in the number of elderly enthrone the senior lobby as an invincible political titan? Or will the young, who must pay for tomorrow's senior benefits, find their political voice while there's still time to do something about it? Averting a de-

3. AGING, HEALTH, AND HEALTH CARE ISSUES

structive conflict between the generations will require a political transformation. But how can the young be encouraged to participate more aggressively in the political process? How do we merge the public interests of young and old and show how dangerous it is for them to become adversaries?

The Global Transformation

I recently asked the head of Japan's Central Bank why Japan has resisted America's requests to cut its budget surplus and stimulate consumer demand. His immediate response was "Because Japan must save so that it can afford its coming retirement wave"—a warning that the abundant and relatively inexpensive supply of foreign capital we have depended on for many years may soon disappear. The banker's reply underscores the high priority that some other industrial nations assign to the economics of aging populations.

Americans have paid little attention, but since 1980 roughly a third of net U.S. domestic investment has been funded by foreign creditors. Although some have expressed concern over how these capital inflows must give rise to a permanent annual debt-service charge on our national income, virtually no one has pointed out a more alarming prospect: not that the inflows will continue but that they could slow substantially as aging populations in other industrial countries consume more of their national income and savings at home. If America cannot boost its domestic savings rate within the next decade, we may enter an era of rising real interest rates, capital rationing, and a forcible curtailment of domestic investment.

There is also the issue of our relation to the less-developed world. When half the population in the United States is over forty, half the population in some emerging markets of Latin America and Asia may still be under twenty-five. Will the current distinction between rich and poor nations gradually come to be seen as a difference between old and young nations? Will the former be characterized by creative consumption, short time horizons, and the defense of the global status quo, while the latter, mainly in Asia and Latin America, become known for energetic investment, long time horizons, and revolutionary changes in the global balance of power? Will the newly democratizing economies of the former Soviet bloc be deprived of the foreign investment they need? Or, alternatively, will a high-saving Third World be exporting capital to a low-saving First World—an ironic turnabout of the policy recommendations of the 1970s? How will these demographic and economic shifts affect global institutions such as the United Nations, the OECD, and the World Trade Organization? Will they effectively address the myriad issues associated with the global age wave and enormous unfunded retirement liabilities?

TURNING AMERICA FROM CONSUMPTION AND DEFICITS TO SAVING AND INVESTMENT: WHAT NEEDS TO BE DONE

To argue in favor of thrift is sometimes enough to earn one the label "declinist"—a person who believes that America's best days may be over. This is not my view. Still, I want to explain why, if we do not face up to the economic and social challenges ahead, America will age prematurely and perhaps enter a precipitous decline. I do not believe it is un-American to suggest that we live in a finite world, that some desires can't be satisfied, and that bad choices can lead to tragic outcomes. On the other hand, some good choices—eminently feasible, gradual, and humane choices—can provide a sound future for all of us.

In an era crowded with social "crises"—from race to class—it may seem presumptuous to say that here we have a "real" problem that deserves our full attention. But let there be no doubt: the economic implications of America's aging population over the next several decades will dwarf, in sheer dollars, any other big issue one might name. Indeed, how we deal with the entitlement and savings crisis may determine how the other issues we face will ultimately play out.

If my analysis so far is correct, we are heading for a major crisis for which our society is unprepared. But our political leaders cannot be expected to take this challenge seriously unless we as individuals do so as well. A program of thrift thus has to work on all fronts, from the halls of Congress to our homes. Here are some workable steps.

1. *Achieve and guarantee long-term budget balance by the year 2002.* A campaign to boost saving must start with the federal budget, which can no longer be a borrower but must be a saver. Of all the policy choices directly available to American voters, none would do a more reliable—and faster—job of raising the national savings rate than eliminating our chronic deficits. I believe that we should achieve budget balance no later than 2002—a date, happily, around which a bipartisan consensus has finally emerged, after considerable Republican pressure. The reforms we make, moreover, should, at least provisionally, guarantee long-term budget balance *after* 2002—not just temporary balance *in* 2002.

The federal deficit is now 2.4 percent of GDP. Over the past thirty years the United States—along with every other major industrial nation—has repeatedly achieved this degree of public-sector deficit reduction in fewer than seven years. Moreover, since the federal deficit is projected to grow rapidly *after* the year 2002, a longer timetable would only make the long-term effort more difficult. Balancing the budget, starting now, is like running to catch a train that's leaving the station. To catch it in two minutes we would have to run harder than we would to catch it in one minute.

13. Will America Grow Up before It Grows Old?

Some experts worry that this is not the right time in the business cycle to initiate a balanced-budget plan. But according to these critics, it may never be the right time. So long as reforms are phased in gradually over seven years, there is little danger that a shift from consumption to saving will seriously depress the economy. Indeed, a credible budget plan might boost the economy if—as many economists, including Alan Greenspan, think likely—the markets react by lowering interest rates, particularly long-term rates, by two percent.

But mere budget balance is too timid a goal. Given the shortage of our national savings, I believe that Congress should aim for a federal budget *surplus* of perhaps one or two percent of GDP through the first two decades of the next century, to make up for our recent profligacy and, more important, to lay up stores during the Boomers' peak earning years for the sudden burden that will accompany their retirement. Or, better, Congress could aim for a smaller surplus but substantially increase spending on targeted public investments in education, worker training, and research and development—the kind of human-infrastructure investment that is essential to an information-age economy, but in which we are now sorely deficient. Either way, we would radically change federal budgeting. We would no longer presume on the good will of our children but would demonstrate our good will toward them by moderating excess consumption, which makes us net takers, in favor of investment, which unites us as net givers.

2. *Reform entitlement programs.* Trying to achieve long-term budget balance without reforming entitlements is like trying to clean out the garage without removing the Winnebago. The following reforms, taken together, would put these programs in long-term sustainable balance well into the twenty-first century.

• *Subject all federal benefits to an "affluence test."* The first sensible step toward long-term budget balance is to scale back entitlement subsidies flowing to people who don't need them. To this end I recommend that we enact a comprehensive "affluence test" that would progressively reduce entitlement benefits to all households with incomes over $40,000—or more than $5,000 *above* the U.S. median household income for 1996. Households with lower incomes would retain all government benefits. The affluence test would be applied *annually*—protecting the elderly in the event of an unexpected loss of income. Higher-income households would lose 10 percent of all benefits that raised their income above $40,000, and 10 percent for each additional $10,000 in income. Thus a household with $50,000 in total income and $10,000 in federal benefits would lose $1,000, or 10 percent of its benefits; a household with $100,000 in income and the same $10,000 in benefits would lose $6,000, or 60 percent; a household with more than $120,000 in income would lose $8,500, or 85 percent—the maximum benefit-withholding rate. (This 15 percent exemption would ensure that even today's most affluent beneficiaries continue to enjoy a respectable tax-free return on their personal FICA contributions.) All income brackets would be indexed for inflation.

Because the test would leave in place all benefits to lower-income households, the original "floor-of-protection" intention of nearly all federal benefits programs would continue to apply. Because such a large share of entitlements now goes to middle- and upper-income Americans (nearly 40 percent of Social Security payments go to recipients with incomes above the U.S. median), savings would be large and would compound as the population aged and the number of beneficiaries grew. Indeed, it is estimated that by 2040 *annual* savings would amount to more than $550 billion. Finally, because the test would also be comprehensive, covering not just Social Security and Medicare but everything from farm aid to federal pensions to veterans' benefits, this plan would not pit one special-interest constituency against another.

Since this affluence test was first proposed, in my book *Facing Up*, it has attracted considerable interest from both Democrats and Republicans. It has also elicited criticism from those who for various reasons don't want entitlements reformed. Some have said that an affluence test would constitute a tax on savings, and thus would discourage thrift. There is no evidence to support this hypothesis. More important, it ignores the larger issue—which is how to increase national savings. Any decline in private saving caused by an affluence test would be dwarfed by the decline in benefit outlays—which in turn would translate dollar for dollar into smaller deficits and greater net national savings.

It has also been said that an affluence test would undermine public support for Social Security and other universal social-insurance programs. The theory seems to be that we must bribe the affluent in order to ensure political support for benefits for the needy. This is dead wrong. Of all major proposals to reform entitlements, affluence testing receives the greatest public support. According to a recent opinion poll by the Concord Coalition—a group that I helped to form in 1992, along with the former senators Warren Rudman and Paul Tsongas—67 percent of those asked would support reductions in Social Security benefits to higher-income households, and 77 percent would support reductions in Medicare benefits. Even majorities of older and of affluent households support such a reform.

• *Raise the eligibility age for full benefits.* Congress has already raised the Social Security full-benefit retirement age from sixty-five to sixty-seven, to be phased in from 2000 to 2027. This is a step in the right direction (although most Americans are not aware of it), but the step is too small and too gradual. My recommendation is that the Social Security retirement age be raised by three months a year until a new eligibility age of seventy is reached in 2015—a phase-in that

3. AGING, HEALTH, AND HEALTH CARE ISSUES

WE MUST PUT ASIDE THE "ENTITLEMENT ETHIC" AND RETURN TO THE "ENDOWMENT ETHIC," ACCEPTING RESPONSIBILITY FOR THE FUTURE.

would leave Boomers plenty of time to plan ahead. In my view, early retirement should still be allowed at age sixty-two, but the benefits extended to early retirees should be reduced commensurately. When this reform had been entirely phased in, workers would still enjoy more years of full benefits than were envisioned when Social Security was founded. As Social Security's full-benefit eligibility age went up to seventy, so should Medicare's. Americans aged sixty-five to sixty-nine could still participate in the program, but only by paying extra premiums.

• *Set limits on federal health-benefit spending.* We must restructure health-care benefits to control federal health-care costs. Currently we offer fee-for-service reimbursement to all eligible comers, with few cost disincentives, and then surround the process with a thicket of regulatory controls. I propose that Medicare, Medicaid, and other health-benefit programs offer three choices: take a fixed-dollar voucher and use it toward the purchase of the health insurance of your choice; enroll in an accredited managed-care program that will then bill the government a fixed annual amount; or remain in the current fee-for-service system and face much greater co-payments and deductibles.

Any reform that seeks to introduce market discipline into our system of federal health benefits must give beneficiaries real incentives to be cost-conscious. Hence the greater co-payments and deductibles for those who choose the expensive fee-for-service option. The Medicare plan that Congress passed last year was all carrot and no stick. It gave beneficiaries the choice of enrolling in new kinds of managed-care plans, but would have imposed no penalty on those who opted to stay in traditional fee-for-service plans.

These measures would shift the task of cost control away from regulators and back to patients and providers, where it belongs. They would also allow Congress to live within a health-benefits budget, like the government of every other industrial country. As for the senior lobby's attachment to a "free choice of doctor" guarantee, voters must be reminded that a declining proportion of today's young workers—whose FICA taxes pay for much of Medicare—enjoy the full freedom of choice that was once common in American medicine. Most young workers consider themselves lucky if their employer pays for any health insurance at all.

Another prime candidate for reform is the unlimited tax deduction for company-paid health-care insurance (which now amounts to a $92 billion annual subsidy from the federal government). This wasteful and regressive deduction should be capped. Federal efforts to establish national health-practice guidelines for doctors, hospitals, and insurers should be encouraged. Although these standards would not be mandatory (patients or providers would still be free to spend their own money for services above the guidelines), they would give everyone a clearer idea of the cost-effectiveness of various treatment options—something all experts agree we lack.

Finally, we need to reduce the huge costs of "defensive medicine," through malpractice reform, and of "heroic" intervention when recovery is highly unlikely. Medicare spends approximately 30 percent of its budget on patients in their last year of life—often when the attempt to prolong life merely prolongs a hospitalized death. Few Americans want to end their lives this way. A recent survey shows that 89 percent of Americans support the concept of living wills. Yet only nine percent actually have them. Until we launch a widespread educational effort, make enforceable living wills widely available at very low cost, and perhaps even provide financial incentives to maintain them, doctors will continue to perform costly and painful procedures on patients who do not (or would not) want them and who will die in a few days or weeks anyway.

We should have no illusions about the future. Whatever reforms we implement, federal health-care costs are going to grow faster than our economy. This is one more reason why we must do everything possible to reduce growth in Social Security and other non-health-related programs.

3. *Extend working lives.* One of the best ways to reduce the crushing burden ahead is to encourage seniors to work longer—and make it easier for them to do so. This would require more than raising the age of eligibility for full benefits under Social Security and Medicare. To encourage longer working lives we should abolish the Social Security "earnings test" for beneficiaries who continue to work. (Let me stress that this reform *must* be implemented along with the affluence test I have described; a stand-alone aberration such as what Congress has recently proposed would be an unearned windfall for senior-citizen CEOs like me.) Yes, there would be a small direct budget cost. But the benefits to the economy and to society, and to seniors themselves, of encouraging later retirement would be far more significant than the small increase in outlays.

The maturity, wisdom, and experience of older adults should not be lost to the workplace. This is a matter not just of combating age discrimination but of unlocking a powerful human resource. The market for jobs for which the elderly might be especially well suited should be explored: for example, full- and part-time service jobs in health care, child care, and various education and training efforts. It is time to do elders the honor of making their phase of life one of ongoing contribution—of genuine "generativity," to use Erik Erikson's classic description—as long as they are willing and able.

13. Will America Grow Up before It Grows Old?

Not everyone, of course, is able to go on working. Richard Trumka, the president of the United Mine Workers, who recently served with me on the Kerrey-Danforth Commission on Entitlement and Tax Reform, warns that later retirement is simply not a realistic option for worn-out industrial laborers in physically demanding occupations. But such workers make up a small and shrinking share of the total labor force. Under my plan they would still have the option of early retirement (though with reduced benefits) and would be protected by federal Disability Insurance and Workers Compensation, not to mention the system of mandatory personal retirement accounts that I propose below. I would also use a small part of the savings achieved by raising the Social Security retirement age to lower eligibility ages and raise benefit levels under Supplemental Security Income, the means-tested floor of protection for the low-income elderly. In sum, we should encourage the elderly to work but not force work on those who are truly incapacitated. In any case, our national retirement policy should not be determined by the miner retiring at age sixty-two any more than by the police officer retiring at fifty-two or the athlete at forty-two.

4. *Establish a system of mandatory pensions or personal retirement accounts*. I have concluded—reluctantly—that a fully funded, privately managed, and portable system of personal retirement accounts should be mandatory. The system I envision would initially supplement Social Security—and over time might increasingly substitute for it. But Social Security would continue to provide a floor of protection to all Americans, albeit one subject to the limits of the affluence test described above. Governments around the world have tried to achieve both these objectives—retirement savings and poverty protection—in a single system. They have achieved neither efficiently.

Why mandatory? In 1993 C. Fred Bergsten, the chairman of the Competitiveness Policy Council (a publicly financed, bipartisan group), asked me to chair a committee on capital formation. An impressive group of the nation's leading economists joined me in this effort. I had expected to hear that certain tax favors for saving (IRAs, for example) would significantly increase net savings—that is, savings beyond the cost of the tax incentive that encourages them. I quickly learned otherwise. The net effect of many of these conventional incentives has been marginal, because much of the money deposited in IRAs is simply shifted out of other investments. When I asked how we might increase net savings significantly, one important area of agreement emerged: mandatory pensions or savings accounts covering the entire work force. In addition to boosting private saving, such plans —by making tomorrow's retirees more self-sufficient—would allow us to reduce traditional Social Security gradually, thus reducing public dissaving as well. I am perfectly well aware of the libertarian argument that decisions about saving should be left entirely to individuals. The melancholy truth, however, is that many Americans are currently too myopic to save for the future unless compelled, and so end up becoming free-riders in the government safety net.

Why fully funded? First, to boost national savings. A funded retirement system would add to America's capital stock; a pay-as-you-go system does not. Second, because the dynamics of pay-as-you-go financing have encouraged politicians around the world to promise benefits that can be paid for only by excessively high taxes on future generations. The only way to avoid that temptation is to make it clear to everyone that above some minimum safety net a worker's future benefits will be determined solely by the resources that have been set aside for that worker, by some combination of employer contributions and the worker's own savings. These pensions must be invested in diversified investment-grade assets and must be the worker's personal property.

Why privately managed? A sound system of mandatory pension accounts must be publicly regulated to maintain fiduciary standards but should be privately managed to maximize returns. The evidence is overwhelming that publicly managed systems, which are often required to invest in low-return government securities, earn far less than privately managed accounts invested in the real economy.

Why portable? The new and fluid global economy, characterized by intense competition, rapid innovation, and relentless technological change, has made "lifetime employment" with one company rare. Instead making several major job changes in one's lifetime—perhaps seven or eight for the average worker now in his or her twenties—is normal, and therefore many workers lack enough years of service in any one job to qualify for a pension. The plan I propose would vest all contributions immediately, and so workers could take their pension savings with them as they moved from job to job.

To provide adequate retirement income, these accounts would require substantial contributions. In my view, all workers (in some combination with employers) should be required to contribute four to six percent of their pay—which, added to FICA, would come to a total contribution of 16–18 percent of pay. As a point of comparison, Australia's new system of mandatory pensions will ultimately result in total contributions of 15 percent of pay. In the scheme I propose, workers would have the option of making additional voluntary tax-free contributions. Employers who currently provide pensions could divert their contributions to workers' savings accounts as well. The primary function of this system would be to finance retirement and survivors' benefits; in time it might also pay for long-term medical care.

Although mandatory pension contributions would be made in addition to current FICA payroll taxes, and thus would decrease the consumable portion of each paycheck, the system would be linked to the Social Security reforms described above—and this would prevent FICA taxes from rising to the alarming levels forecast for the next century.

3. AGING, HEALTH, AND HEALTH CARE ISSUES

Eventually workers would be paying no more (and maybe substantially less) in combined FICA and savings contributions under my plan than they would be paying in FICA taxes alone in a status quo future. By putting more of our income into genuine savings today, we could relieve the crushing payroll-tax rates that unfunded public transfers will otherwise exact on workers tomorrow.

The reform I propose would also require that any current-year Social Security or Medicare cash surplus be transferred, on a pro rata basis, to workers' personal retirement accounts. This provision would be consistent with pay-as-you-go accounting. Meanwhile, workers would have a direct stake in reforms that constrain future growth in federal benefits. To the extent that Social Security declines as a share of payroll, a growing share of FICA taxes would automatically be transferred to workers' savings accounts. Let me repeat: My proposal is for a two-tiered system under which everyone would continue to receive Social Security benefits. But over time my proposal would also allow us to go a step further. As the savings in private retirement accounts built, the current universal Social Security system could be converted into a purer and much less costly floor of protection that paid out benefits only to the truly needy.

A mandatory savings plan would generate substantial net gains in household (and national) savings—and thus ultimately gains in productivity and living standards. For middle- and upper-income workers subject to the affluence test this system would at least make up for reduced government retirement benefits—and probably go much further. For lower-income workers, who are the least likely to save (either on their own or through pensions), it would vastly reduce the chances of a destitute retirement. Seniors who were beneath the affluence-test threshold would receive their private pension on top of full federal benefits. True, the deduction from wages would be a burden, but it's worth noting that because of the Earned Income Tax Credit, the existing FICA tax on many of the working poor is now entirely borne by the federal government.

Dismissed until recently as too "radical," "privatization" of Social Security has burst upon the scene over the past year. Major proposals are under development at half a dozen think tanks—left, right, and center. Privatization has been featured on the cover of *Time* and embraced by the presidential candidate Steve Forbes, and, in one form or another, is endorsed by seven out of thirteen members of the Administration's official Social Security Advisory Council.

My plan has elements in common with many of these proposals. Where it differs from most is that it would fully pay for the transition to a funded Social Security system—and would do so *without adding to the national debt and without new general-purpose taxes*.

The challenge is that a single generation must somehow pay for two retirements—its own and that of its parents. Some proposals simply ignore the challenge. Take Steve Forbes's plan to keep all benefits for current retirees intact and yet permit younger workers to shift a substantial share of their FICA contributions into personal retirement accounts. What his plan would add to private savings it would cancel out dollar for dollar by increasing the federal deficit. Other proposals would issue Treasury debt directly to Social Security beneficiaries in the amount of the system's accumulated liabilities. This, too, is a zero-sum game that will leave tomorrow's workers no better off than if we had never reformed the system. A few proposals, like that of the Social Security Advisory Council, are more honest. But to pay for the transition they would resort to large general-purpose tax increases.

My plan would pay for transition costs the old-fashioned way: not with smoke and mirrors but by taking the essential step of asking current beneficiaries and current workers to give something up—the former by forgoing some benefits, the latter by saving more. This would not be painless. The magic of compounded returns from the stock market and other long-term investments cannot solve all our problems. To save more, we must consume less, at least temporarily. This "transition cost" is the price of escaping the generational chain letter we have so far depended on.

5. *Shift our tax base from income to consumption*. In an aging society taxpayers should be penalized for what they take out of the economy (consume) as opposed to what they put in (save). I therefore propose that only "consumed income"—spending, that is—be taxed. It is true that by exempting savings from taxation this reform would narrow the tax base. On the other hand, it would also widen the base, by rendering taxable various forms of government-financed and -subsidized consumption—from Social Security benefits and the insurance value of Medicare to employer-paid health care—which today are partly or fully tax exempt. Tax rates thus need not be any higher than they are today.

Many will object that consumption taxes are regressive; but the consumed-income tax plan introduced by Senators Sam Nunn and Pete Domenici—in which the more one spends, the higher one's tax rate—demonstrates that consumption taxes need not sacrifice the principle of progressivity, which I support. Moreover, without increased saving we cannot expect the real income of the typical American household to grow again—and without such growth the distribution of incomes will continue to widen. The zero-sum politics of economic stagnation will overwhelm whatever weak contribution to economic equality we might continue to derive from our current system of progressive income-tax rates.

We are currently bombarded with tax-reform proposals of every variety—from sales taxes to flat taxes. Some want to get rid of the Internal Revenue Service. Some want tax returns that can be filled out on a postcard. Who doesn't want simplicity? And who likes the IRS? But I would suggest a more important criterion for evaluating tax reforms: Which is most likely to increase net national savings?

6. *Mount a broad-scale public-education effort to promote saving.* National leaders must help to mobilize citizens by articulating a sense of moral imperative. A thrift plan needs a bully pulpit.

Can the right kind of education and exhortation make a difference? Consider Japan. Until the 1950s, when the country rallied behind a campaign to promote thrift, the Japanese were poor savers. Since then they've become famous for their saving. Or consider Singapore, whose Central Provident Fund has furnished much of the investment capital that has fueled Singapore's legendary economic growth—not to mention the savings that have enabled nine out of ten households to become homeowners. Or consider Chile and Australia, which have also established national pension systems based on the principles of full funding and portability. In each instance public education was crucial to securing public support. In Chile, for instance, Jose Piñera, then the Minister of Labor, went on national television, often weekly, to explain why the mandatory pension plan was such good news for Chileans.

In a society like our own, where grassroots consensus is so important to governance, public discussion and debate are all the more important. The problem is that for at least three decades leaders have been telling us that consumption, not savings, is the key to prosperity. The campaign in favor of consumption has worked—all too well. Now it's time for a different kind of campaign—one in which not only our political leaders but also our businesses, our universities, and our public-policy institutions must persuade Americans to adapt to the realities of our aging society.

What we need most of all is a moral vision, a Middle-Class Bill of Responsibilities—not a gaggle of leaders falling over one another in their rush to propose a Middle-Class Bill of Rights, or the middle class silently rehearsing the mantra "We are not part of the problem and we need not be part of the solution." Instead we must be encouraged to ask, What do we expect individuals and families to do for themselves, and what do we expect federal, state, or local governments to do for them? What are our responsibilities to our own children and grandchildren? How can we strengthen families so as to provide support for older people? What are our obligations as a nation to our collective progeny?

The manual for German's social-security system looks, at first glance, a lot like our own—page after page describing the benefits due if one retires, is widowed, or loses one's job. The most obvious difference is the generous benefits to German children. But there is a more striking contrast. For each benefit, alongside a box describing "Your Rights" is a box describing "Your Duties." Citizens are thus reminded that society must always balance the payer against the payee, the future against the present. We need to find that balance again in our culture.

Why can't the President call for a White House Conference on Aging different from the one held last year—not one that panders to the senior lobby but one that encourages serious dialogue between old and young? Why can't the President call for a global summit at which the leading economies focus on reducing their tremendous and unsustainable unfunded liabilities, and at which developing economies with younger populations concentrate on avoiding the mistakes the industrial countries have made in providing old-age security?

Companies also have a major educational responsibility. With their human-resources and accounting departments, they are able to educate workers on the basics of saving—why they should save more, the power of compound interest, how to invest. They can also make it easier for their employees to save—through automatic salary deductions, 401K plans, stock-purchase and dividend-reinvestment plans.

Bringing our youth into the savings crusade is another key. John F. Kennedy once challenged us to ask not what our country can do for us but what we can do for our country. Today's youth see the most conspicuous interest groups in our political system busily asking what the country can do for them. But who represents the future and the general interest? The young, alas, are the new silent majority. The demographer Samuel Preston once remarked apropos of the relentless growth in senior entitlements that the political system would behave a lot differently if people were forced to live their lives backward—that is, if they had to look forward to the burdens imposed upon youth as their own future.

I suggest that young people embark on dual careers—a private career and one as a citizen. As citizen lobbyists in behalf of the future, they are responsible for becoming informed about the debts they are going to assume, the unfunded liabilities they are going to pay for, and the unsustainable taxes they are going to bear. Once they are informed, perhaps America's youth will initiate an honest dialogue with their parents and grandparents, without assuming that their elders are greedy old fogies who don't care. My generation may be uninformed and even misinformed, but we do

THE FIRST STEP TOWARD LONG-TERM BUDGET BALANCE IS TO REDUCE ENTITLEMENTS FOR PEOPLE WHO DON'T NEED THEM.

care about our children, our grandchildren, and our collective future. But if anyone is to create a general-interest lobby in behalf of the future, youth must lead the way.

If we expect our leaders to lead, the voters must make it safe for them to do so. The Concord Coalition is a bipartisan grassroots "lobby for the future," dedicated to breathing new life into the American Dream. The warm reception we have received from countless concerned citizens has rekindled my faith that we can still build a special interest in behalf of the general interest.

CONCLUSION

AND what of the special role for geezers like me? Pessimists say, "Forget it"—Americans will not reform senior benefits until a severe crisis is actually upon us, but will persist in viewing them as contractual obligations that by definition are always affordable. After all, an America that acknowledges limits is an America that has lost the one illusion that makes it unique and creative. According to this view, America must always be an unteachable force of nature that can never back away from any promise or expectation, no matter how extravagant. This, pessimists say, is why American voters repeatedly elect leaders who promise lower taxes, higher benefits, rejuvenated economic growth, and a magic bullet for every social problem—without caring how the pieces fit together.

But I have a more optimistic view. Two years ago I was interviewed by *60 Minutes* about the need to enact gradual but far-reaching structural reforms in federal entitlements for the elderly. The show's producers, after patiently taping my arguments, invited me to join them at a middle-class retirement community. Here, they said (with a few wry smiles), I could explain my suggestions to those who would be immediately affected.

Standing before this group of retired grandparents, I began by showing photographs of my own grandchildren. I explained my concerns about their future and the world they would inherit. I then reminded the retirees how much of our national affluence today rests on the willingness we had to make collective sacrifices during the Great Depression and the Second World War. Back then we felt that we were "all in this together" for the sake of tomorrow. I told them that the German theologian Dietrich Bonhoeffer said it best for us when he observed, at the height of the Second World War, "The ultimate test of a moral society is the kind of world that it leaves to its children."

Sooner or later, I told the retirees, we will have to prepare for the future. We will have to balance our public budgets, trim back benefits to those who need them least, save more as households, retire somewhat later from the work force, explore innovative means of economizing on health care, take a more effective public interest in the welfare of children, and offer the rising generation some tangible evidence that we are willing to make sacrifices in their behalf. If we do so sooner, we still have time to plan for a gradual and humane transformation. If we do so later, the changes are likely to be forced upon us, suddenly and painfully, in the midst of an economic, political, and family crisis that will leave the eventual outcome much in doubt.

Given all that, I asked them, if everything else were also put on the table and it really would lead to a balanced budget, how many of you would be willing to give up some share of your federal benefits, above what you need to live on, in order to ease the deficit burden on younger generations? To the visible surprise of the *60 Minutes* producers nearly everyone raised a hand.

The generation I was speaking to survived the Depression and fought and won the Second World War. After the war this generation provided its returned veterans with college educations, built the interstate highway system, eradicated polio, took us to the moon, and won the Cold War against communism. Against these monumental accomplishments what it would take to solve our current crisis seems small. I believe that this generation is capable of doing the right thing, and that politicians might well discover that it is better to appeal to their nobler instincts than to pander to their baser ones.

A people who have made a tradition of quick gratification must now be asked to focus on the requirements of a society graced with the patina of age—on saving rather than consumption, on prudence rather than desire, on collective restraint rather than individual satisfaction. As Americans grow older, they will have to recognize that the live-for-today attitude that may be endearing or at least understandable in youth is not just unseemly but ruinously dysfunctional at the far end of life. They would do well to heed the eighteenth-century French moralist Joseph Joubert, who warned, "The passions of the young are vices in the old."

The Cruelest Choice

Medicine is forcing some parents to decide the fate of their severely premature babies

Sharman Stein
Sharman Stein is a Tribune staff writer.

In her sunny, North Side Chicago apartment, Ellen Baren Skowronski feeds raisins one by one to her son.

"Chew, Leon, chew," she tells the 4-year-old. Lying on a pillow at the center of a living room crowded with toys, Leon smiles hugely at his mother and his 2-year-old sister, Phoebe.

Leon, who weighed 677 grams at his birth—about a pound and a half—was born at Evanston Hospital 24 weeks into his mother's pregnancy. He was nearly four months early. Due Nov. 16, 1990, he was born July 27.

The doctor held him in one hand. It would be nearly 5½ weeks before he would be strong enough to be held in his mother's arms. Wide-eyed and sweet-tempered, fair-haired Leon now weighs just 27 pounds, the weight of an average 2-year-old. He cannot walk, talk, sit up or stand. He can hardly hear. He has cerebral palsy—moderate in his arms, severe in his legs. He is brain-damaged. It takes him nearly 45 minutes to chew his food at each meal.

"He can't hug me, even though I know in his mind he would like to," says Baren Skowronski, 36, a dark-haired, outspoken woman who studied photography at The School of the Art Institute and now waitresses nights at Second City Comedy Club. "He can't even kiss me. He has never said 'Mom.' When he's hungry, he can't tell me what he wants to eat. Sometimes when I'm not giving him what he wants, he just starts to cry. At 4 years old, he should be able to say, 'Mom, I want a peanut butter sandwich.'"

Born at the very threshold of viability, Leon would have died at birth if he had been born just a few years ago. Medical technology has advanced to the point of saving babies as premature as he was; doctors can save 90 percent of infants born as early as 28 weeks.

What the doctors cannot do is to predict with certainty which of these premature children will turn into so-called "miracle babies" without lingering disabilities and which will live severely handicapped lives. It is akin to a surgeon who might capably remove your appendix but could not give you any reliable projection of how you would fare after the operation or whether you would survive.

The younger and smaller the babies are, the greater the possibility of severe disability and death. Leon was born in the so-called "gray zone" between 23 and 27 weeks, the size and weight of five or six sticks of butter, his eyes still closed, his skin covered with lanugo, the fine, soft hair that covers the infant inside the mother's womb.

With the ability to save ever-tinier babies—a Ft. Worth girl weighing just 12 ounces at birth went home from the hospital a few months ago—parents and doctors confront a Solomon-size dilemma.

Under the stark fluorescent lights of the hospital and the glare of attention from doctors and nurses and strangers, parents who had anticipated the joyful delivery of a healthy baby now end up facing the spectre of death or long-term disability for their child.

Baren Skowronski says she now wishes the doctors had told her how grim the future could be while Leon, the baby she had very much wanted, lay there in the neonatal intensive care unit. "He was

3. AGING, HEALTH, AND HEALTH CARE ISSUES

just born too early," she says, her tone alternating between sadness and anger. "They didn't have any business saving Leon. They were experimenting with a fetus, keeping it alive outside his mother's womb. The doctors say they care for humanity, but they're not here taking care of him now."

Helen Harrison, author of "The Premature Baby Book," whose severely disabled 19-year-old son was born prematurely, hears frequently from parents around the country who claim their physicians did not give them the opportunity to have a say about their baby's future. "So many of these babies are saveable [but] for lives of incredible handicap and impairment," Harrison says. "They're not going to be healthy, functioning citizens. They're in incredible pain for months, sometimes years.

"Some doctors say babies weighing 800 grams or less shouldn't be treated because their outcome is so bad; some say they have to treat every baby with a heartbeat," she says. "In some hospitals, lots of babies are not resuscitated to begin with because of their gestational age. There is no standard procedure."

Nor is there any case law codifying what the procedures are or should be. Much of what happens—how much the parents' wishes are honored and what doctors do after the birth of a severely disabled child—depends largely on the policies of the individual doctors and hospitals.

A Michigan case due to go to trial in January has called the question of parental-infant rights into sharp focus. Gregory Messenger, a dermatologist on staff at the East Lansing hospital where his wife, Traci, had a premature delivery, requested that doctors take no actions to save the infant after birth. At the hospital, in the hours before the 1-pound-11-ounce boy was born, a neonatologist had informed the couple that the baby's prospects were very doubtful.

But after the delivery, a physician's assistant placed the infant boy on a respirator anyway. As doctors and nurses watched but did nothing to stop him, Messenger disconnected the machine, and his wife held the infant until he died.

Messenger will go on trial for involuntary manslaughter in mid-January. The case is being closely watched by parents, neonatologists and ethicists, even though, because of its dramatic and unusual nature, it may not set a precedent.

Society's devotion to the sanctity of life demands that separate attention be paid to the interests of the child—a person, even when handicapped, with the right to life.

The so-called Baby Doe laws passed in the 1980s required doctors to try their utmost to save children with Down's syndrome and spina bifida. No laws have been passed to address the issue of premature infants, of whom more than 29,000 are born severely premature—more than three months before term—each year. Nearly 40 percent of those infants born at less than 28 weeks gestation die.

Neonatology advances can now save preterm babies as early as 22 weeks gestation (normal gestation is 39 to 40 weeks). At 28 weeks, the infant is considered to have an excellent chance of survival, and many doctors believe they should give these babies their best chance at survival, all the more so as technology is making it more and more possible to do so.

It is the parents, however, who have to live with the consequences of medical intervention at such a stage, and on this basis, Frank Reynolds, Gregory Messenger's attorney, argues: "The real important issue here is parental rights. These parents made a medical decision that was within their rights as parents to make. They did not want to prolong their child's death. Their concern was to provide care and comfort and to bond with the baby until he died."

Michigan law indicates that parents do have the right to make decisions affecting their children. But Ingham County, Mich., prosecutor Donald Martin says Messenger made his decision without waiting for standard medical tests on the baby, which determine, among other things, how well oxygen is being processed through the blood and the development of the baby's lungs.

"What information did the good doctor-father have to justify his actions that whatever he was doing was in the best interest of his child?" Martin asks. "He made a unilateral decision to end his infant son's life. And we believe that in so doing, he did not act in the child's best interest. It was a reckless act that brought about the child's death."

Dr. Mark Siegler, a professor of internal medicine and director of the University of Chicago's MacLean Center for Clinical Medical Ethics, calls Messenger's actions "outside the bounds of a civilized response to a tragic situation." A great ethical divide occurs once the baby leaves the woman's womb, and "comes under the protection of our society," Siegler says. Neonatologists eventually would have backed the parents' decision to remove the baby from life support if he were truly beyond help, he adds. "What the father did sets a terrible precedent."

In Baren Skowronski's case, she has had plenty of time to think about the freakish turn her life took with Leon's birth. That she and her husband, Mike Skowronski, love him devotedly, there is no doubt. But they also agonize over their ability to care for Leon as he gets bigger. In a year or two, they will have to move from their second-floor apartment to something more wheelchair-accessible.

Mike, a tall, burly man, insists that Ellen immediately fell in love with the baby and could not bear to think of letting him go, and he says his impulse at the time was to support her. But he is frank about how difficult life has become. Other men he has met through support groups also complain, he says, and many men end up walking out, overwhelmed by this kind of fatherhood. "The big thing with these guys is that that their wives are never the same; the fun is no longer there. They're always going, going, going with the child. They don't get back to themselves."

The Skowronskis are sadly acquainted with some of the perils in Leon's future. The Johnsons, of Oak Brook, are at the beginning of their journey with an extremely premature infant.

Karen Johnson gave birth to a daughter three months early in September at University of Chicago Hospitals. She and her husband, Dwain, named the baby Faith. "We're rather religious, we have a strong belief in God, and we believe it's a miracle so far," says Karen Johnson, 32, an accountant.

Born weighing 1 pound 6 ounces, Faith appeared to be doing well from the start, and the couple never considered letting her die, despite her extreme prematurity and the risks ahead. "We don't want her to suffer for the rest of her life. If she can be healthy and normal, we want her to live. If she had to suffer, God would do the best thing."

Each day, sometimes twice a day, she and Dwain visit Faith at the hospital. They were hoping she would be well enough to come home at Christmas, but an infection and a couple of other setbacks may delay that homecoming.

Actually, many infants with a poor chance of remotely healthy survival are being allowed to die quietly in hospitals throughout the country, including the University of Chicago, Northwestern, Evanston and other local hospitals that care for severely premature infants.

But such situations are handled quite differently in each hospital, depending on the doctor, the prognosis for the infant and the hospital's general policies.

One physician who strongly favors saving any baby with a chance of survival is Dr. Sheldon Korones, professor of pediatrics at the University of Tennessee College of Medicine in Memphis. "When I look at a baby, he's the whole world," Korones says. "I don't know where that baby is going to fall in the statistics of outcomes.

14. Cruelest Choice

Preterm babies can now be saved as early as 22 weeks gestation out of the normal 39 to 40 weeks.

I can't predict. So I opt for support. I can't go by parents' fears of outcome. If they ask me to let the baby die and the baby is responding to treatment, I tell them this is murder, that's what it is."

Many of the highly premature infants being saved by Dr. Korones and others like him face a highly risky future. In the hospital, many suffer brain bleeds, seizures, difficulty in breathing and swallowing and require painful operations. Afterward, they often suffer cerebral palsy, mental retardation, respiratory problems, deafness, blindness and a host of other disabilities.

The seriousness of some disabilities sometimes are not readily apparent, says author Harrison, who lives in Berkeley, Calif. "Our son, who tests as mildly retarded, can't leave the house by himself. He can read and count, but he can't understand what he's reading or counting. He can never live independently."

The financial costs of prematurity are enormous. The Center for Risk Management and Insurance Research at Georgia State University in 1992 estimated that cost at $5.6 billion a year nationally. The average amount that the Illinois Department of Public Aid spends for its clients for a normal vaginal delivery and a hospital stay of about two days is $2,814. For extremely premature deliveries, which involves an average hospital stay of 46 days, its average bill is nearly $60,000.

Blue Cross/Blue Shield of Illinois estimates that doctor and hospital costs for a premature infant are nine times as expensive as a normal delivery. Doctors' fees start at $1,250 the first day, compared to $200 a day for normal births.

The costs of caring for a very premature baby who has been in the hospital for many months can easily total hundreds of thousands of dollars.

Each additional week of gestation makes an enormous difference in the ability of a premature infant to survive intact. In the gray area, between 22 and 26 weeks of gestation, the chances are greater for an infant to survive, but with serious disabilities.

A recent study by the Johns Hopkins University School of Medicine showed that 39 percent of 142 infants born between 22 and 25 weeks of gestation survived six months; nearly 100 percent of the babies at the lowest gestational age, 22 or 23 weeks, either died or had severe brain damage. Among those born at 25 weeks gestation, or about four months early, 26 percent had severe brain abnormalities.

Joanne Bregman, a developmental psychologist at Evanston Hospital, says most studies at neonatology wards in hospitals across the country show that between 15 and 20 percent of infants born between 24 and 26 weeks gestation are severely disabled, and about half of them exhibit a wide range of learning disabilities.

"The best interests of the child may not always mean saving the child, if the child is going to suffer a lifetime of pain and hospitalization," Harrison says. Treating every newborn, no matter how severely premature, is a step back to Baby Doe laws, she says, which required every newborn to be treated until death is unavoidable. "I hear from parents every day who are being put through the mill on this," Harrison says.

Some 800 to 1,000 very premature babies weighing less than a pound and a half are born at the University of Chicago Hospitals every year. About twice a week, parents and doctors decide to stop treatment and let the baby die.

Dr. William Meadow, associate professor of pediatrics at the U. of C., says some of these infants, especially as they get closer to weighing about 2 pounds, are "obviously viable," and in those cases, there is less of an issue about whether or not the babies should be given every chance of surviving.

But when babies are considered "previable," Meadow says, "the most compassionate thing to do is to put it in the arms of its loving parents rather than on a ventilator and prolonging the dying."

Most very small newborns, depressed by the rigors of labor, are fairly inactive at delivery. They are scrawny, and their eyes are sometimes still fused shut. What usually happens, Meadow says, is that the child is stabilized initially as more information is gathered. "Many options are perfectly reasonable," he says. "I never really *know* what to do next."

Many of the smallest and sickest babies die very quickly—70 to 80 percent of them within the first 48 hours. "No matter what the doctors do, nature takes its course," says Dr. John Lantos, a pediatrician and ethicist at the U. of C.

For most of the remaining 20 percent, doctors and parents reach an agreement without much trouble, Lantos adds. And contrary to the impression left by the most controversial cases, like that of the Messengers, most decisions to stop treatment are initiated by doctors, not parents, he says.

When parents resist stopping treatment and ask for more time, a meeting is held with additional family members and hospital staff present. Usually there is a consensus after that second meeting. Sometimes a third meeting is needed, but eventually treatment is stopped, Lantos says.

"It's a well-kept secret among neonatologists that the problem is not with parents begging us to stop treatment," Lantos says. "In the majority of problem cases, it's the parents who are hoping for a miracle. They don't trust the doctors; they feel that stopping treatment is killing the baby."

Indeed, a Georgia doctor has been charged with murder after she allowed a 39-day-old newborn to die because the baby had lung disease and failing kidneys and, in the doctor's judgment, its vital signs had irreversibly dropped. But the parents said the doctor did not ask for their consent before stopping treatment.

In a minority of cases, the roles are reversed, with parents opting to stop intervention and doctors wanting to continue it, Lantos says. "The neonatologists get pressure from the parents not to be so aggressive, to think about quality of life, to give parents the right to participate."

In the thorniest situations, the choices—even when the ethics committee and a host of professionals are consulted—can be difficult, Meadow says.

In a recent case at the U. of C., a tiny preemie had had a very bad episode of bleeding in his brain, the most common neurological abnormality in premature infants and which often leads to significant degrees of mental retardation. Eight neonatologists who considered the case could not agree on whether or not to treat the baby. Meadow, the attending neonatologist, explained the options to the parents. They elected to discontinue treatment.

"I believe the families should be empowered to make these decisions," Meadow says. "Many times, I would recommend one thing, and the parent chooses another. If a parent asks what I think they should do, I try to give them the sense of how strongly *I* feel one way or another, but I also say that there are many options that are perfectly reasonable."

Dr. Elaine Farrell, a neonatologist at Evanston Hospital, says there is usually enough time after the expectant mother is admitted to the hospital to fully inform the family about the baby's chances, the risks and potential problems.

"If the families say, 'Don't do this, don't try to save them,' I support that," Farrell says. "I feel that, for them, it is the right decision. You have to have a committed family to take that on. If they think that the risks are too high, that they have three other kids at home, and if they consider the emotional, the physical and the financial costs, all of which are prohibitive, they could conclude that nature

3. AGING, HEALTH, AND HEALTH CARE ISSUES

did not intend this child to be there."

In that case, Farrell says, the nurses help to make the baby comfortable. They wrap the infant in a blanket, put a hat on it and let the parents hold it. "We encourage them to take pictures, and when the baby dies, we give them the baby's wrist band. These mementos are the only ones these parents are going to have. They grieve, they say goodbye. They made the best decision they could. They made a decision in the child's best interests."

The most difficult cases for her, Farrell says, are the full-term infants who suddenly run into problems, whether it's something congenital, an infection or some other unexpected trauma. "Our complacency and our sense of entitlement in the United States is that if you get to full term, you will go home with a [healthy] baby. That doesn't always happen. With preterm babies, everyone knows we're doing the best we can, pushing the frontiers forward, doing amazing things, and everybody knows what's on the line. With full-term babies, parents' expectations are so high, that when the baby is hurt, or dies, that's the hardest."

The parents' grief at their babies' deaths is so strong, Farrell says, it would be classified as "psychotic" if it were occurring in any other instance. "The bond is so deep, and the parents are so bound up in the sense of responsibility and expectations . . . they have bonded with the baby since the time they first thought they might be having a baby."

The first group of very tiny premature children born in the mid-1980s are just now reaching school age. And the news is not good.

A new study published Sept. 22 in the New England Journal of Medicine tracked 68 children born five to nine years ago weighing less than 1 pound 10 ounces. Researchers found that the children had a greatly increased chance of mental retardation, cerebral palsy and severe vision problems and were below average in cognitive ability, psychomotor skills and academic achievement.

About half of the children in the study had borderline-to-normal IQs or lower; many of the rest were having a lot of trouble in school trying to pay attention, follow directions and learn some subjects, especially mathematics. "These children are at serious disadvantage in every skill required for adequate performance in school," the study said.

"I talk about three categories of treatment," says Dr. Robert Nelson, associate professor of pediatrics and bioethics at the Medical College of Wisconsin in Milwaukee. "The first category is what's clearly beneficial to the child; the next is the clearly useless, where continuing it could be seen as medical abuse; and the third is a midrange category, where it's not clear if treatment is beneficial; and usually under those circumstances, people err on the side of continuing therapy until the uncertainty is resolved."

Dr. William Silverman, a retired California obstetrician who is credited with helping to create modern neonatal intensive care, calls the outcome of infants weighing below 750 grams (a little over a pound) a "no man's land, with no set guidelines," and where treatment is still largely experimental.

Silverman recalls realizing that "the timid notion of a 'natural limit of viability' vanished" in the mid-1950s, shortly after the Apgar score was introduced to systematically appraise the health of newborns. Requiring that Apgar scores be taken for every newborn eliminated the previous practice of allowing marginally viable infants to die quietly and labeling them stillborns.

"We don't know enough to guarantee results," Silverman says. "We don't know how much oxygen to give them, what kind of feeding; every aspect of their care is controversial. Every aspect convinces me that the parental view has to be taken into account. The professional view is so weak, so iffy, so difficult to defend."

Others disagree. Dr. Korones of Memphis says he is not an extremist; he regularly stops treatment for babies when it is clear that they are going to die soon anyway. But he admits that he is one of those physicians decried by parents because he does not believe that parents should have the ultimate decision over the baby's life. He insists on trying to save every infant with even the slightest chance of living, whatever the problems they may have to face later on. "Can I ask parents, 'Do you want this baby or not?' I can't do that," he says.

Korones, 70, says he frequently sees the severely disabled children he has saved and is aware of the pain they and their families are bearing. "I know I'm causing misery, but I don't know if I'm causing any more misery than they are," Korones says, referring to other physicians who may be more willing to give up on babies when the prognosis is poor and the parents want to let them die. "I don't know how many otherwise normal, gratifying human beings are being killed in that process. The basic difficulty is, my mistakes you see; the other guy's mistakes you bury."

The United States is considerably more aggressive in its efforts to save very small premature infants than most other countries, except perhaps Canada.

In Denmark, doctors generally would not try to save infants below 25 or 26 weeks gestation. In Australia, they would not treat infants smaller than 650 grams, or about 25 weeks. In Sweden, doctors withhold artificial ventilation from the most premature babies, those weighing less than 750 grams.

In Japan and other countries, doctors consider it cruel to let the parents make final decisions about the baby's fate.

Many states have laws that permit parents and other surrogates to act on behalf of their sick children. A 1991 Illinois law gives parents the right to make life-or-death decisions when children suffer from a terminal, incurable or irreversible condition. The law does not specifically address prematurity.

"Parents are supposed to obtain medical care for their children, to protect their children from harm, to relieve suffering, and failure to do so is neglect," says Dr. Norman Fost, professor of pediatrics and director of the program in medical ethics at the University of Wisconsin. "Our society is based on a pro-life assumption; we assume life is better than death, even handicapped life, but undertreatment has been replaced with overtreatment."

Neonatologists say that parents with more formal education choose to stop treatment more frequently than those with less education, probably because they are better informed about possible problems.

At Evanston Hospital, where 86 percent of those who come to its maternity ward are from the middle and upper-middle class, Farrell says parents have very high expectations for their children, and when the child's outcome is very risky, "they generally say we should stop treatment. But I have had families who look me right in the eye and say, 'We've been trying for 15 years [to have a child], and we can't stop now.' And then we go on. And usually those children turn out to be what we say they are.'"

Doctors say, however, that there are many good reasons not to allow parents to make unilateral decisions and instead lead doctors to consider themselves the representative of the child. Farrell recalls a father—a doctor but with no professional experience with newborns—who requested that attending doctors stop feeding and treating his daughter, born at 28 weeks. He was concerned about the effects of prematurity on the child's brain development. With doctors convinced that the child would thrive, the hospital went to court for authority to treat the child against the parents' wishes. And the girl did very well, Farrell says.

Between 15 and 20 percent of infants born between 24 and 26 weeks gestation are severely disabled.

14. Cruelest Choice

A 1992 Georgia State University study estimated the financial cost of prematurity at $5.6 billion a year nationally.

Farrell could not comment on Leon Skowronski's case, but she says Evanston Hospital physicians provide parents with abundant information about the possible handicaps children could suffer. "There are a lot of angry parents, what with all their sorrow and loss and pain and work, who don't remember what they were told," Farrell says. "They only know what they're dealing with now."

"These are agonizing decisions," says Dr. Arthur Kohrman, president of LaRabida Children's Hospital in Chicago. "These are issues about the nature of futility, the likelihood the child will live and the question of if they do survive, as what? You can't standardize this. We're all mucking around in the unknown together."

Ellen Baren Skowronski well remembers the feeling of euphoria she felt after giving birth to Leon. "I remembered thinking of all the things a son represents. I had all those feelings."

She has thought about it a lot. She has heard the well-meaning "but condescending" comment that "God does not give you more than you can handle." As she considers the hardships Leon and her family have already been through—the heart surgery, the seizures, the spasms, the frustrations—and her fears for a future of many more miseries and difficulties, Baren Skowronski says she believes her son was not intended to live.

"But he has helped me put my life into perspective," she says. "He renewed my faith in humanity." She's particularly touched by the committed professionals who have dedicated their lives to helping handicapped children such as Leon.

"If I had another child born prematurely, I wouldn't go to the hospital. There's no need. I would just stay at home. When Leon was born, I needed somebody to say there's a really good chance this baby will be severely disabled and that we'll have to take care of him for the rest of his life."

Even then her words belie the uncertainty that must always be there. "If I had known what was ahead, if they had told me, I probably would have decided to let him go."

Probably.

A New Look at Health Care Reform

COMPETITIVE MARKET FORCES

Delivered as the Keynote Address to the Annual Health Policy Conference of the Quincy Foundation for Medical Research, San Francisco, California, February 9, 1995

MURRAY WEIDENBAUM

MURRAY WEIDENBAUM, *Mallinckrodt Distinguished University Professor and Director of the Center for the Study of American Business, Washington University*

THE time is ripe for taking a new look at health-care reform. Conventional approaches have bogged down in the legislative process. A fresh start is necessary. Truly reforming the health-care delivery system of the United States requires developing a sensible and sensitive mechanism to balance the demand for health care with its supply. That is the only effective way of dealing simultaneously with the powerful demand for medical services, the limited resources available, and the pressures of rising costs and prices.

I put aside the question of lack of universal health-insurance coverage. My justification for doing so is that most public discussions equate lack of insurance with lack of medical care. That is erroneous. A large array of health-care providers do give medical services — at low or no cost — to those without insurance. To be sure, often the result is inefficient, such as the excessive use of emergency rooms. But, that is just a special case of a problem that I will be dealing with — people demanding expensive health care without paying the full cost.

One complication is curable. At present, employees — or employers acting in their behalf — cannot buy a modest health-care plan. State insurance commissions dictate the composition of these plans and they are very amenable to lobbying by special interests. In many states, the plans must include hair transplants, acupuncture, and other optional items. The purchaser of health insurance cannot buy a Ford. It must be a Lincoln — or nothing. As Voltaire said it, the best is the enemy of the good. Of course, this is not a federal case. Each state insurance commission should shift its focus from serving the special interests among health-care providers to meeting the needs of the patients.

The Two Basic Alternatives

Let us begin with the fundamentals. There is a spectrum of possible responses to the health-care dilemma, each with its own set of advantages and disadvantages.

At the free market end of the policy spectrum is an approach based on each family or unattached individual making their own choices on what type of medical outlays they will request — and pay for. This means a general elimination of third-party payments and a restoration of the traditional producer-consumer relationship which is found in most other product and service markets.

The primary reliance on third-party payments is a relatively recent phenomenon — which reminds us that the present pattern can be changed. Third-party payments have become important only in the last several decades. Back in 1960, people paid 49 percent of their health-care costs, while government agencies paid 24 percent and insurance companies paid 22 percent. A complete reversal has occurred in the intervening years. By 1993, people paid less than 18 percent of their medical costs. The lion's share was borne by government (44 percent) and insurance (34 percent). For hospital service, the patient now pays only 3 percent. For doctor bills, the average patient payment is 15 percent of the total.

The implication of the shift to "third-party" financing of health care cannot be overestimated. Important evidence comes from an experiment by the non-profit Rand Corporation. Thousands of families were given one of four health insurance plans. The difference between the plans was the co-payment rate, the portion of health expenses paid by the family. The co-payment rate varied from 0 to 95 percent. Under all the plans, if a family's out-of-pocket expenses reached $1,000, the insurance paid for all additional expenses.

The main finding was that the higher a family's co-payment rate, the less often members of that family went to a doctor and the less often they incurred medical expenses generally. In the words of my colleague David R. Henderson,

> "People do consume more health care when they are spending other people's money."

Rand found no substantial improvement in health outcomes for the higher spending by the families with low co-payment rates.

Relying on the marketplace is the self-policing way to control medical costs. When patients pay the bills directly, they become cost conscious — and so do the people and organizations serving them. The market approach differs fundamentally from the typical "third-party" payments so widely used in the United States. Under this latter method, patients usually do not know the prices and costs of their medical care before hand, if ever.

Third parties that pay the bills have effectively removed the patient from the traditional consumer role of watchdog.

15. Health Care Reform

Rarely are prices of physician and hospital services or goods such as prescription drugs advertised to consumers.

Of course, there always were exceptions to the operation of the free market in health care. Modern society has never been willing to accept the full consequences of allocating medical care solely on the desire and ability to pay. However, in this approach, market forces are supplemented, not supplanted. Poor people receive free or low-cost medical treatment, although sometimes of a lower quality than the rest of the society and usually at greater inconvenience.

Primary reliance on the market means that the price system rations the amount of health care produced and consumed. That amount is likely to be less than the results of current policy. A sensible step toward the free-market approach is to reduce the governmental subsidies which increase people's demand for the "best" health-care service. A good place to begin is to eliminate the tax advantage now given to health care over other consumer expenditures. Employer-financed health insurance should be included in taxable employee compensation along with direct payments of wages and salaries. Employer-financed insurance plans became popular during World War II as a loophole to get around wage controls. The special tax treatment is not justified by any canon of efficiency or fairness and should be eliminated.

Much of the formal effort to "economize" on health-care costs by departing from marketplace competition is illusory. A major example is the cost shifting under Medicaid and Medicare. That does little to reduce the nation's total medical outlays. That procedure forces other patients to pay for a portion of health care for the poor and the elderly.

To some significant extent, private health plans — goaded by employers who are unhappy at the steady stream of premium increases — can try to weed out high-cost providers, to limit the use of expensive specialists, to monitor closely the performance of health-care providers, and to emphasize preventative care. Such pressures can be reinforced by giving employees a similar stake in controlling health-insurance premiums.

At the other end of the policy spectrum is the notion that the society should finance whatever level of health care is required by each citizen. This general notion is embodied in the "single payer" plan, whereby government simply pays everyone's health bills. Practical problems abound. When health care becomes a free good, the individual response quickly becomes "Nothing's too good for me if I don't have to pay for it."

Because human wants are insatiable, the notion that each of us is entitled to all the medical care that we ask for exhausts the ability of even the most generous source of financing. Therefore, in practice, each single payer plan adopts some form of rationing. One of the most widely used means of limiting care is indirect. It is the bureaucratic technique of delay and inconvenience — forcing people to wait longer than they now do before they receive medical services, including having to go through a variety of reviews or "gatekeeper" approvals. It has been said of some high-risk surgical procedures under the Canadian system that the patient is more likely to die while waiting his or her turn than on the operating table.

Rationing by delay appeals to the bureaucratic instinct. It does not require making many difficult decisions. It is easy to administer. The queue even sounds fair: first come, first served. But, rationing by delay distributes the benefits of limited care arbitrarily.

A safety valve often accompanies the queue approach. It favors upper income individuals or at least people who value health care highly enough to pay for it. Wealthy Canadians, for example, come to the United States for serious surgery when they are not content with the quality or the time availability of the health care provided in Canada.

One of the claimed benefits of the single payer approach could be achieved without resorting to a massive expansion of the government's role. A standard medical card for each person with the vital personal and insurance information would avoid the repetitious collection of the same data by each health-care provider. The transcription errors which occur so frequently would be avoided, as well as the delays bedeviling patients and medical offices alike.

Surely, in this electronic age, the paperwork burden could be reduced substantially. Voluntary cooperation on the part of key private associations — the American Medical Association, the American Hospital Association, the American Pharmaceutical Association, etc. — should be able to accomplish this useful change.

Along these lines, the Quincy Foundation for Medical Research has proposed the establishment of a network of computer terminals located at care delivery sites. Each participant in the program would receive a code card containing his or her social security number and basic personal and medical data. We can endorse this proposal without embracing the notion of using the card to administer eligibility for a variety of governmentally imposed benefits.

All in all, it seems unlikely that public policy will adopt either of the two extremes. Yet, it is useful to view the various individual proposals in terms of whether they move the health-care system toward the governmental pole or toward the individual choice pole.

An Upbeat Outlook

It is pertinent to acknowledge a separate and noteworthy development. While the Congress and the Clinton Administration have been debating inconclusively how to provide and finance better health care, the institutions that actually provide medical care have been undergoing an unprecedented but voluntary restructuring. The health-care delivery system is being reformed. The marketplace is transforming itself and is delivering health care at reduced costs or at a slower rate of price increase.

The voluntary changes being made in health care are taking many forms. By the end of 1994, a majority of privately insured Americans were enrolled in managed-care plans that limit choice of doctors and treatments. In California, three-fourths of all privately insured patients are now in Health Maintenance Organizations. Three-fourths of all physicians had signed contracts, covering at least some of their patients, to reduce their fees and to accept oversight of their medical decisions. About nine out of every ten doctors who work in group practices have agreed to managed-care arrangements.

Large insurance companies are setting up "community care" networks. They are acquiring hospitals and clinics, so that they can offer a full spectrum of treatment for a fixed price. In suburban Atlanta, Aetna has opened six primary care centers. In the same area, another large insurance company, Cigna, has acquired medical practices and is recruiting doctors for its own clinics.

The Michigan health care network is a good example of the voluntary changes taking place. The network is vertically integrating the Henry Ford Health System, Mercy Health Services, and Michigan Blue Cross/Blue Shield. The net-

3. AGING, HEALTH, AND HEALTH CARE ISSUES

work of 13 hospitals offers health care to groups of 100 employees or more. It requires a fixed monthly payment for an individual or a family. New York Hospital has established a regional alliance with seven other non-profit hospitals, two nursing homes, and four walk-in clinics.

Three large hospital alliances, created in the last two years, now care for three-fourths of the hospital patients in the St. Louis area. Each alliance is actively buying up the practices of primary-care physicians.

In many communities, hospitals have been hiring or buying out the practices of primary care doctors — family practitioners, general internists, and pediatricians — to assure a stream of patient referrals and to increase their bargaining power with insurance companies. The South Carolina Medical Association has been developing an alternative approach. It is forming a statewide network of doctors to negotiate contracts with employers and take responsibility for controlling their health costs.

Health-care networks already dominate Southern California. Hospitals, physicians, and insurance companies all have established health-care networks. Mullikin Medical Enterprises, which is owned by 200 physicians in Southern California, is acquiring the practices of other medical groups around the state. Solo practitioners are becoming rare.

On a national scale, an unprecedented wave of mergers and acquisitions is occurring among major health-care providers. Columbia/HCA Healthcare, the country's largest for-profit hospital chain, has bought out Medical Care America, the largest chain of surgery centers. In contrast, Surgical Care Affiliates, which operates a chain of outpatient surgery centers, is luring patients away from hospitals. These centers provide a lower-cost setting for many of the less critical operations, such as removal of cataracts, tonsillectomies, and laparoscopic gallbladder removals.

The large pharmaceutical companies — squeezed by national policy and regional health-care providers — have been actively diversifying within the health-care sector. Merck acquired Medco, the managed-care drug distributor. SmithKline Beecham merged with Diversified Pharmaceutical, another managed-care drug marketer. Eli Lilly bought PCS Health Systems, the largest processor of payments for prescription drugs. Zeneca Group, a manufacturer of cancer drugs, acquired 50 percent of Salick Healthcare, an operator of cancer care centers. Thrifty Drug Stores bought the Payless drugstores of Kmart. Revco acquired Hook-Supe Rx and Rite Aid purchased Perry Drug Stores.

Meanwhile, many individually owned pharmacies are finding that they lack the resources to compete for managed-care business and are becoming members of chains, franchises, and other group efforts. In the future, perhaps insurance companies and hospitals will get together. Between them, they have the large organizational skills and recordkeeping that are necessary. Hospitals have the patients and insurance companies have the market — the willingness of employers to pay for the health care of the employees.

Ultimately, these conglomerates may include, in addition to insurance companies and hospitals, some of the following — outpatient clinics, doctors' offices, nursing homes, hospices, home health-care services, pharmacies, drug treatment centers, and medical equipment suppliers.

Conclusions and Recommendations

The operation of market forces often proceeds more rapidly and more effectively in responding to serious problems than do the more ponderous decisionmaking mechanisms of the public sector. Often the reduction of governmental impediments to competition represents the most efficient and least costly solution. Medical care is no exception to that basic proposition.

The most effective driving force to slow the rapid rise in health-care costs is now the business firms who find that this special expense reduces their competitiveness in an increasingly global marketplace. The pressure they exert on their health-insurance carriers, in turn, is transmitted to health-care providers. As we have seen, hospitals, physicians, and pharmaceutical firms have been engaged in an unprecedented effort to restructure, streamline, diversify, and otherwise reduce their costs — while they maintain or expand their share of a rapidly and radically changing marketplace for health care.

There is an important role for public policy in this important adjustment process, but it is not the role usually envisioned. To continue the movement to greater efficiency while meeting the needs of the patient, it is necessary to further reduce the impediments to the fuller operation of competitive market forces.

The most fundamental change needed is to reduce the dependence on third-party reimbursements. To the extent that patients view medical care as a "free" or low-cost good to them, the ability to contain costs will be greatly limited.

For the typical middle class patient/consumer, it makes no sense to go through an insurance/reimbursement system for routine office and out-patient hospital visits and procedures. What is required is to stop looking at health insurance as a benefit or, worse yet, as an entitlement. Rather, each of us must consider health insurance as a form of insurance protecting us from chance but potentially devastating circumstances. The implication of that seemingly simplistic change is profound.

Take automobile insurance as a basis for comparison. Each vehicle owner chooses a form of deductible. This means that many fender benders or paint scratches (the equivalent of the routine office visit) are not covered by insurance. There is no massive outcry that this approach is "unfair" to poor people. Motorists generally understand that a deductible is necessary to avoid swamping the insurance system with the paperwork that would push up premiums very sharply. As a result, of course, many paint scratches and dented fenders go unfixed — but that is considered choice of the owners who would rather spend their money on something else.

Indeed, one company in Virginia has moved in this direction. It treats its health program like true insurance, reimbursing for insurable events rather than for routine medical expenditures. The plan is structured so that employees are reimbursed for a small number of large claims rather than a large number of small claims. Savings from shifting away from traditional health-care coverage are shared equally between the employer and the employees.

Under the present array of public policies, primary reliance on third-party reimbursement strikes most taxpayers as highly desirable. Not many citizens are sophisticated enough to understand that such fringe benefits as employer-paid health insurance are a substitute for wages in the employee's compensation package. But even among the growing minority that comprehend the process, the status quo is still considered to be a good deal because wages and salaries are taxable income, while fringe benefits are not.

The answer is to make the entire compensation package taxable, including employer-paid health-insurance premiums. That will not eliminate the demand for such fringes, even among the most sophisticated, for a variety of reasons. Some of these are eminently sensible, such as the desire to obtain the economies of scale that result in lower group rates. A level playing field in the taxation of compensation would not constitute a panacea but it surely will help.

Increasing the knowledge available to consumers will enable them to make more informed choices. In the purchase of pharmaceutical products, government policy now restricts or prevents the patient from acquiring information concerning the prices charged by different providers for the same or similar products. Many states prohibit advertising the price of prescription drugs. Such restrictions make it difficult for consumers to shop for the best price. Every state which has enacted such anti-consumer legislation should promptly repeal it.

At the federal level, the Food and Drug Administration should reduce the barriers it has set up that inhibit advertising prescription medicines. Because consumers must obtain a prescription from a physician in order to acquire prescription drugs, there is little reason to fear deception in advertising in this market. Experience shows that direct advertising can reduce the prices that consumers pay. Such evidence was cited by the Supreme Court in the decision overturning state bans on advertising of eyeglasses.

The current FDA rules on advertising are needlessly bureaucratic. The agency should reconsider the requirement for the misnamed "brief summary" which must accompany any ad that both mentions a health condition and the name of a drug which can be used for the condition. The "brief summary" is actually a lengthy statement in small print listing side effects and contraindications associated with a prescription drug. Such information is essential for physicians, for whom the brief summaries were originally designed. But, for the average patient, the technical language is incomprehensible.

The FDA also discourages prescription drug ads from being shown on television, a major source of information for many consumers. The high cost of ads in the print media — resulting from the FDA requirements — also reduces their use. Like so much government regulation, the result is the opposite of what the FDA says it wants. Due to the restraint on advertising, consumers may not be aware that a treatment exists for a certain condition and so they will not consult a physician. In other circumstances, consumers may suffer some symptoms (e.g., thirst) without realizing that these are symptoms of a treatable disease (e.g., diabetes). Alternatively, a new remedy with reduced side effects may become available, but patients are not aware of it and do not visit their physicians to obtain a prescription.

If there is any single conclusion that emerges from this presentation, it is that no single solution — no silver bullet — is available to cure all the ailments besetting the American health-care system. What will help in a fundamental way is to acknowledge that difficult choices have to be made among imperfect alternatives. I trust that the package of alternatives I propose, based primarily on the free market, is less imperfect than the others.

GUNS, MONEY & MEDICINE

The proliferation of powerful new weapons has sent the cost of crime spiraling. Here's why you pay

One glance in the rearview mirror of his 1978 Cadillac Eldorado and 21-year-old Dewayne Bellamy knew that his evening was over. Approaching the car near a decaying corner of the nation's capital was the teenage son of a woman with whom Bellamy was having an affair. The boy had a gun. Before Bellamy could draw his own arsenal of semiautomatic weapons, he heard the familiar pop of a 9-millimeter pistol. There was no pain, no blood. Only after he awoke from a coma three days later did Bellamy receive two pieces of news. The first was that he had been shot 13 times. The second was that he would never walk again.

From the moment paramedics lifted him into the ambulance, Bellamy became the charge of the nation's taxpayers. And for the next eight months, the meter would never stop ticking. Covering everything from $3 scalpels to $2,283 CT scans, Bellamy's hospital bills would ultimately total $562,561. Doctor's fees would add tens of thousands more to the tab. For Bellamy, a one-time car thief who used to earn $5,000 a day selling crack cocaine, that's big money. But he doesn't worry about it. After all, he's not paying the bills.

In emergency rooms and rehabilitation centers across the country, Bellamy's is a depressingly familiar tale. By the year 2003, according to the federal Centers for Disease Control and Prevention, gunfire will have surpassed auto accidents as the leading cause of injury deaths in the United States. In seven states, it already has. But unlike victims of car crashes, who are almost always privately insured, 4 out of 5 gunshot victims are on public assistance or uninsured. That means taxpayers bear the brunt of medical costs that have spiked nearly ninefold in the past decade, to a stunning $4.5 billion a year.

Nationwide, the number of violent crimes has held steady for the past four years, yet gun sales continue to soar. While most gun owners buy their weapons legally, keeping them for self-protection and recreation, a flourishing illegal-drug trade has caused a dramatic rise in the number of powerful semiautomatic weapons used to commit crimes. The result is a flood of new gunshot victims to the nation's emergency rooms.

Multiple wounds. Although injuries from military-style assault weapons are rare, multiple wounds inflicted by semiautomatics such as 9-millimeter pistols are becoming so common as to make some trauma specialists practically nostalgic for the days of the cheap Saturday night special. "It seems like we never see just one shot anymore," says orthopedic surgeon Andrew Burgess of the University of Maryland's shock-trauma center in Baltimore. The increased firepower means doctors are saving fewer patients — and seeing greater damage to those who do survive.

Today's gunshot victims are a distinctive breed. Headlines highlight shootings of innocent bystanders, but the fact is that probably half of gun homicide victims — in some cities as many as 70 percent — are offenders themselves. They are due no less care, doctors say, but they confront modern medicine with an unsettling paradox: Physicians invest countless hours at huge expense to bind wounds and even heal their gunshot patients, only to return them to the streets, where many promptly resume a life of crime. "About 20 percent of our gunshot victims are what we call our 'frequent fliers,' " says Burgess. "It's not as if they leave here and find Jesus."

Criminals or bystanders, those shot by semiautomatic weapons can test the limits of even the best emergency care. Lamarr Wilson of Newark, N.J., was one such victim. Shot seven times with a semiautomatic, the 23-year-old was riddled with so many holes that doctors in the trauma unit of the University of Medicine and Dentistry of New Jersey

Gunshot violence in the United States costs $20 billion a year, a fifth of it in medical expenses. That's $200 per household.

DEWAYNE BELLAMY

The former crack dealer was shot Dec. 12, 1993, while sitting in his car watching television.

■ INJURIES: Hit 13 times, Bellamy had several surgeries to repair damage to his liver and lungs. Two years later, he was shot twice more. Some expenses:

■ SURGICAL ROOM:	$38,295
■ BLOOD:	$11,142
■ INTRAVENOUS DRUGS:	$19,052
■ SURGICAL SUPPLIES:	$72,593
■ NURSING CARE:	$46,878
■ X-RAYS:	$11,901
■ LAB TESTS:	$21,510
■ TIME IN HOSPITALS:	8 months
■ TOTAL ACUTE CARE:	$334,029
■ TOTAL COSTS:	over $500,000

16. Guns, Money & Medicine

The proportion of spinal cord patients injured by gunfire has more than doubled in the past 20 years, from 14% to 30%.

couldn't treat them fast enough. "We'd plug up one hole, only to find two more," says Tonni Glick, an emergency room nurse. The perforations caused the contents of Wilson's bowels to spill into his lacerated vital organs. Wilson's abdominal skin eroded so badly it had to be replaced with a sheet of plastic wrap. Altogether, he endured 14 different surgical procedures. "This one, we never thought he'd make it," says Glick. "But these young guys are tough. We saved his life." A Medicaid patient, Wilson spent 61 days in the hospital. The bottom line: $268,181.

In the seemingly endless debate over gun control, one fact is unassailable: Gunshot patients are far more expensive to take care of than are victims of other kinds of crime. A typical stab wound, for example, cost $6,446 to treat in 1992; the average gunshot case cost $14,541. Although gunshot wounds account for fewer than 1 percent of injuries in hospitals nationwide, they generate 9 percent of injury treatment costs. That's because more than half of all gunshot victims require expensive emergency surgery. Typical are laparotomies (average cost at one urban hospital: $41,000), thoracotomies (average cost: $26,000) and procedures on the neck and extremities. And that's often just the beginning: About a fifth of all gunshot victims require additional surgery later on.

"Disruption." One reason for the higher treatment costs is physics. A bullet causes trauma to human tissue by transmitting energy beyond the capacity of the tissue to absorb and dissipate it. That causes what doctors call "disruption." The extent of the damage depends on the size and speed of the bullet and the type of tissue affected. A bullet can stretch human tissue, creating an opening that in the most severe cases may expand to many times the size of the bullet. Whether the cavity is temporarily or permanently damaged depends on the body area affected. Elastic tissue like that of a bowel wall is more resistant to permanent damage; inelastic tissue like that of the liver and brain is less so. "If a rubber ball and a raw egg of equal weight are dropped on a cement floor from the same height, these two missiles of equal kinetic energy will sustain different degrees of damage," explains Dr. Jeremy Hollerman of the Hennepin County Medical Center in Minneapolis. "The rubber ball behaves like skeletal muscle or lung, the raw egg like the brain or liver."

At higher velocities, bullets pack more destructive force, causing more extensive damage to soft tissue. Bullets fired at high velocity also tend to create a kind of suctioning action when they strike human tissue, carrying external bacteria deep into internal wounds. (Contrary to popular belief, bullets are not sterilized in the heat of firing.) Slugs are often left in the body when their removal poses a greater danger to a victim, but they can cause lead poisoning and degenerative arthritis if lodged in a joint. Bullets fired at high velocity are also more likely to shatter when they strike bone or metal, producing multiple and even more destructive

---

EDDIE MATOS

A former crack dealer, Matos was shot once in the neck in September 1990. His assailant, Matos says, "got his."

■ INJURIES: Matos is paralyzed from the neck down and dependent on a ventilator. He is subject to lung and bladder infections.

■ TIME IN HOSPITALS: nearly 6 years

■ MEDICAL COSTS: over $1 million

---

projectiles. Says Dr. Kenneth Swan of the University of Medicine and Dentistry of New Jersey: "In the face, these secondary (bullets) often cause more damage to the brain and eyes than the primary bullet."

"T10 complete." When they survive, victims of multiple gunshots almost always go on to live more complicated—and more expensive—lives. Nestor Cantor, 22, of Brooklyn, N.Y., took seven shots in the small of his back from a 9-millimeter semiautomatic fired by a hit man in Richmond Hill, Queens. The bullets exploded, driving lead fragments deep into his spinal cord. Extensive operations repaired lacerations to his bladder and liver and drained fluid from his lungs. The doctors call Cantor a "T10 complete"—paralyzed from the waist down. Two weeks in the intensive care unit, 3½ months at Bellevue Hospital and 1½ years in a public rehabilitation facility have generated a Medicaid bill in excess of $300,000. "I never see what it costs," says Cantor. "I haven't paid anything out of my pocket."

At George Washington University Medical Center in Washington, D.C., former Medical Director Keith Ghezzi, an emergency room physician, totes up the financial toll of a weekend of violence in the nation's capital. A typical gunshot patient spent 16 days in the intensive care unit at $1,487 per day. The patient required drugs costing $13,580, X-rays at $2,738, and bandages, tubes and miscellaneous supplies totaling $16,280. Nursing care, physical therapy and other services added thousands more to the bill. By the time the man was discharged from the hospital, he had racked up a bill of $100,838, not including doctor's fees. Medicaid will pay about 70 percent of the bill; the patient will pay nothing.

The story is repeated every few days. Last year, a homeless man who had served time for armed robbery and assault was taken to George Washington after he was shot while wielding a knife outside the White House. In just two days, the man received more than $70,000 in medical care. He died. The hospital ate the cost of his treatment.

Cost shifting. Such cases show how handgun violence affects Americans who have never even seen a gun or heard one fired in anger. Like most institutions, George Washington covers the costs of treating uninsured and underinsured patients by increasing the bills of those who do pay. Such cost shifting, a recent report to Congress estimated, forced private patients to pay an average of 29 percent above the actual costs of their care in 1993. According to one study, the University of California-Davis Medical Center, despite incurring three-year losses of nearly $2.2 million on gunshot victims, actually made a profit on its trauma center, so heavily did it shift the burden to patients who could pay.

As health maintenance organizations demand more and more savings, however, hospitals are finding it more difficult to pawn off on anyone the costs of the uninsured. The consequences for trauma units are dire. Once sure-fire moneymakers, more than 60 urban trauma centers have closed in the past 10 years, leaving less than one quarter of the nation's population residing anywhere near topflight trauma care. In a study by the General Accounting Office for members of

3. AGING, HEALTH, AND HEALTH CARE ISSUES

The share of gunshot victims without private health insurance: 80% The share of the general public without health insurance: 17%

Congress, all the shuttered trauma centers blamed their troubles on the growing burden of uncompensated services—millions of dollars of which resulted from treating indigent victims of handgun violence.

For every patient who dies from a gunshot wound—and there were 39,720 in 1994—three others are injured seriously enough to be hospitalized. Of those, one on average suffers from a disabling, lifelong injury. The worst injuries are to the spinal cord, and the higher on the cord the blow, the greater the area paralyzed. If a patient is injured anywhere between the first and third cervical vertebrae, for instance, he may lose all feeling from the neck down. Most spinal-cord-injured gunshot victims are paraplegics, paralyzed only from the waist down.

Eddie Matos was unluckier than most. In the past six months, the 21-year-old former drug dealer has not moved from his room at New York's Goldwater Memorial Hospital, where he keeps the shades pulled tight and watches soap operas and videos all day. He could motor around the grounds in the $5,000 electric wheelchair he operates by puffing on a straw. But why bother? he says. He sees the same old patients, and they all look like him. Before his accident, Matos was a prospering businessman. He had four "spots": three for crack, one for cocaine. One spot could make $11,000 on a weekend; Matos kept $2,000. The money bought cars—a Cadillac, a Pathfinder, a Mustang and a Volvo. It bought jewelry and his own apartment. It also paid for a 9-millimeter semiautomatic pistol. "My favorite," Matos says. "It does damage."

He should know. One night in September 1990, another man with a 9 millimeter jumped Matos outside a grocery store and shot him once in the neck. The gunman has since "gotten his," Matos says. But his own life is shattered. Lying in the quadraplegic ward of the aging city-run hospital, his only movements are the painful spasms that convulse his muscles every so often. He cannot feed himself or breathe without a ventilator. He must clench a wand in his teeth to turn the pages of a book. Matos has stayed at Goldwater longer than any other gunshot victim. His treatment has cost the public well over $1 million.

Aiming to maim? For patients paralyzed by gunfire, bills like Matos's are not uncommon. Quadraplegics, paralyzed from the neck down, require round-the-clock care. They need aides to change catheters, tracheotomy tubes and bladder bags; to feed, bathe and clothe them; to help wean them, if possible, from their ventilators. Unable to cough, their lungs must be suctioned several times a day to prevent pneumonia, which threatens lives already shortened by ventilator dependency. Bladder infections, which strike with troubling frequency, must be attacked aggressively or they will spread. Beyond medical care, there is arduous physical therapy

---

TALMADGE CONOVER

The former cocaine dealer was shot while in a car in Newark, N.J., at 2 a.m. on March 2, 1994.

■ INJURIES: A paraplegic, Conover suffered from internal bleeding and a collapsed lung. Infected bedsores required surgery.

■ TIME IN HOSPITALS: 6 months

■ MEDICAL COSTS: over $134,000

---

to prevent muscle atrophy and occupational therapy to help patients function in a nonhandicapped world.

All in all, a bullet in the spinal cord is an expensive proposition. In 1992 dollars (the most recent figures available), the National Spinal Cord Injury Statistical Center estimated first-year medical costs for a high quadraplegic (injured in the uppermost cervical vertebra) at $417,067, plus $74,707 for each year thereafter. The first-year costs for a paraplegic were $152,396, plus $15,507 for each year thereafter. For a 25-year-old quadraplegic, that would amount to lifetime medical costs of $1.3 million; for a paraplegic, $427,700.

So common are spinal cord injuries among gunshot victims today that some health care providers suspect gunmen are deliberately aiming for the neck. "It's as if the gunmen are saying, 'We don't want to kill you; we just want to paralyze you,'" says Glick of the University of Medicine and Dentistry of New Jersey. "We want to keep you alive so you will always remember what happened to you." In Los Angeles, at least half of all spinal cord injuries are caused by gunshots. Since most insurance plans have lifetime benefit caps, even those patients with private health insurance eventually end up on Medicaid. Roughly 75 percent of all gunshot victims are under 30, as are half of all spinal cord victims. That means better survival rates, of course—and many costly years ahead.

At the Kessler Institute for Rehabilitation in West Orange, N.J., whose stellar reputation for treating head- and spinal-cord-injured victims has attracted celebrities like dancer Ben Vereen and actor Christopher Reeve, gunshot survivor Talmadge Conover improved steadily under a rehabilitation program that costs $1,000 a day. But once the 18-year-old paraplegic returned to his drab third-floor apartment in a fading section of Newark, N.J., with three bullets still in his abdomen, he found it harder to keep doing the pull-ups that flipped his skinny body from side to side. The result: bedsores so infected they started eating away at his bone. Now, Conover is recovering from a successful skin-graft operation, studying for a high school equivalency degree and working the phones from a $30,000 Clinitron bed, a sort of heated hammock of delicate silicone balls. He says he has stopped dealing cocaine. Estimated cost of his treatment: more than $134,000.

Carrying a nine. That Conover was shot with a 9-millimeter semiautomatic weapon would come as no surprise to anyone who has spent time in an urban trauma center. Introduced in the early 1980s to revive a sagging gun industry, "nines" are now the weapon of choice on city streets. They are cheap and concealable, and, with extended magazines, they allow the shooter to fire up to 36 rounds without reloading. "You

By the year 2003, gunshots will have surpassed car accidents as the leading cause of injury death in the United States.

carry [a nine] to get a rep," explains Matos, "to get respect."

The Treasury Department's Bureau of Alcohol, Tobacco and Firearms lists two brands of 9 millimeters—the $410 Ruger P89 and the $609 Glock 17—among the top 10 guns found at crime scenes. There are now more than 3 million 9 millimeters on America's streets, and while many of those are arming law enforcement officers, the number of 9 millimeters used by criminals has nearly doubled since 1987. In Philadelphia in 1987, 9 millimeters sent 57 victims to local trauma hospitals; by 1993, the number of victims hospitalized by 9 millimeters had soared to 351.

Vernon Parker, a 31-year-old Brooklyn man, still carries nine bullets in his right thigh from the 17 rounds of an Intratec TEC-9 semiautomatic fired into him outside a housing project in the Bedford-Stuyvesant section on Oct. 19, 1993. (The manufacture of TEC-9s, along with certain magazines, was banned under the 1994 assault weapons law, but thousands made before the ban remain in circulation.) Slugs from the TEC-9 struck Parker's groin, buttocks and shoulder, necessitating three operations and two years in the hospital. The cost: well over $500,000. Today, there is little hope that Parker will walk again. "It used to be that just flashing a gun was enough," says Parker, a convicted drug dealer who speaks from experience. "But these young guys today, they'll shoot a whole crowd in broad daylight just to get one dead."

To doctors after a while, the entries on emergency-room-admissions forms start to look the same: *GWS, BL, M, 1976, MA*—gunshot wound, black, male, 20 years old, medical assistance. Only the faces change. "there is a lot of frustration and angst about these injuries," says Stephen Hargarten, an emergency room physician at the Medical College of Wisconsin in Milwaukee. It is no longer enough, he says, for emergency room doctors to simply treat gunshot victims and release them. "Doctors must leave the bedside," he says, "and go to the legislatures."

Solutions? And so they are. Physicians are lobbying for restrictions on U.S. handguns as strict as those for imports. They want childproof guns, a heavier tax on ammunition and other reforms.

In their more discouraging moments, however, doctors admit the prognosis is poor. Nestor Cantor, after all, says he knows seven people who have been shot, six of them killed. Eddie Matos counts at least five. Talmadge Conover says he knows more than a dozen victims of handgun violence, three of them dead. He has had days when he wanted to join them. But in a country where there is one handgun for every other household, even those relegated to wheelchairs show no inclination to disarm. The phenomenon, says Cantor, "is just too big. It's out of control."

BY SUSAN HEADDEN

NESTOR CANTOR

Shot seven times on July 25, 1994, Cantor says he was mistaken for a crack dealer.

■ INJURIES: His spinal cord severed, Cantor is paralyzed from the waist down. He had several operations on his bladder and liver.

■ TIME IN HOSPITALS: 2 years

■ MEDICAL COSTS: over $300,000

THE LINE ON SEMIAUTOMATICS

■ SPEED AND POWER: Most gunshot injuries are caused by small- and medium-caliber revolvers, but emergency room specialists point to an alarming increase in multiple wounds caused by high-powered semiautomatic pistols.

■ POPULARITY: Semiautomatics are popular with the young. In one survey of inmates in four states, 55 percent of juveniles admitted to carrying a semiautomatic pistol.

■ FIREARMS IN CIRCULATION: 216 million

■ HANDGUNS IN CIRCULATION: 72 million

■ 9-MM SEMIAUTOMATICS IN CIRCULATION: 3 million

■ PREVENTION: Emergency room doctors are urging policy makers to focus on gun design. There can be safer guns, they argue, just as there are safer cars.

Mental Illness Is Still a Myth

Thomas Szasz

Thomas Szasz is professor emeritus of psychiatry at the State University of New York in Syracuse, New York. He is author of The Myth of Mental Illness; Our Right to Drugs: The Case for a Free Market; *and* A Lexicon of Lunacy: Metaphoric Malady, Moral Responsibility, and Psychiatry (*the latter published by Transaction Publishers*). *His most recent book is* Cruel Compassion: Psychiatric Control of Society's Unwanted.

In a memorable statement C. S. Lewis once remarked, "Of all the tyrannies a tyranny sincerely exercised for the good of its victims may be the most oppressive.... To be 'cured' against one's will and cured of states which we may not regard as disease is to be put on a level with those who have not yet reached the age of reason or those who never will; to be classed with infants, imbeciles, and domestic animals." These words still apply to psychiatry today.

Anyone with an ear for language will recognize that the boundary that separates the serious vocabulary of psychiatry from the ludicrous lexicon of psychobabble, and both from playful slang, is thin and permeable to fashion. This is precisely wherein lies the richness and power of language that is inexorably metaphoric. Should a person want to say something sensitive tactfully, he can, as the adage suggests, say it in jest, but mean it in earnest. Bureaucrats, lawyers, politicians, quacks, and the assorted mountebanks of the "hindering professions" are in the habit of saying everything in earnest. If we want to protect ourselves from them, we had better hear what they tell us in jest, lest the joke be on us.

As far back as I can remember thinking about such things, I have been struck by the analogic-metaphoric character of the vocabulary of psychiatry, which is nevertheless accepted as a legitimate medical idiom. When I decided to discontinue my residency training in internal medicine and switch to psychiatry, I did so with the aim of exploring the nature and function of psychiatry's metaphors and to expose them to public scrutiny as figures of speech.

During the 1950s, I published a score of articles in professional journals, challenging the epistemological foundations of the concept of the mental illness and the moral basis of involuntary mental hospitalization. In 1958, as my book *The Myth of Mental Illness* was nearing completion, I wrote a short paper of the same title and submitted it to every major American psychiatric journal. Not one of them would accept it for publication. As fate would have it—and because the competition between psychologists and psychiatrists for a slice of the mental health pie was then even more intense than now—*The American Psychologist* published the essay in 1960. The following year, the book appeared. I think it is fair to say that psychiatry has not been the same since.

Responses to my work have varied from lavish praise to bitter denunciation. American psychiatrists quickly closed ranks against me. Official psychiatry simply dismissed my contention that (mis)behaviors are not diseases and asserted that I "deny the reality that mental diseases are like other diseases," and distorted my critique of psychiatric slavery as my "denying life-saving treatment to mental patients." Actually, I have sought to deprive psychiatrists of their power to involuntarily hospitalize or treat competent adults called "mental patients." My critics have chosen to interpret this proposal as my trying to deprive competent adults of their right or opportunity to seek or receive psychiatric help. By 1970, I had become a non-person in American psychiatry. The pages of American psychiatric journals were shut to my work. Soon, the very mention of my name became taboo and was omitted from new editions of texts that had previously featured my views. In short, I became the object of that most effective of all criticisms, the silent treatment—or, as the Germans so aptly call it, *Totschweigetaktik.*

In Great Britain, my views elicited a more favorable reception. Some English psychiatrists conceded that not all psychiatric diagnoses designate *bona fide* diseases. Others were sympathetic to the plight of persons in psychiatric custody. Regrettably, that posture rested heavily on the misguided patriotic belief that the practice of psychiatric slavery was less common in England than in the United States.

Not surprisingly, my work was received more favorably by philosophers, psychologists, sociologists, and civil libertarians, who recognized the merit of my cognitive challenge to the concept of mental illness, and the legitimacy of my questioning the morality of involuntary psychiatric interventions. I thus managed to set in motion a controversy about mental illness that is still raging.

When people now hear the term "mental illness," virtually everyone acts as if he were unaware of the distinction between literal and metaphoric uses of the word "illness." That is why people believe that finding brain lesions in some mental patients (for example, schizophrenics) would prove, or has already proven, that mental illnesses exist and are "like other illnesses." This is an error. If mental illnesses are diseases of the central nervous system (for example, paresis), then they are diseases of the brain, not the mind; and if they are the names of (mis)behaviors (for example, using illegal drugs), then they are not diseases. A screwdriver may be a drink or an implement. No amount of research on orange juice and vodka can establish that it is a hitherto unrecognized form of a carpenter's tool.

Such linguistic clarification is useful for persons who want to think clearly, regardless of consequences. However, it is not useful for persons who want to respect social institutions that rest on the literal uses of a master metaphor. In short, psychiatric metaphors play the same role in therapeutic societies as religious metaphors play in theological societies. Consider the similarities. Mohammedans believe that God wants them to worship on Friday, Jews that He wants them to worship on Saturday, and Christians that He wants them to worship on Sunday. The various versions of the American Psychiatric Association's (APA) *Diagnostic and Statistical Manual* rest on the same sort of consensus. How does behavior become illness? By the membership of the American Psychiatric Association reaching a consensus that, say, gambling is an illness and then issuing a declaration to that effect. Thereafter "pathological gambling" is a disease.

Obviously, belief in the reality of a psychiatric fiction, such as mental illness, cannot be dispelled by logical argument any more than belief in the reality of a religious fiction, such as life after death, can be. That is because, *inter alia*, religion is the denial of the human foundations of meaning and of the finitude of life; this authenticated denial lets those who yearn for a theo-mythological foundation of meaning and who reject the reality of death to theologize life and entrust its management to clerical professionals. Similarly, psychiatry is the denial of the reality of free will and of the tragic nature of life; this authenticated denial lets those who seek a neuro-mythological explanation of human wickedness and who reject the inevitability of personal responsibility to medicalize life and entrust its management to health professionals. Marx was close to the mark when he asserted that religion was "the opiate of the people." But religion is not the opiate of the people. The human mind is. For both religion and psychiatry are the products of our own minds. Hence, the mind is its own opiate; and its ultimate drug is the word.

Freud himself flirted with such a formulation. But he shied away from its implications, choosing instead to believe that "neuroses" are literal diseases, and that "psychoanalysis" is a literal treatment. As he wrote in his essay "Psychical (or Mental) Treatment":

> Foremost among such measures [which operate upon the human mind] is the use of words; and words are the essential tool of mental treatment. A layman will no doubt find it hard to understand how pathological disorders of the body and mind can be eliminated by 'mere' words. He will feel that he is being asked to believe in magic. And he will not be so very wrong. . . . But we shall have to follow a roundabout path in order to explain how science sets about restoring to words a part at least of their former magical power.

I took up the profession of psychiatry in part to combat the contention that abnormal behaviors are the products of abnormal brains. Ironically, it was easier to do this fifty years ago than today. In the 1940s, the idea that every phenomenon named a "mental illness" will prove to be a bona fide brain disease was considered to be only a hypothesis, the validity of which one could doubt and still be regarded as reasonable. Since the 1960s, however, the view that mental diseases are diseases of the brain has become scientific fact. This contention is the bedrock claim of the National Alliance for the Mentally Ill (NAMI), an organization of and for the relatives of mental patients, with a membership in excess of one hundred thousand. Its "public service" slogan, intoned like a mantra, is: "Learn to recognize the symptoms of Mental Illness. Schizophrenia, Manic Depression, and Severe Depression are Brain Diseases."

Diagnoses are social constructs which vary from time to time and from culture to culture.

Psychiatrists and their powerful allies have thus succeeded in persuading the scientific community, the courts, the media, and the general public that the conditions they call "mental disorders" are diseases—that is, phenomena independent of human motivation or will. This development is at once curious and sinister. Until recently, only psychiatrists—who know little about medicine and less about science—embraced such blind physical reductionism.

Most scientists knew better. For example, Michael Polanyi, who made important contributions to both physical chemistry and social philosophy, observed: "The recognition of certain basic impossibilities has laid the foundations of some major principles of physics and chemistry; similarly, recognition of the impossibility of understanding living things in terms of physics and chemistry, far from setting limits to our understanding of life, will guide it in the right direction." It is no accident that the more firmly psychiatrically inspired ideas take hold of the collective American mind, the more foolishness and injustice they generate. The specifications of the Americans With Disabilities Act (AWDA), a federal law enacted in 1990, is a case in point.

Long ago, American lawmakers allowed psychiatrists to literalize the metaphor of mental illness. Having accepted fictitious mental diseases as facts, politicians could not avoid specifying which of these manufactured maladies were covered, and which were not covered, under the AWDA. They had no trouble doing so, creating a veritable "DSM-Congress," that is, a list of mental diseases accredited by a congressional, rather than a psychiatric, consensus group. Thus, the AWDA covers "claustrophobia, personality problems, and mental retardation, [but does not cover] kleptomania, pyromania, compulsive gambling, and...transvestism." It is reassuring to know that the Congress of the United States agrees with me that stealing, setting fires, gambling, and cross-dressing are not diseases.

Thus, the various versions of the APA's *Diagnostic and Statistical Manual of Mental Disorders* are not classifications of mental disorders that "patients have," but are rosters of officially accredited psychiatric diagnoses. This is why in psychiatry, unlike in the rest of medicine, members of "consensus groups" and "task forces," appointed by officers of the APA, make and unmake diagnoses, the membership sometimes voting on whether a controversial diagnosis is or is not a disease. For more than a century, psychiatrists constructed diagnoses, pretended that they are diseases, and no one in authority challenged their deceptions. The result is that few people now realize that diagnoses are not diseases.

Diseases are demonstrable anatomical or physiological lesions, that may occur naturally or be caused by human agents. Although diseases may not be recognized or understood, they "exist." People have hypertension and malaria, regardless of whether or not they know it or physicians diagnose it.

Diagnoses are disease names. Because diagnoses are social constructs, they vary from time to time, and from culture to culture. Focal infections, masturbatory insanity, and homosexuality were diagnoses in the past; now they are considered to be diagnostic errors or normal behaviors. In France, physicians diagnose "liver crises"; in Germany, "low blood pressure"; in the United States, "nicotine dependence."

These considerations raise the question: Why do we make diagnoses? There are several reasons: 1) Scientific—to identify the organs or tissues affected and perhaps the cause of the illness; 2) Professional—to enlarge the scope, and thus the power and prestige, of a state-protected medical monopoly and the income of its practitioners; 3) Legal—to justify state-sanctioned coercive interventions outside of the criminal justice system; 4) Political-economic—to justify enacting and enforcing measures aimed at promoting public health and providing funds for research and treatment on projects classified as medical; 5) Personal—to enlist the support of public opinion, the media, and the legal system for bestowing special privileges (and impose special hardships) on persons diagnosed as (mentally) ill.

It is no coincidence that most psychiatric diagnoses are twentieth-century inventions. The aim of the classic, nineteenth-century model of diagnosis was to identify bodily lesions (diseases) and their material causes (etiology). The term "pneumococcal pneumonia," for example, identifies the organ affected, the lungs, and the cause of the illness, infection with the pneumococcus. Pneumococcal pneumonia is an example of a pathology-driven diagnosis.

Diagnoses driven by other motives—such as the desire to coerce the patient or to secure government funding for the treatment of the illness—generate different diagnostic constructions and lead to different conceptions of disease. Today, even diagnoses of (what used to be) strictly medical diseases are no longer principally pathology-driven. Because of third-party funding of hospital costs and physicians' fees, even the diagnoses of persons suffering from *bona fide* illnesses—for example, asthma or arthritis—are distorted by economic considerations. Final diagnoses on the discharge summaries of hospitalized patients are often no longer made by physicians, but by bureaucrats skilled in the ways of Medicare, Medicaid, and private health insurance reimbursement—based partly on what ails the patient, and partly on which medical terms for his ailment and treatment ensure the most generous reimbursement for the services rendered.

As for psychiatry, it ought to be clear that, except for the diagnoses of neurological diseases (treated by neurologists), no psychiatric diagnosis is, or can be, pathology-driven. Instead, all such diagnoses are driven by non-medical, that is, economic, personal, legal, political, or social considerations and incentives. Hence, psychiatric diagnoses point neither to anatomical or physiological lesions, nor to disease-causative agents, but allude to human behaviors and human problems. These problems include not only the plight of the denominated patient, but also the dilemmas with which the patient, relatives, and the psychiatrist must cope and which each tries to exploit.

My critique of psychiatry is two-pronged, partly conceptual, partly moral and political. At the core of my conceptual critique lies the distinction between the literal and metaphorical use of language—with mental illness as a metaphor. At the core of my moral-political critique lies the distinction between relating to grown persons as responsible adults and as irresponsible insane persons (quasi-infants or idiots)—the former possessing free will, the latter lacking this moral attribute because of being "possessed" by mental illness. Instead of addressing these issues, my critics have concentrated on analyzing my motives and defending psychiatric slavery as benefiting the "slaves" and society alike. The reason for this impasse is that psychiatrists regard their own claims as the truths of medical science, and the claims of mental patients as the manifestations of mental diseases; whereas I regard both sets of claims as unwarranted justifications for imposing the claimants' beliefs and behavior on others. Because the secret to unraveling many of the mysteries of psychiatry lies in distinguishing claims from assertions, descriptions, suggestions, or hypotheses, let us briefly examine this concept.

Psychiatrists have the power to accredit their claims as scientific facts and rational treatments.

Advancing a claim means seeking, by virtue of authority or right, the recognition of a demand—say, the validity of an assertion (in religion), or entitlement to money damages (in tort litigation). To use my previous example, Muslims, Jews, and Christians all claim that God created the world in six days and on the seventh He rested. However, each faith names a different day of the week as the day of rest. Similarly, (some so-called) psychotics assert that they hear voices that command them to kill their wives or children; psychiatrists assert that such persons suffer from a brain disease called "schizophrenia," which can be effectively treated with certain chemicals; and I claim that the assertions of psychotics and psychiatrists alike are claims unsubstantiated by evidence. The point, however, is that psychiatrists have the power to accredit their own claims as scientific facts and rational treatments, discredit the claims of mental patients and psychiatric critics as delusions and denials, and enlist the coercive power of the state to impose their views on involuntary "patients."

The difference between a description and a claim is sometimes a matter of context rather than vocabulary. For example, the adjective "schizophrenic" may describe a man who asserts that his wife is trying to poison him (assuming that she is not); but it functions as a claim when, after shooting his wife, the killer's court-appointed lawyer, desperate to "defend" him (perhaps against his nominal client's wishes), claims that the illegal act was caused by schizophrenia and that the killer should therefore be "acquitted" and treated in a mental hospital, rather than punished by imprisonment. Because psychiatrists view mental diseases and their treatments as facts rather than as claims, they reject the possibility that the words "illness" and "treatment" may, as all words, have a literal or metaphorical usage. Although some psychiatrists now concede that hysteria is not a genuine disease, they are loath to acknowledge that it is a metaphorical disease, that is, not a disease at all. Similarly, many psychiatrists acknowledge that psychotherapy—that is, two or more persons listening and talking to one another—is radically unlike surgical and medical treatment. But, again, they do not acknowledge that it is a metaphorical treatment—that is, not a treatment at all.

Psychiatry is a branch of the law and a secular religion rather than a science or a therapy.

Finally, psychiatrists, who potentially always deal with involuntary patients, delight in the doubly self-serving claim that their patients suffer from brain diseases and that these (psychiatric) brain diseases (unlike others, such as Parkinsonism) render their sufferers incompetent. This claim lets psychiatrists pretend that coercion is a necessary, yet insignificant, element in contemporary psychiatric practice, a claim daily contradicted by reports in the newspapers. Understandably, psychiatrists prefer to occupy themselves with the putative brain diseases of persons called "mental patients" than with the proven social functions of psychiatric diagnoses, hospitals, and treatments.

Lawmakers do not discover prohibited rules of conduct, called crimes, they create them. Killing is not a crime; only unlawful killing is—for example, murder. Similarly, psychiatrists do not discover (mis)behaviors, called mental diseases, they create them. Killing is not a mental disease; only killing defined as due to mental illness is; schizophrenia thus "causes" heterohomicide (not called "murder") and bipolar illness "causes" auto-homicide (called "suicide").

My point is that psychiatrists, who create diagnoses of mental diseases by giving disease names to personal (mis)conduct, function as legislators, not as scientists. It was this sort of diagnosis making alienists engaged in when they created masturbatory insanity; that Paul Eugen Bleuler engaged in when he created schizophrenia; and that the task force committees of the American Psychiatric Association engage in when they construct new psychiatric diagnoses, such as body dysmorphic disorder, and deconstruct old ones, such as homosexuality.

I am not arguing that rule making, such as politicians engage in, is not important. I am merely insisting on the differences between phenomena and rules, science and law, cure and control. Treating the sick and punishing criminals are both necessary for maintaining the social order. Indeed, breakdown in the just enforcement of just laws is far more destructive to the social order than the absence of equitable access to effective medical treatment. The medical profession's traditional social mandate is healing the sick; the criminal justice system's, punishing the lawbreaker; and the psychiatric profession's, confining and controlling the "deviant" (ostensibly as diseased, supposedly for the purpose of treatment). This is why I regard psychiatry as a branch of the law and a secular religion, rather than a science or therapy.

I want to add a brief remark here on the so-called anti-psychiatry movement with which my name is often associated. As detailed elsewhere, I consider the term anti-psychiatry imprudent and the movement it names irresponsible. As a classical liberal, I support the rights of physicians to engage in mutually consenting psychiatric acts with other adults. By the same token, I object to involuntary psychiatric interventions, regardless of how they are justified. Psychiatrists *qua* physicians should never deprive individuals of their lives, liberties, and properties, even if the security of society requires that they engage in such acts. In adopting this view, I follow the example of the great Hungarian physician Ignaz Semmelweis who believed that obstetricians, *qua* physicians, should never infect their patients, even if the advancement of medical education requires that they do so.

I do not deny that involuntary psychiatric interventions might be justified vis-à-vis individuals declared to be legally incompetent, just as involuntary financial or medical interventions are justified under such circumstances. Individuals who are disabled by a stroke or are in a coma cannot discharge their duties or represent their desires. Accordingly, there are procedures for relieving them—with due process of law—of their rights and responsibilities as full-fledged adults. Although the persons entrusted with the task of reclassifying citizens from moral agents to wards of the state might make use of medical information, they should be lay persons (jurors) and judges, not physicians or mental health specialists. Their determination should be viewed as a legal and political procedure, not as a medical or therapeutic intervention.

I have sought to alert the professions as well as the public to the tendency in modern societies—whether capitalist or communist, democratic, or totalitarian—to reclassify deviant conduct as (mental) disease, deviant actor as (psychiatric) patient, and activities aimed at controlling deviants as therapeutic interventions. And I have warned against the dangers of the destruction of self-discipline and criminal sanctions which these practices create—specifically the replacing of penal sanctions with psychiatric coercions rationalized as "hospitalization" and "treatment." To describe the confusion arising from the use of the metaphorical term "mental disease," I have suggested the phrase "the myth of mental illness." For a political order that uses physicians and hospitals in place of policemen and prisons to coerce and confine miscreants and which justifies constraint and compulsion as therapy rather than punishment, I have proposed the name "therapeutic state."

The personal freedom of which the English and American people are justly proud rests on the assumption of a fundamental right to life, liberty, and property. This is why deprivations of life, liberty, and property have traditionally been regarded as punishments (execution, imprisonment, and the imposition of a fine), that is, legal and political acts whose lawful performance is delegated to specific agents of the state and is regulated by due process of law. No physician *qua* medical healer has the right to deprive another of life, liberty, or property. Formerly, when the clergy was allied with the state, a priest had the right to deprive a person of life and liberty. In the seventeenth century, the state began to transfer this role to psychiatrists (alienists or mad-doctors), who eagerly accepted the assignment and have served as state agents authorized to deprive persons of liberty under medical auspices. Now, we are witnessing a clamor for granting physicians the right to kill persons—an ostensibly medical intervention euphemized as "physician-assisted suicide."

It is a truism that the interests of the individual, the family, and the state often conflict. Medicalizing interpersonal conflicts, that is, disagreements among family members, the members of society, and between citizens and the state, threatens to destroy not only respect for persons as responsible moral agents, but also for the state as an arbiter and dispenser of justice. Let us never forget that the state is an organ of coercion with a monopoly on force—for good or ill. The more the state empowers doctors, the more physicians will strengthen the state (by authenticating political preferences as health values), and the more the resulting union of medicine and the state will enfeeble the individual (by depriving him of the right to reject interventions classified as therapeutic). If that is the kind of society we want, that is the kind we shall get—and deserve.

Poverty and Inequality

It is not clear whether poverty is the result or the cause of inequality, but where one is found, so is the other. Most individuals, regardless of how little or how much they have, agree that poverty is bad, but they do not necessarily agree that inequality is bad. To those raised in capitalistic societies, inequality is seen as the driving force behind the American success story. The ability to improve one's economic position, the chance to move up through corporate hierarchies, and the opportunity to have access to the best that life offers are what has made America great. Conversely, the lack of any real upward mobility is the direct cause of the fall of communism. It is not inequality that is bad, but the degree of inequality. It is when the gap between the top and bottom becomes extreme, when the number of individuals at the bottom greatly exceeds the number at the top, and when the opportunity of improving oneself is removed that questions of inequality emerge as a social problem.

President Clinton vowed to end welfare as we know it, a declaration that appeals to many Americans. The Republicans' "Contract with America" contained similar rhetoric. But this goal is extremely difficult to achieve. This unit explores the extent to which America is an open society, the implications of corporate downsizing, what is happening to the middle class, and various strategies for making welfare work.

Robert A. Rosenblatt reports in " 'Glass Ceiling' Still Too Hard to Crack" that although significant progress has been made in opening doors for women and minorities to many areas of employment, the doors to the boardrooms remain closed. He asserts that until the "glass ceiling" is cracked, affirmative action programs must persist.

John Cassidy argues that the "Death of the Middle Class" is due in part to anticapitalistic education, global competition, technological programs, the demise of the trade union movement, and inflation. Since 1973, the bottom two-fifths of American families experienced a drop in their incomes, and the middle incomes were stagnant. Only the top groups enjoyed any growth in income. Cassidy concludes that the "Golden Era" that followed World War II is gone forever and, with it, the great middle class.

To maintain their corporate advantage, many corporations have resorted to downsizing. They often survive fiscally, but leave in their wake "On the Battlefields of Business, Millions of Casualties," as the New York Times headline put it. "The Downsizing of America" creates extreme job insecurity, "diluting a sense of self-worth, splintering families, fragmenting communities, altering the chemistry of workplaces, roiling political agendas, and rubbing salt on the very soul of the country," warns Louis Uchitelle and N. R. Kleinfield.

The most significant factor determining whether a teenager from a poverty-stricken family will find stable employment or wind up hanging around street corners is the presence or absence of an adult mentor (other than parents). Big Brother/Big Sister programs provide this mentor for millions of kids, among them those from poverty. Though not professionals, these mentors help to guide many young people into productive lives, as described in "Social Change One on One."

Is the answer to America's social welfare mess "Dismantling The Welfare State"? D. Stanley Eitzen says no! He argues that the problem is that not enough is being invested in the poor and suggests enhancing programs designed to eliminate poverty, protect the family, anticipate corporate downsizing, and reduce extreme income/wealth disparities.

Adam Wolfson, in "Welfare Fixers," takes the opposite stance. He sees the existing welfare system as the problem, not the solution to poverty. Wolfson evaluates three dominant theories focusing on the causes of poverty and probable solutions. He concludes that none of the three can fix the problem singly, but that together they provide a coherent solution.

"Its Not Working: Why Many Single Mothers Can't Work Their Way Out of Poverty," shows how difficult it is for single mothers to escape poverty. Most of these mothers work less than full-time, experience job discrimination, and lack adequate child and health care.

Chuck Collins, in "Aid to Dependent Corporations: Exposing Federal Handouts to the Wealthy," claims that many large corporations receive substantial handouts

UNIT 4

from the federal government, which he labels "wealthfare." These handouts include major tax benefits, accelerated depreciation, subsidies, and funded research (which allows the benefits to be reaped by private firms).

Looking Ahead: Challenge Questions

How is the distribution of poverty changing?

What are the major problems facing a president who wants to eliminate or reduce poverty significantly?

What societal-level problems are ensuring that the poor stay poor?

Can welfare programs be fixed, or should they be eliminated? Defend your answer.

Why has volunteerism thrived in Big Brother/Big Sisters programs while declining elsewhere?

What is meant by "the glass ceiling," and what are its implications for affirmative action programs?

Is the middle class dying? If so, what is killing it? If not, why not?

What implications does "downsizing" have for corporate integrity? What moral obligations do major corporations have to their employees?

What is "wealthfare," and does it hurt the middle-class taxpayer?

How would sociologists differ in the ways they go about studying problems of inequality and poverty?

What are the conflicts in rights, values, obligations, and harms that underlie each of the issues covered in this unit?

'Glass Ceiling' Still Too Hard to Crack, U.S. Panel Finds

■ **Careers:** *Few women and minorities are even in line for top jobs, report to say. Biased notions, fear of change cited.*

ROBERT A. ROSENBLATT

TIMES STAFF WRITER

WASHINGTON—The informal "glass ceiling" blocking women and minorities from the president's chair in American corporations won't be shattered any time soon because few of them hold the sales, marketing and production jobs that eventually lead to the top of the business world, a bipartisan federal commission will report today.

"Serious barriers to advancement remain—such as persistent stereotyping, erroneous beliefs that 'no qualified women or minorities are out there,' and plain old fear of change," according to the first major report by the Glass Ceiling Commission, a panel of legislators and business officials selected by President George Bush in 1991.

The report found that among the top 1,000 U.S. industrial firms and the 500 biggest firms of all types, ranked by Fortune Magazine, 97% of the senior managers are white and an estimated 95% to 97% are male.

Although the "glass ceiling" was a phrase coined to describe the problems of women in reaching the executive suite, the figurative barricade blocks minority men too, the report says.

"At the highest levels of business, there is indeed a barrier only rarely penetrated by women or persons of color," the commission said. The 20-member group carefully avoided making any policy suggestions, which could be viewed as taking sides in the growing national debate over affirmative action. Their report was limited to presenting information. Recommendations for action, if any, won't be issued until later in the year.

But the figures and analysis will be closely studied by all participants in the debate as they seek ammunition. The commission talks about women and minorities frustrated by the lack of promotions. And it describes white males filled with anxiety about losing jobs or promotions.

The political debate intensified Wednesday as supporters of affirmative action, including members of Congress and California state legislators, met with President Clinton at the White House to urge him to resist political pressure to retreat on the issue.

Rep. Maxine Waters (D-Los Angeles) said minorities and women will no longer stand with Clinton or his party if affirmative action is abandoned.

"No party is so important that we will belong to it if it undermines us on this issue. No President is so important that we will belong to him if he undermines us on this issue," she said.

"The President is trying in a sensible way to have a national conversation on this issue which results in good, common-sense policy," White House spokesman Mike McCurry said.

On the Republican side, Senate Majority Leader Bob Dole (R-Kan.) said the Administration's affirmative action policies are "fatally flawed." Dole, who is becoming increasingly critical of the Administration on the issue, said the Justice Department in some of its decisions has been supporting reverse discrimination.

During the Bush Administration, Dole played a key role in persuading the President to establish the Glass Ceiling Commission. Other key findings of the report:

• Most women and minorities work in the government of the "third sector": schools and health and social welfare agencies. However, Asians are more likely to work in the private sector.

• Latinos, with a college completion rate below the general population, are less likely to have the "advanced degrees that are now considered a prerequisite for climbing the corporate ladder." Latinos who become managers and administrators are more likely to work in government and the nonprofit sector. In the business world, they are "relatively invisible in corporate decision-making positions."

• African American men with professional degrees earn 79% of the salaries of their white male counterparts. African American professional women earn 60% of what white professional males make.

• Asians and Pacific Islanders, although almost twice as educated as the general population—38% have bachelor's degrees or more, compared with 20% of all Americans—are much less likely to become managers or executives.

"I didn't realize how bad the situation was," said commission member Maria Contreras-Sweet, vice president of public affairs at the 7-Up Royal Crown Bottling Co. in Vernon, Calif.

As I talked to some of my business colleagues, they said they were getting a sense women and minorities are in the pipeline and it's only a matter of time," she said. "I don't think this is accurate. Based on the numbers I'm seeing, there is not a critical mass yet. We have to be actively engaged in programs aggressively reaching minorities and women," she said.

Contreras-Sweet, who was concerned because the commission had no white male members other than its chairman, labor Secretary Robert B. Reich, organized a series of meetings with white male corporate leaders, including sessions in Los Angeles with top executives from such firms as Southern California Edison, Atlantic Richfield Co., Bank of America and IBM.

The commission's activity "could never be positioned as affirmative action," she said.

Instead, she said, the commission moved toward recognizing the issue of diversity in executive ranks "as an economic imperative" for business profitability. "That was the tap dance I did at every meeting so the report looked and sounded like a business plan."

The meetings and focus groups with the white male executives, she said, demonstrated the common pattern of picking familiar types to fill top posts. "If you have a job need, you call someone you know. If I have a job, I call a woman. But we all have to say, 'How can I find somebody who enhances the group I have and reflects the marketplace? I will do that because I am a smart businessman,'" she said.

She expressed concern that many firms, while tracking the numbers and career development of female and African American workers, did not make similar efforts for Latinos and Asians. The commission report said it takes 25 or 30 years of work in a corporation, in a variety of jobs, to become a contender for the top executive positions.

"The critical career path for senior management positions requires taking on responsibilities most directly related to the corporate bottom line," the report said.

"But the relatively few women and minorities found at the highest levels tend to be in staff positions, such as human resources or research or administration, rather than line positions, such as marketing or sales or production. Similarly, most companies require broad and varied experience in core areas of the business to advance—experience of the sort that, even now, too few women or minority men are in a position to develop."

Another commission member, Judith L. Lichtman, president of the Women's Legal Defense Fund, said the commission report shows that "the world of discrimination hasn't ended."

The report's figures "reveal that while women and men of color have come very far, how far we have to go," she said. "To link this report and the debate over affirmative action is inescapable."

The report was based on five public hearings including a session in Los Angeles; a survey of chief executive officers; special research studies; focus groups, and analysis of census data.

Death of the middle class

The ideal that once defined how Americans lived and dreamed has gone. **John Cassidy** looks at what has replaced it.

John Cassidy writes for *The New Yorker* magazine, where a longer version of this article originally appeared.

WILLIAM J McDONOUGH, the President of the Federal Reserve Bank of New York, watches his words as closely as a Savile Row tailor watches his stitches, and with good reason. Any mis-statement on his part, or even an intentional but slight deviation from the previous official line, can send the financial markets into a multi-billion-dollar tizzy.

McDonough shuns press conferences, but in November 1994 he invited 35 academics, executives and journalists to a day-long conference at the New York Fed's headquarters in lower Manhattan. When they had assembled, some from as far away as Los Angeles and London, he addressed them as follows: 'I am very pleased that all of you are here today to discuss what I feel is a critical issue facing our country. The issue is, of course, the growing disparity in wages earned by different segments of our labor force. It is deeply troubling that during the 1980s the real wages of low-skilled workers in the United States have fallen sharply, both in absolute terms and relative to the wages of highly-skilled workers. These dramatic wage developments raise profound questions for the United States, issues of equity and social cohesion issues that affect the very temperament of the country.'

For a pillar of capitalism like McDonough to express concern about low wages is surprising. For him to then question, as he did, whether America 'will be able to go forward together as a unified society' is virtually unprecedented.

Until recently, it was an empirical law of American economics that the majority of citizens, including virtually all those who considered themselves middle class, received steadily rising wages. In the three decades after the Second World War in particular, the American dream of moving to the suburbs, buying a house and even sending the kids to college was no mere election slogan. Home ownership soared and the living standards of the middle class – idealized in television sitcoms – were the envy of the world.

Today that image is as dated as the television shows it spawned. Falling wages and rising prices have transformed the home economics of tens of millions of Americans. The trend is best illustrated with the help of a mental experiment. Imagine lining up the entire population of the US in order of ascending income, with the poorest on the extreme left and the richest on the extreme right. The person smack in the center of the line – the meridian – would be, by definition, the most middle-class American alive.

In September of 1979 this person was earning (in constant, inflation-adjusted dollars) $498 a week, or $24,700 a year. By 1995 he or she had suffered a wage cut of about a hundred dollars a month, or 4.6 per cent.

The citizens on the right of the income line-up fared very differently. In 1979 the typical full-time worker in the top third of the income distribution was earning $890 a week, or $46,280 a year. By September of 1995 his or her pay-check had swelled to $960 a week, or $49,920 a year – an increase of 7.9 per cent.

The fortunate souls on the extreme right of the income line-up were doing best of all. In 1979 the richest five per cent of American families earned, on average, $137,482, according to Census Bureau data. By 1993 their income had risen to $177,518, an increase of $770 a week, or 29.1 per cent. The top one per cent of families have made spectacular gains. According to the Congressional Budget Office, between 1977 and 1989 their average income rose from $323,942 to $576,553 – a gain of $252,611, or 78 per cent.

The numbers prove what many Americans have suspected for a long time: living standards have fallen or stagnated for the majority, while a small minority have enjoyed a bonanza. Taken together, recent wage and income trends suggest an unavoidable conclusion: America is no longer a middle-class country; indeed, the term 'middle class' has lost its meaning.

Vague meaning

The idea that the United States is a middle-class country is at least as old as de Tocqueville ('The whole country seems to have melded into one middle class') and Matthew Arnold ('That which in England we call the Middle Classes is in America virtually the nation'). By the 1950s opinion polls showed that the vast majority of Americans referred to themselves as middle class, regardless of their income. Despite this consensus, the exact meaning of 'middle class' remained vague. In contrast to what it meant in Europe, it did not mean the bourgeoisie, who were clearly defined by Engels as 'the class of modern capitalists, owners of the means of social production and employers of wage labor'.

19. Death of the Middle Class

Few inhabitants of California's Orange County or New York's Suffolk County owned factories or speculated on Wall Street. Most were regular employees of major corporations like McDonnell Douglas, Grumman or Hughes Aircraft. If they didn't go to work they risked losing their livelihoods, their houses and their cars. They were, in fact, not middle class at all in the Marxian sense of the word. They were working class but, unlike similar people in Britain or Germany, they called themselves middle class.

Most commentators let them get away with the theoretical confusion, and it is easy to see why. From 1945 until 1973 – a period that economic historians now refer to as the Golden Era – the American economy resolutely refused to conform to the pattern predicted by the left-wing graybeards. Yes, the rich got richer, but almost everybody else got richer with them, and at roughly the same pace. The spoils of economic growth were divided remarkably evenly. Broadly speaking, all Americans' incomes doubled—secretaries', factory workers' and bank executives'.

A chart that is particularly worth examining divides the nation's 68.5 million families into fifths, or quintiles, with the poorest fifth on the left and the richest on the right, and covers two periods: from 1947 to 1973 and from 1973 to 1993.

As Paul Krugman, an economist at Stamford University, has noted, the 1947-73 graph looks like a picket fence, which is entirely fitting: during the Golden Era income growth, like the picket fence, was an icon of middle-class America. The annual growth rate of family income was between 2.4 and 3.0 per cent, regardless of where the family stood in the income distribution.

A glance at the 1973-93 chart, however, shows that the picket fence has been replaced by a small staircase – and some of the steps are underground. The bottom two-fifths of American families saw their income fall, while the average family in the middle saw its income basically stagnate. Only the top 40 per cent enjoyed any income growth, and only the very rich enjoyed growth comparable with that of the Golden Era.

Some conservative economists have attempted to challenge these findings, but with little success. However the figures are shuffled, the basic picture remains the same: the staircase has replaced the picket fence and the country has experienced an unprecedented redistribution of income towards the rich.

The staircase graph and the changing distribution of income suggest that the country has now split into four groups. At the top there is an immensely wealthy élite, which has never had it so good. At the bottom there is an underclass which is increasingly divorced from the rest of society. And between these extremes there are, instead of a unified middle class, two distinct groups: an upper echelon of highly-skilled, highly-educated professionals who are doing very well; and a vast swathe of unskilled and semi-skilled workers who are experiencing falling wages, stagnant or declining living standards, and increased economic uncertainty. To label this group 'middle class' doesn't make sense. That phrase implies two things – rising living standards and a high degree of economic security – that no longer apply.

The dramatic rise in inequality has had one beneficial side effect: it has provided gainful employment for hundreds of economists who have been burrowing away in universities and research institutes trying to solve the mystery of who killed the middle class. Sad to report, they have yet to come up with a single murderer.

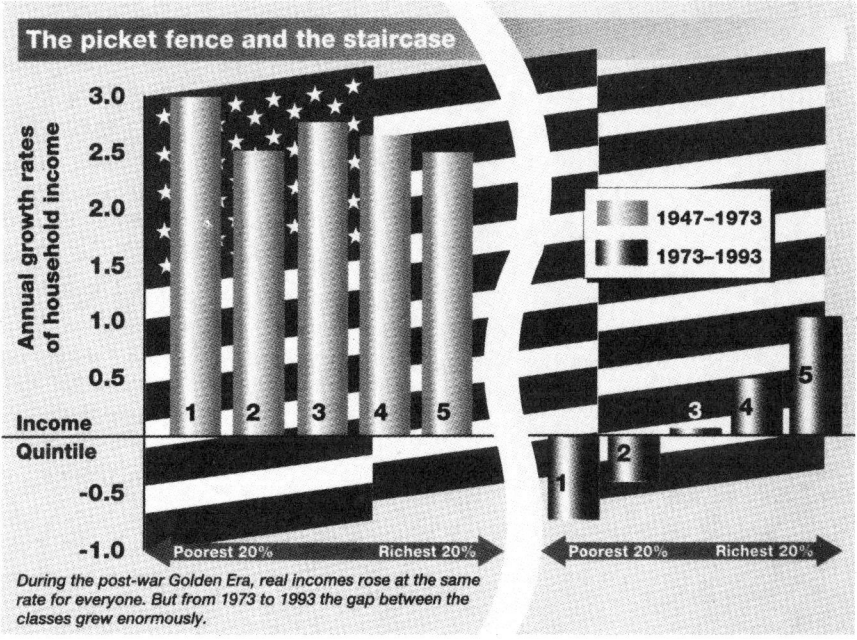

During the post-war Golden Era, real incomes rose at the same rate for everyone. But from 1973 to 1993 the gap between the classes grew enormously.

One thing we do know is that the murder has nothing to do with taxation. This bit of knowledge will disappoint both conservatives who lay the blame for the middle-class squeeze on a growing tax burden, and liberals who link rising wealth concentration to regressive tax policies. But it is indisputable: the fact is that the rise in equality happened before the Internal Revenue Service got its hands on anybody's paycheck. And, contrary to popular myth, the over-all burden of taxation has remained remarkably constant. In 1973 federal taxes swallowed 19.5 per cent of the gross domestic product and in 1993 the figure was 19.9 per cent. It was the free market that decreed that most people's wages should fall or stagnate after 1973. And it was the free market that handed out pay bonanzas at the top of the income distribution.

Anti-capitalism

Education, global competition, technological progress and the demise of the trade-union movement have all been identified as possible suspects – some of the factors probably interacted and reinforced each other. The rise of global competition may have encouraged managers to break unions and invest in computer technology. Similarly, the threat of corporate relocation and the growth of cheap immigrant labor may have contributed to the weakness of labor unions. But what nobody has yet explained satisfactorily is the explosion of incomes at the top of the distribution. The gap between the shop floor and the executive suite has turned into a chasm.

Labor Secretary Robert Reich, somebody who understands what is happening, says that the fallout from the decline of the middle class 'could be very divisive in this country'. He goes on: 'You end up pitting working American against working American, or against the poor, for portions of an ever-decreasing slice of the national pie.' The process is already visible in the rise of anti-immigration, anti-affirmative-action and anti-incumbency rhetoric. At least one Wall Street observer thinks he has spotted a new populist movement on the horizon: anti-capitalism. Stephen Roach, the chief economist at Morgan Stanley, told his clients in a circular that 'worker backlash could be one of the key issues of the 1996 presidential campaign'.

At some point politicians are going to have to level with the voters and tell them the truth: that the postwar Golden Era is gone forever, and the great middle class has gone with it.

On the Battlefields of Business, Millions of Casualties

**LOUIS UCHITELLE and
N. R. KLEINFIELD**

Drive along the asphalt river of Interstate 95 across the Rhode Island border and into the pristine confines of Connecticut. Stop at that first tourist information center with its sheaves of brochures promising lazy delights. What could anyone possibly guess of Steven A. Holthausen, the portly man behind the counter who dispenses the answers?

Certainly not that for two decades he was a $1,000-a-week loan officer. Not that he survived three bank mergers only to be told, upon returning from a family vacation, that he no longer had a job. Not that his wife kicked him out and his children shunned him. Not that he slid to the bottom step of the economic ladder, pumping gas at a station owned by a former bank customer, being a guinea pig in a drug test and driving a car for a salesman who had lost his license for drunkenness. Not that, at 51, he makes do on $1,000 a month as a tourist guide, a quarter of his earlier salary. And not that he is worried that his modest job is itself fragile, and that he may have to work next as a clerk in a brother's liquor store.

That, however, is his condensed story, and its true grimness lies in the simple fact that it is no longer at all extraordinary in America. "I did not realize on that day I was fired how big a price I would have to pay," Mr. Holthausen said, in a near whisper.

More than 43 million jobs have been erased in the United States since 1979, according to a New York Times analysis of Labor Department numbers. Many of the losses come from the normal churning as stores fail and factories move. And far more jobs have been created than lost over that period. But increasingly the jobs that are disappearing are those of higher-paid, white-collar workers, many at large corporations, women as well as men, many at the peak of their careers. Like a clicking odometer on a speeding car, the number twirls higher nearly each day.

Peek into the living rooms of America and see how many are touched:

• Nearly three-quarters of all households have had a close encounter with layoffs since 1980, according to a new poll by The New York Times. In one-third of all households, a family member has lost a job, and nearly 40 percent more know a relative, friend or neighbor who was laid off.

• One in 10 adults—or about 19 million people, a number matching the adult population of New York and New Jersey combined—acknowledged that a lost job in their household had precipitated a major crisis in their lives, according to the Times poll.

• While permanent layoffs have been symptomatic of most recessions, now they are occurring in the same large numbers even during an economic recovery that has lasted five years and even at companies that are doing well.

• In a reversal from the early 80's, workers with at least some college education make up the majority of people whose jobs were eliminated, outnumbering those with no more than high school educations. And better-paid workers—those earning at least $50,000—account for twice the share of the lost jobs than they did in the 1980's.

• Roughly 50 percent more people, about 3 million, are affected by layoffs each year than the 2 million victims of violent crimes. But while crime bromides get easily served up—more police, stiffer jail sentences—no one has come up with any broadly agreed upon antidotes to this problem. And until Patrick J. Buchanan made the issue part of the Presidential campaign, it seldom surfaced in political debate.

Yet this is not a saga about rampant unemployment, like the Great Depression, but one about an emerging redefinition of employment. There has been a net increase of 27 million jobs in America since 1979, enough to easily absorb all the laid-off workers plus the new people beginning careers, and the national unemployment rate is low.

The sting is in the nature of the replacement work. Whereas 25 years ago the vast majority of the people who were laid off found jobs that paid as well as their old ones, Labor Department numbers show that now only about 35 percent of laid-off full-time workers end up in equally remunerative or better-paid jobs. Compounding this frustration are stagnant wages and an increasingly unequal distribution of wealth. Adjusted for inflation, the median wage is nearly 3 percent below what it was in 1979. Average household income climbed 10 percent between 1979 and 1994, but 97 percent of the gain went to the richest 20 percent.

The result is the most acute job insecurity since the Depression. And this in turn has produced an unrelenting angst that is shattering people's notions of work and self and the very promise of tomorrow, even as the President proclaims in his State of the Union Message that the economy is "the healthiest it has been in three decades" and even as the stock market has rocketed to 81 new highs in the last year.

Driving much of the job loss are several familiar and intensifying stresses bearing down upon companies: stunning technological progress that lets machines replace hands and minds; efficient and wily competitors here and abroad; the ease of contracting out work, and the stern insistence of Wall Street on elevating profits even if it means casting off people. Cutting the payroll has appeal for the sick and the healthy—for gasping companies that resort to it as triage and to the soundly profitable that try it as preventative medicine against a complicated future.

The conundrum is that what companies do to make themselves secure is precisely what makes their workers feel insecure. And because workers are heavily represented among the 38 million Americans who own mutual funds, they unwittingly contribute to the very pressure from Wall Street that could take away their salaries even as it improves their investment income.

The job apprehension has intruded everywhere, diluting self-worth, splintering families, fragmenting communities, altering the chemistry of workplaces, roiling political agendas and rubbing salt on the very soul of the country. Dispossessed workers like Steven Holthausen are finding themselves on anguished journeys they never imagined, as if being forced to live the American dream of higher possibilities in reverse.

Many Americans have reacted by downsizing their expectations of material comforts and the sweetness of the future. In a nation where it used to be a given that children would do better than their parents, half of those polled by The Times thought it unlikely that today's youth would attain a higher standard of living than they have. What is striking is that this gloom may be even more emphatic among prosperous and well-educated Americans. A Times survey of the 1970 graduating class at Bucknell University, a college known as an educator of successful engineers and middle managers, found that nearly two-thirds doubted that today's children would live better. White-collar, middle-class Americans in mass numbers are coming to understand first hand the chronic insecurity on which the working class and the poor are experts.

All of this is causing a pronounced withdrawal from community and civic life. Visit Dayton, Ohio, a city fabled for its civic cohesion, and see the detritus. When Vinnie Russo left his job at National Cash Register and went to another city, the 85 boys of Pack 530 lost their Cubmaster, and they still don't have a new one. Many people are too tired, frustrated or busy for activities they used to enjoy, like church choir.

The effects billow beyond community participation. People find themselves sifting for convenient scapegoats on which to turn their anger, and are adopting harsher views toward those more needy than themselves.

Those who have not lost their jobs and their identities, and do not expect to, are also being traumatized. The witnesses, the people who stay employed but sit next to empty desks and wilting ferns, are grappling with the guilt that psychologists label survivor's syndrome. At Chemical Bank, a department of 15 was downsized to just one woman. She sobbed for two days over her vanished colleagues. Why them? Why not me?

The intact workers are scrambling to adjust. They are calculating the best angles to job security, including working harder and shrewder, and discounting the notion that a paycheck is an entitlement. The majority of people polled by the Times said they would work more hours, take fewer vacation days or accept lesser benefits to keep their jobs.

Even the most apparent winners are being singed. A generation of corporate managers have terminated huge numbers of people, and these firing-squad veterans are fumbling for ways to shush their consciences. Richard A. Baumbusch was a manager at CBS in 1985 when a colleague came to him for advice: Should he buy a house? Mr. Baumbusch knew the man's job was doomed, yet felt bound by his corporate duty to remain silent.

4. POVERTY AND INEQUALITY

The man bought the house, then lost his job. Ten years have passed, but Mr. Baumbusch cannot forget.

A Question of Expectations

One factor making this period so traumatic is that since World War II people have expected that their lives and those of their children would steadily improve. "It's important to recall that throughout American history, discontent has always had less to do with material well-being than with expectations and anxiety," David Herbert Donald, a social historian at Harvard, said. "You read that 40,000 people are laid off at AT&T and a shiver goes down your back that says, 'That could be me,' even if the fear is exaggerated. What we are reacting against is the end of a predictable kind of life, just as the people who left the predictable rhythms of the farm in the 1880's felt such a loss of control once they were in cities."

As the clangor from politicians over the jobs issue has begun to be heard, aspirants to public office may find an audience in that group of households in which a lost job produced a major crisis.

The Times poll revealed something of their signature. Only 28 percent, versus 44 percent of the entire population, say they are as well off as they imagined at this juncture of their lives. The vast majority feel the country is going in the wrong direction, and they are more pessimistic about the economy. They are more likely than the overall population to be divorced or separated. They are better educated. Politically, they are more apt to label themselves liberal. They are more likely to favor national health insurance, and to say that curbing government programs like Medicare, Medicaid and welfare is a misguided idea. And more than 63 percent, compared with 47 percent in the whole population, want the Government to do something about job losses.

Wherever one turns one encounters the scents and sounds of this sobering new climate. Ask Ann Landers. Last year, when she adopted a stone-hearted view in her column to a laid-off worker, lecturing him that he had a "negative attitude," she was swamped by 6,000 venomous letters, one of the largest outpourings to any of her columns. "They were really giving me the dickens," Ms. Landers said. "This is the real world, girl. Now I am trying to be supportive."

People run into acquaintances and don't ask how their job is, but whether they still have it. Surf the internet or flick on the comedy channels and take in the macabre jokes: Sales clerk: "What size are you?" Customer: "I'm not sure. I used to be a 42 Regular. But that was before I was downsized." Wife: "But why'd they fire you?" Husband: "They said something about the company making too much money. If the business tanks, they said they'd call me back."

Such graveyard humor is pervasive in Scott Adams's popular comic strip, Dilbert, about a 1990's computer engineer who quakes under a gruff and hectoring boss. In one strip, Dilbert competes with Zimbu, a monkey, for a job, and loses. In another, the boss informs Dilbert that he is about to become involved in all aspects of the company's production. "Dear Lord," Dilbert realizes. "You've fired all the secretaries." Raw material arrives daily in the form of E-mail from demoralized workers.

In an effort to somehow cauterize the emotional damage of the dismissals, managers have introduced a euphemistic layoff-speak. Employees are "downsized," "separated," "severed," "unassigned." They are told that their jobs "are not going forward." The word downsize didn't even enter the language until the early 1970's, when it was coined by the auto industry to refer to the shrinking of cars. Starting in 1982, it was applied to humans and entered in the college edition of the American Heritage Dictionary.

Meanwhile, the word layoff has taken a fresh meaning. In the past, it meant a sour but temporary interruption in one's job. Work was slow, so a factory shift would be laid off. But stay by the phone—the job will resume three weeks or three months from now when business picks up. Today, layoff means a permanent, irrevocable goodbye.

A Portrait of the Victims

Imagine the downsized posed shoulder to shoulder for an annual portrait, some sort of dysfunctional graduation picture. Mostly young, male, blue-collar workers dominated the glossies of the 1980's. Now, white-collar people stare out from every row. Many more of them are women and those whose hair flashes with gray. Instead of factory clothes, far more wear adornment appropriate for carpeted offices.

At his office in the Labor Department's Bureau of Labor Statistics, Thomas Nardone, an associate commissioner, keeps a chart that tracks income and layoffs. In the 1980's, the chart shows, the higher the income, the less frequent layoffs. Now the two lines rise in tandem.

The job insecurity reaches beyond corporations. Government is also scaling back, although not as drastically as corporations, erasing many of the jobs that historically elevated the poor. Between 1979 and 1993, 454,000 public service jobs vanished.

Academia is contributing to the dislocation by paring its rolls and increasingly leaving college teachers in jeopardy by denying them tenure. Doctors, once leading the way along the smug path to American bounty, are succumbing to the cost-containment convulsions in health care.

What so many middle-class workers are experiencing for the first time is achingly familiar to poorer people. Job security never seemed to apply to them. Indeed, those at the lower end of the economic ladder are slipping even further. Rene Brown is a thrice-downsized woman who is still in her 40's. Since the start of the 1980's she has been downsized out of an $8.50-an hour job at a meatpacking plant, a $7.25-an hour job in a bank mailroom and a $4.75-an-hour job loading newspapers. Presently she earns $4.25 cleaning office buildings in Baltimore.

Ms. Brown has done this menial work for three years, without a raise. She is annoyed that, despite a high school diploma and a year of community college, she cannot find a way back up the income ladder.

The poor are losing out in another way. The newly pinched middle class has grown increasingly intolerant of having its tax dollars applied to social programs benefiting the disadvantaged.

What Changed

People, of course, always lost their jobs. In the 19th and early 20th centuries, it didn't take much; job security was not yet an American concept.

World War II, however, ushered in an unprecedented era of economic growth. Demand for workers soared. The post-war years led many people to the succoring belief that they had an almost divine right to a very particular American dream entailing a home, a secure job and a raise every year. An unwritten social contract, codified in part by strong labor unions, came into being, under which managers and workers pledged their loyalty to one another.

The booming economic growth that made this possible slackened in the 1970's, and the economy has remained stuck at a lower volume. The steady and pronounced progress of technology has kept taking tasks from human beings and giving them to machines, undermining the bedrock notion of mass employment. Whereas the General Motors Corporation employed 500,000 people at its peak in the 1970's, today it can make just as many cars with 315,000 workers. Computer programs rather than lawyers prepare divorce papers. If 2,000 movie extras are needed, the studio hires 100 and a computer spits out clones for the rest. Behind every A.T.M. flutter the ghosts of three human tellers.

By the late 1970's, the convergence of these trends prompted companies to sanction large-scale layoffs. At first, the job losses occurred largely in beleaguered smokestack industries. Now the most modern and prosperous industries like telecommunications and electronics are shedding jobs regularly—companies like Sun Microsystems, Pacific Telesis and I.B.M. Media companies, including The New York Times, are also doing so.

Labor Department statistics show that more than 36 million jobs were eliminated between 1979 and 1993, and an analysis by The New York Times puts the number at 43 million through 1995. Many of the jobs would disappear in any age, when a store closes or an old product like the typewriter yields to a new one like the computer. What distinguishes this age are three phenomena: white-collar workers are big victims; large corporations now account for many of the layoffs, and a large percentage of the jobs are lost to "outsourcing"—contracting out work to another company, usually within the United States.

Far more jobs are being added than lost. But many of the new jobs are in small companies that offer scant benefits and less pay, and many are part-time positions with no benefits at all. Often, the laid off get only temporary work, tackling tasks once performed by full-timers. The country's largest employer, renting out 767,000 substitute workers each year, is Manpower Inc., the temporary-help agency.

In this game of musical jobs, people making $150,000 resurrect themselves making $50,000, sometimes as self-employed consultants or contractors. Those making $50,000 reappear earning $25,000. And these jobs are discovered often after much time, misery and personal humiliation.

The Rationale for Cutting

Most chief executives and some economists view this interlude as an unavoidable and even healthy period during which efficiency is created out of inefficiency. They herald the downsizings, messy as they are, as necessary to compete in a global economy. The argument is that some workers must be sacrificed to salvage the organization.

4. POVERTY AND INEQUALITY

Sears, Roebuck and Company felt its very existence threatened in a world of too many stores and too many ways for people to buy what Sears sold for less. Cost cutting, in the form of 50,000 eliminated jobs in the 1990's, was part of the response. "I felt lousy about it," Arthur C. Martinez, Sears's chairman, said. "But I was trying to balance that with the other 300,000 employees left, and balance it with the thousands of workers in our supplier community, and with 125,000 retirees who look to Sears for their pensions, and with the needs of our shareholders."

At the Newport News, Va., shipyard of Tenneco Inc., a diversified manufacturer, 11,000 of 29,000 workers have been shed since 1990, largely because of technological efficiencies like automated welding. It's also true that the Pentagon is buying fewer ships. Dana G. Mead, Tenneco's chairman, boasts that Newport News is now as efficient as any shipyard in the world. Four workers operating robots can cut all the ribs of a tanker, a task that had required 21 and took longer.

"We put in automation to get more competitive," Mr. Mead said, adding that the change won important tanker and submarine contracts. "Then how many workers you build back depends on the rate the commercial business grows, and what the Navy decides to build."

Robert E. Allen, the AT&T chairman who has recently been turned into something of a symbol of corporate avarice for authorizing the elimination of 40,000 jobs, said that intensifying competition left him without choices. He said that with the Baby Bells free to invade AT&T's long-distance stronghold, AT&T's bloated staff of middle managers is no longer affordable. "The easy thing would be to rest on our laurels and say we are doing pretty well, let's just ride it out," he said. "The initiative we took is to get ahead of the game a little bit."

Also intrinsic to the new message is that the lion's share of raises and bonuses must be channeled to those judged most talented and diligent. This new standard of "pay for performance" has made a growing divide among incomes a hallmark of the layoff era. In essence, a new notion of growth and job creation has emerged in which, rather than an expanding economy benefiting all, only the stellar performers—or those providentially in the right careers—come out ahead.

At the same time, some layoffs seem rooted in economic fashion. An unforgiving Wall Street has given its signals of approval—rising stock prices—to companies that take the meat-ax to their costs. The day Sears announced it was discarding 50,000 jobs, its stock climbed nearly 4 percent. The day Xerox said it would prune 10,000 jobs, its stock surged 7 percent. And thus business has been thrust into a cycle where it is keener about pleasing investors than workers.

How this all plays out is a matter of debate. Some contend that through these adjustments American companies will recapture their past dominance in world markets, and once again be in a position to deliver higher income to most workers. Others predict that creating such fungible workforces will leave businesses with dispirited and disloyal employees who will be less productive. And many economists and chief executives think the job shuffling may be a permanent fixture, always with us, as if the nation had caught a chronic, rasping cough.

The Hardest Hit

The tally of jobs eliminated in the 1990's—123,000 at AT&T, 18,800 at Delta Airlines, 16,800 at Eastman Kodak—has the eerie feel of battlefield casualty counts. And like waves of strung-out veterans, the psychically frazzled downsized workers are infecting their families, friends and communities with their grief, fear and anger. The metabolic changes taking place in the country are only beginning to be understood, but there is no missing the deep imprint on the life of Steven Holthausen, the loan officer turned tourist guide.

His high-velocity slide has caused him to go into his soul with calipers. He is suffused with anger, much of it toward himself. Why, he berates himself over and over, did he give so many evenings and weekends to his employer? Why didn't he see that his job was doomed? And then when the dismal news came that August day in 1990, he took it as he felt an executive should, coolly accepting the unfeeling reality of modern economics. Accepting it, that is, until he learned that some of his duties had been assumed by a 22-year-old at a fraction of his pay.

Once laid off, he not only withdrew from work, he withdrew from sight. He had been co-chairman of the trustees of a church in Westbrook, Conn., as well as vice chairman of the police advisory board. He quit both posts. No longer a banker, he felt he had lost the requisite dignity to participate in civic activities. "You feel the community has lost its respect for you," he said.

For almost a year, Mr. Holthausen scraped by on severance pay, on meager commissions earned as a freelance mortgage broker and on unemployment insurance. The fact that this income was taxed made him resent the government. If the Federal budget were balanced by scaling back spending, he reasoned, less of his skimpy income would be taken from him.

Accordingly, Mr. Holthausen voted for Ross Perot in 1992, warming to his pledge to cure the deficit. He now considers himself a budget-balancing Republican, although he has yet to settle on a candidate.

He lives alone with his torments in a humble apartment owned by a brother. He sat stock-still as he ruminated on the tatters of his family. Even while he was a banker, tensions underlay the marriage. When he was fired, the couple sought therapy. At the sessions, he beseeched his wife to help him regain his shattered confidence. He found her unsympathetic. Six months later, she ordered him out. Soon after, she filed for divorce and, after years of not working, found a job as a medical secretary. His two teen-age children avoided him. Their view, he felt, was that he must have shortcomings or he would not be jobless.

"The anger that I feel right now is that I lost both my family and my job," he said. "That is not where I wanted to be at this point in my life."

In a society in which identity is so directly quantified by work, the psychological fall involved in losing a job is leading many to stress-induced illnesses. "What makes it so hard for people is very often these situations come about very suddenly," said Dr. Gerd Fenchel, the head of the Washington Square Institute for Psychotherapy and Mental Health in New York, who has seen his caseload swell with downsized workers. "We have a diagnosis called post-traumatic stress syndrome that applies to this."

The impact of job loss on marriages varies. The divorce rate, according to several studies, is as much as 50 percent higher than the national average in families where one earner, usually the man, has lost a job and cannot quickly find an equivalent one. On the other hand, many families where both husband and wife are employed seem to be drawing closer to muster their energies against the common enemy of job insecurity.

The effect on community unity seems more straightforward. In city after city, downsized people are withdrawing from the civic activities that held communities together. Sociologists report that involvement has tumbled at P.T.A.'s, Rotary clubs, Kiwanis clubs, town meetings and church suppers. Bowling leagues are unraveling, even though more people are bowling than ever. The reason is they are visiting alleys not as part of corporate or community leagues, but singly or with a friend. "The 'we', has become a 'me,' or at least a narrower 'we,' " Robert D. Putnam, a Harvard professor who has documented this contracting participation, said. He fingers downsizing as a culprit, although not as insidious as television.

Social Change One on One

The New Mentoring Movement

Gary Walker
and Marc Freedman

Gary Walker is president of Public/Private Ventures. **Marc Freedman**, vice-president of Public/Private Ventures, is author of *The Kindness of Strangers*.

Past the video games, the 11th Frame bar, and the orange formica shoe-rental island, occupying half the Saturday afternoon lanes at Mel's Bowl in Oakland is a scene to warm the civic heart: 25 lanes packed with bowlers, impervious to the brilliant spring day outside, immersed in spontaneous sociability.

The crowd is at Mel's today to participate in Bowl for Kids' Sake, a fundraiser for the Big Brothers/Big Sisters program that has generated nearly $125 million in donations since 1981. This year the Oakland program expects to net close to $150,000. Nationally, with endorsements from both the Professional Bowlers Association and the Bowling Proprietors Association of America, two million Bowl for Kids' Sake participants will produce more than $15 million, second only to United Way contributions as a source of financial support for the mentoring work of Big Brothers/Big Sisters organizations.

The remarkable feature of Bowl for Kids' Sake is that it is fundamentally a fundraiser for social capital, helping to underwrite a program that constitutes one of the most important forms of connectedness that political scientist Robert Putnam and other observers fear is disappearing [see Robert Putnam, "The Strange Disappearance of Civic America," *TAP*, Winter 1996, and the correction from Putnam and John F. Helliwell (*The American Prospect*, July–August 1996), page 18.]

As the late sociologist James Coleman—who first introduced the concept of social capital to an American audience—explained nearly a decade ago, "Social capital in the community exists in the interest, even the intrusiveness, of one adult in the activities of someone else's child." For Coleman this was most vividly expressed in relationships that crossed generations and were characterized by "attention, personal interest and intensity of involvement" on the part of the adults.

Coleman might as well have been reciting the mission statement of Big Brothers/Big Sisters of America. For nearly 100 years Big Brothers/Big Sisters has defined the enterprise of face-to-face connecting so completely that it owns the trademark, One to One. It is also one of the best-known youth programs in America. A 1994 Gallup poll found that 78 percent of Americans are aware of Big Brothers/Big Sisters, which matches 75,000 children from single-parent homes ("Littles") with volunteer mentors ("Bigs") through more than 500 chapters nationwide. The activities encouraged by the program are primarily informal and friendly. Depending on a youngster's interests, a mentor takes the child out to eat, to watch a ballgame, to go to a concert, or just to talk—on average three times a month, three and a half hours each time. That amounts to 126 hours a year, or about three 40-hour work weeks, which is no small commitment.

The case of Big Brothers/Big Sisters and the new mentoring movement it has helped spawn offer a concrete opportunity to reexamine the sweeping thesis that social capital in this country is plummeting, without organic prospects for replenishment. By focusing on one of its most important and elemental expressions—the attempt to create one-to-one relationships between adults and youth—it is also possible to explore if and how public social policy can help rebuild the civic infrastructure of neighborhoods.

THE PERSONAL TOUCH

Big Brothers/Big Sisters began its long crusade to foster adult-youth connections in 1904, launched by Ernest Coulter, a New York newspaperman who left journalism to work in the city's first children's court. Appalled by the harsh juvenile justice and social welfare systems he encountered in his new job, Coulter appealed for help to his friends at the Men's Club of the Central Presbyterian Church of New York.

Recounting the story of one promising child relegated to the custody of an uncaring system, Coulter told the professionals and businessmen of the Men's Club: "There is only one possible way to save that youngster, to have some earnest, true man volunteer to be his big brother, to look after him, help him to do right, make the little chap feel that there is at least one human being in this great city . . . who cares whether he lives or dies."

The experience of the 39 volunteers who signed up that night was far from glorious. Coulter himself spent eight years unsuccessfully trying to rehabilitate a member of the Fagins street gang. Nevertheless, the movement caught on, largely by contrasting itself to the emerging bureaucracies of early-twentieth-century urban America.

At a revival-like rally for the new movement held at the Casino Theater in New York in 1916, a multiracial, ecumenical crowd of 2,000 recruits listened to Rabbi J.L. Magnes and a parade of clergy lambast these institutions and their "social machinery." "In this day of cold efficiency—efficiency in business, efficiency in charity," Magnes charged, "it is a miserable small justice our great organized charities do. . . . The personal touch is absent."

Almost 100 years later, the circumstances that so outraged Coulter and Magnes and launched the Big Brothers/Big Sisters movement are dramatically worse. British child care expert Penelope Leach warns that if we want to socialize our children, we must socialize with them. Yet increasingly, many of this country's youth have few responsible adults to lean on for nurturance and support. Market forces, individual decisions, and public policies have all converged to produce a situation for kids that everyone decries.

For one, there are fewer adults in families. Today more than one in four children is born into a single-parent household; among African Americans, it's two of three. The 1980s alone saw a doubling of *no-parent* households (that is, households where kids are being raised by grandparents, other relatives, or foster parents), and nearly 40 percent of our youth grow up in fatherless homes. And regardless of the number of adults, the time famine afflicting so many working parents makes it difficult for them to spend sufficient time with their kids.

The settings that might compensate for these shifts—neighborhood streets and schools—frequently don't offer much help. One poll finds that nearly three of four Americans do not know the person living next door; meanwhile, in many urban neighborhoods, the fear of violence has driven community adults, once an important source of socialization and support, behind locked doors. The problem is particularly acute in schools that are now often impersonal teaching factories, reeling in the wake of budget cuts. Student-counselor ratios exceed 500 to 1 in most urban districts. Class size is 30 or even 40 students, and the average teacher will often face 200 students in a single day. Even the most caring find it hard to connect with more than a few young people.

These figures are all the more troubling given the level of stress so many young people confront. Today's youth face a bewildering array of hazards, with those living in the inner city particularly besieged. Seventy percent of these children, by the age of 15, have witnessed someone being beaten, while nearly a third have watched someone being shot. A recent survey of 2,000 teenagers by Louis Harris and Associates found that one in eight youth carries a weapon to school for protection. In high crime areas the number jumps to almost two in five—with one out of three cutting classes or staying away from school regularly out of fear for their safety.

In short, many youth are without adult support at precisely the juncture in their lives when they need it most. And while common sense tells us this is an unhappy situation, social science—through an array of studies on "resilient" children who overcome the odds of poverty—suggests that it is also a missed opportunity.

In the most substantial of these studies, developmental psychologist Emmy E. Werner followed 500 Hawaiian children growing up in poverty on Kauai. Examining their lives from birth to adulthood, over a 30-year period, Werner found that the youth who managed to make it, against the

4. POVERTY AND INEQUALITY

odds, all could count on the support of an adult mentor other than their parents.

Anthropologists William Kornblum and Terry Williams followed 900 children in urban and rural poverty across the U.S., concluding that "the most significant" factor determining whether teenagers would end up on the corner or in a stable job "is the presence or absence of adult mentors." Earlier this year, Arthur Levine, the president of Teachers College at Columbia University, reached the same conclusion in a book of case histories chronicling the experience of 24 disadvantaged children who had made it to college.

MAKING A DIFFERENCE

Against this backdrop, the busy lanes at Mel's Bowl are reassuring. Better yet, Mel's patrons are not alone. While overall volunteering is on the sharp decline (dropping from 54 to 48 percent of adults from 1989 to 1993, according to a Gallup survey), the number of volunteers in Big Brothers/Big Sisters programs continues to grow steadily. Indeed, Big Brothers/Big Sisters has helped to create the new mentoring movement by providing support to other initiatives and establishing the legitimacy of matching youth with unrelated adult mentors.

Over the past decade hundreds of corporations, universities, youth organizations, and religious and civic groups have hopped on the mentoring bandwagon. Nationally, a wide range of groups have joined Big Brothers/Big Sisters in promoting the mentoring cause, from Proctor and Gamble to the Rainbow Coalition. California, New York, and Rhode Island have established statewide mentoring campaigns, with citywide efforts launched to considerable fanfare in Kansas City, Newark, Oakland, Baltimore, Milwaukee, and dozens of other locations around the country. A recent volume, *Nurturing Young Black Males*, published by the Urban Institute, suggests a particularly rich flowering of these efforts in the African-American community, sponsored by church groups, fraternities, sororities, and networks like Concerned Black Men and One Hundred Black Men.

We also now have powerful evidence that mentoring works. In 1992 the organization for which we work, Public/Private Ventures, undertook an independent evaluation of Big Brothers/Big Sisters. (Public/Private Ventures is a nonprofit social policy development and evaluation firm in Philadelphia that focuses much of its work on youth and young adults; the evaluation of Big Brothers/Big Sisters was financed by the Pew Charitable Trusts, Commonwealth Fund, Lilly Endowment, and an anonymous donor.) To carry out the research, Public/Private Ventures studied nearly 1,000 10- to 16-year-olds who applied to Big Brothers/Big Sisters in 1992 and 1993 but were still on a waiting list. More than 60 percent of the sample were boys; more than half were members of minority groups, mostly African Americans. Over 80 percent came from impoverished families, and almost all were being raised by a single parent, usually the mother. Approximately 40 percent were from homes with a history of drug or alcohol abuse and nearly 30 percent came from families with a record of domestic violence.

Half these young people, randomly chosen, were matched with a Big Brother or Big Sister, while the rest stayed on the waiting list. Eighteen months later, the differences between the two groups were dramatic. The involvement of a Big Brother or Big Sister in a young person's life for a single year reduced first-time drug use by 46 percent (at a time when drug use is mounting among teenagers), cut school absenteeism by 52 percent, and lowered violent behavior by 33 percent. Youth with a Big Brother or Big Sister were more likely to perform well in school, much more likely to relate well to family and friends, less likely to assault somebody, and much less likely to start using alcohol. The effects were sustained for both boys and girls and across races.

What's especially startling about these findings is that the mentors were not trained in drug prevention, remedial tutoring, antiviolence counseling, or family therapy: Their instructions were to gain the kids' trust and become their friends. A companion study that looked in-depth over 18 months at 82 Big Brothers/Big Sisters relationships concluded that those adults who could carry out those instructions—not those determined to "straighten these kids out"—were far more likely to gain the trust and time necessary to have an influence on youths' lives.

In short, while many argue that "nothing works" for poor kids—especially if you don't reach them well before adolescence—the evidence says powerfully otherwise. At a time when "social engineering" of any kind has reached a nadir in the public's confidence, we find that the most delicate of social forms—human relationships—can be created, and can accomplish important results.

OBSTACLES TO EXPANSION

With common sense and powerful evidence of effectiveness now lined up behind mentoring, you might think that the road ahead is now significantly clearer: We can floor the accelerator, and get on with it. But two obstacles stand in our way. They are the numbers of adults available to serve as mentors, and the organizational resources necessary for carrying out a successful program. Their repair is inextricably linked to a deeper structural flaw in the route to a more civil society, namely the widespread American belief that public policy is a necessary evil, only to be used after individual and social crises have occurred, and not as an integral, dynamic part of building and maintaining the social fabric.

Available adults. For all the success of Big Brothers/Big Sisters, the program is too small—far too small—in comparison to the number of young people who want or need a mentor. At present, the waiting list for Big Brothers/Big Sisters equals nearly half the 75,000 youth it matches in a year, a list comprised disproportionately of African-American boys. According to one Big Brothers/Big Sisters official, between 5 million and 15 million children could benefit from being matched with a mentor.

The modest number of Big Brothers/Big Sisters mentors results in part from a careful and lengthy screening process; only about one in four people who show initial interest actually become

> While many argue that "nothing works" for poor kids—especially if you don't reach them well before adolescence—the evidence suggests otherwise.

mentors. But that selectivity is critical, not simply to screen out pedophiles but because relationships that don't work can be damaging to kids. It's important to know if an interested, well-meaning adult really has the time to mentor.

There are a multitude of smaller mentoring programs around the country, and there is no current and reliable estimate of the number of mentors they deploy or the time the mentors put in. But if the findings of an earlier Public/Private Ventures survey hold up, a generous estimate is that these many smaller efforts triple the Big Brothers/Big Sisters number. Given the estimate of the kids who could use mentoring, it seems reasonable to conclude that at best a few percent of kids who could benefit from mentoring are getting it.

Infrastructure and resources. Though some Americans would like to believe that doing good springs simply from the heart, the Big Brother/Big Sister experience suggests that, at least in the case of mentoring, making a genuine difference requires a great deal more. It takes persistent, consistent involvement, and, as already noted, that necessitates substantial care in recruiting, screening, matching, and supporting the volunteers. Paid caseworkers carry out these critical functions for Big Brothers/Big Sisters; as a result, the program costs on average $1,000 per year per match.

Cheaper programs may not have the same positive results. In other foundation-supported research, we have studied several mentoring programs that were much less structured, and thus much less costly, than Big Brothers/Big Sisters. In these programs many more adult volunteers were ill-prepared for the commitment and empathy required for mentoring a young stranger. On average the relationships did not last as long, and far fewer of them were successful, judging by interviews with both young people and mentors.

Perhaps the Big Brothers/Big Sisters results can be achieved for something less than $1,000 per year—but the amount needed won't be zero. If the new, streamlined mentoring were to cost only $500 per year and we used the lower estimate of five million kids in need, we would still be about $2.5 billion short.

Big Brothers/Big Sisters and the many smaller mentoring agencies do a good job of raising money through bowling fundraisers, United Way contributions, black-tie affairs, and foundation grants. They may also be able to squeeze 2 to 3 percent more money from private sources. But even if they are spectacularly successful at fundraising, there is still insufficient monetary support for the programs. Thus if a few million new volunteers suddenly appeared, there would be inadequate infrastructure to do anything with them. And there is plenty of sobering experience with poorly administered volunteers.

These are, remember, children.

Facing shortages of volunteers and dollars, Big Brothers/Big Sisters programs across the country are busy innovating and adapting. They are trying to recruit more older adults as mentors. They now involve high school students as "Bigs" and work with schools and other partners in an attempt to lower supervisory costs. Indeed, this willingness to adapt is what separates Big Brothers/Big Sisters from many civic organizations that have become passé. Still, these measures are insufficient.

Until we accept the integral role of the public sector in scaling up effective private initiatives, the potential of mentoring will remain unfulfilled. Unfortunately, for some time public debate has been marooned over ideological opposition between voluntary action and public involvement. (George Bush was fond of contrasting the "good" and the "Great" societies, highlighting the superiority of the former over the latter, and this tendency continues among many right-wing champions of voluntarism today.)

Nevertheless, there are hopeful signs. In Congress, for example, Democrat Frank Lautenberg of New Jersey successfully shepherded the Juvenile Mentoring Program Act (JUMP) to passage, providing federal support for grassroots mentoring programs. JUMP has also won the hearts of some conservative Republicans. Dan Coats of Indiana, who—along with Connecticut liberal Christopher Dodd—is one of two former Big Brothers in the Senate, announced that he would include a measure based on JUMP, the Character Development Act, as one of 18 bills in his Project for American Renewal. Coats's act will offer federal grants to link public schools and local mentoring programs. Republican Governor Pete Wilson of California has proposed putting $15 million in state funding directly into local mentoring programs to address teenage alcohol and drug abuse, pregnancy, violence, and school failure.

But these hopeful signs are only that; on balance, most policy efforts remain small, scattered, and symbolic.

An Army for Youth

Blending public and voluntary contributions would help mentoring programs to address the needs of young people for adult contact. Yet it would be a mistake to underestimate these needs or to overestimate the capacity of volunteers to meet them. The isolation of young people is a structural problem resulting from fundamental, corrosive social changes. Volunteer mentoring is a critical step in the right direction, but rather than being a sufficient response, it highlights an unmet need and calls out for reinforcements.

From where might these reinforcements come? For one, we need to expand the number of paid youth workers available to connect with children, as mentors, in the hours kids spend outside of school. In this vein, David Liederman, director of the Child Welfare League of America, urges establishment of a corps of inner-city youth workers, "able to hit the streets and work directly with kids in their own neighborhoods." He argues that these are the role models kids really need—not famous athletes on television, but caring adults they can "see, touch, and talk to." Hugh Price of the National Urban League estimates that we could support 500,000 such youth workers for the crucial afternoon and early evening hours for the price of the 100,000 new cops that were so central to the debate over the 1994 Crime Act.

In fact, this vision of a "small army" of adults committed to youth is being partly realized through national service, another example of the role that public policy can play in rebuilding the social capital available to kids. A significant portion of the 20,000 national service participants in AmeriCorps are working in direct and intensive one-to-one roles with youth, many of them helping to expand grassroots mentoring projects.

Friends of the Children in Portland, Oregon, offers a compelling example. Created by a local financier, the program employs "full-time caring, loving adults" who each work intensively with eight young children identified by teachers as destined for trouble. The adult Friends spend time in the classroom, serve as a bridge between school and home, and act like surrogate family to the kids. Their goal is to stick with them from second grade to high school. Until last year, however, budget restrictions limited this promising program to four adult friends. With AmeriCorps dollars, Friends of the Children has moved from a complement of 4 to a corps of 24 mentors, and the number of children served has increased dramatically.

The Foster Grandparent Program, a little-known product of the War on Poverty that is now run by the Corporation for National Service, is another potential platoon in this much-needed army. Foster Grandparents work one-on-one with 90,000 children a year, making it the biggest one-to-one program in the U.S., bigger even than Big Brothers/Big Sisters. These Grandparents are low-income women and men over the age of 60 who serve 20 hours a week in schools, Head Start centers, and youth organizations for a stipend of about $200 a month.

As with Friends of the Children and other youth workers, Foster Grandparents are able to maintain relationships with as many as ten children each, providing consistent, weekly (if not daily) attention, while dramatically expanding the number of young people that can be served. At this ratio, 500,000 youth workers might eventually be able to reach as many as five million young people at a level of intensity equivalent to that provided by Big Brothers and Big Sisters.

A GREAT TRANSITION?

When James Coleman argued that social capital was declining ten years ago, he presented his case in historical terms extending back to the Industrial Revolution. When men left the household in droves to work in factories in the nineteenth century, an extensive public investment was required "in a new form of social capital, mass public schooling." Now that women are leaving home, we must think again, he suggested, about institutional changes. These reforms should focus not on classroom instruction, but on caring for children—as Coleman put it, "all day; from birth to school age; after school, every day till parents return home from work; and all summer."

In this context, mentoring appears as a transitional movement, part of a much larger working through of potential institutional and social changes. Indeed, in the latter part of the nineteenth century, the limitations of the Friendly Visiting movement—built, like mentoring, around middle-class volunteers, a belief in the transformative power of personal relationships, and a focus on improving the lives of poor children—led to the establishment of the social work profession.

However, the potential implications of mentoring extend further, past the creation of new roles, to the reform of institutions—in particular, transformation of those faceless bureaucracies that Rabbi Magnes and the early proponents of Big Brothers/Big Sisters so despised. As we seek to develop more powerful strategies for supporting kids, we cannot afford simply to rail against these institutions or to write them off. Rather, we must strive to combine the efforts of volunteers and staff in ways capable of transforming the two settings where young people spend the majority of their time outside the home: schools and youth organizations.

Today, little of the money being spent on reform of schools, foster care, and other youth-serving institutions leads to more sustained adult contact with youth. Our goal, in contrast, should be to work through schools and youth organizations to construct a web of support for children, sustained by adults acting together, as partners, to help out with the long-term, complex process of developing young people. This is not just because these changes seem kinder and more gentle, but because a caring climate is the key to successful results, according to research on a wide range of children's initiatives.

Ultimately, this approach suggests a new approach to schools and youth programs. We need to fill these places with interested adults—not only with volunteer mentors, but also with youth workers, teachers, coaches, counselors, and others with the time and inclination to establish close ties with young people. Young people in these settings

> Little of the money spent on institutions like schools leads to more sustained contact between youth and adults.

would find ample opportunities to develop natural connections and to select the right mentor at the right time. As we go about "stocking the pond," so to speak, we should also help youth to "fish," to make best use of the adults they find in their path. Doing so builds off a central lesson from the study of resilient children, who typically overcome poverty by actively *recruiting* mentors from the surrounding community. More kids can learn how to reach out to adults, too.

ELEMENTARY CARING

How will we ever develop the constituency for such sweeping reforms—and the dollars they will require? The issue provokes a final point about mentoring's potential to rebuild social capital and revitalize civic engagement. While the isolation of youth is a primary reason for mentoring programs, many of the middle-class adult volunteers are just as isolated from the realities of poverty as participating youth are from middle-class life.

At a time when statistics no longer shock, mentors are brought face-to-face with the unfair impact of poverty on innocent children. Many wonder

4. POVERTY AND INEQUALITY

how their own children would fare under such circumstances. This education can build not only empathy, but also advocacy. In other words, mentoring can be every bit as much a social program for adults as for kids, a vehicle for developing their civic instincts while building bridges between communities.

Investment banker Felix Rohatyn's involvement as a sponsor of the "I Have a Dream" project illustrates this process. While mentoring young people in a New York City public school class he adopted, Rohatyn concluded that the youth worker he hired to connect with the children was *the* critical ingredient in the program's success. As a result, he became an influential advocate for reducing the ratio of students to guidance counselors in New York elementary schools. He even attempted to hold up approval for the city's budget in the late 1980s until the school system began making changes to create more caring climates for kids.

The anthropologist Mary Catherine Bateson observes that every adult needs a relationship with a flesh-and-blood child to imagine what the future will be like as that child's life unfolds. For her, this amounts to "the elementary school of caring." The new mentoring movement offers vivid examples of such a school—an elementary school of caring for other people's children, particularly children of the poor.

But by serving as a catalyst to broader institutional changes, the movement holds the potential to graduate beyond elementary contributions. In this process, social policy can offer critical assistance, if it is savvy about identifying the opportunities for rebuilding social capital and willing to blend public and voluntary support.

This may well be something that Democrats and Republicans, liberals and conservatives, can agree upon. The forces pulling us apart today seem almost inexorable, yet Big Brothers/Big Sisters and the mentoring movement are timely reminders that our social instincts continue to function, offering not only a basis of hope but also an opportunity to act.

Dismantling The Welfare State

IS IT THE ANSWER TO AMERICA'S SOCIAL PROBLEMS?

Address by D. STANLEY EITZEN, *Professor Emeritus of Sociology, Colorado State University*
Delivered at the College of Liberal Arts and Sciences Alumni Week Lecture, University of Kansas, Lawrence, Kansas, April 25, 1996

The welfare state is under attack by Republicans, the New Democrats (including President Clinton), and various pressure groups, most notably corporate America and the Christian Coalition. The emerging ideological consensus among these groups has three related propositions. First, government subsidies exacerbate social problems rather than solving them. Second, individuals must rely on their own resources and motivations if they are to succeed. And, third, individuals who fail are to blame for their failure. In a most telling irony, these propositions are assumed for individuals but not for corporations. My task this evening is to present the facts of two representative social problems, to assess the current public policy debate surrounding them, to examine the consequences of dismantling the welfare state, and to propose an alternative vision.

REPRESENTATIVE SOCIAL PROBLEMS AND PUBLIC POLICY

Poverty. Using the government's official definition of poverty, which understates the actual numbers, 15.1 percent of the population (39.3 million people) were officially poor in 1993. If the cost of basic necessities, especially housing, were included in the formula instead of just multiplying food costs by three, the number in poverty would be approximately 60 million.

Some social categories are over represented among the poor. In terms of race, 9.9 percent of whites are officially poor, compared to 30.6 percent of Latinos and about one-third of African Americans and Native Americans. Two out of three impoverished adults are women. Slightly more than one-fourth of children under age six are poor. Comparatively, poor American children don't do very well. In a study of 18 industrialized nations, our poor children ranked 16th in living standards.

The outlook for these poor children is grim. Because they had the misfortune to be born to disadvantaged parents they are denied, for the most part, adequate nutrition, decent medical and dental care, and a safe and secure environment. As a result, by age 5 or so, they are less alert, less curious, and less effective at interacting with their peers than more privileged children. Thus, they begin school already behind. They, most likely, attend the most poorly staffed, overcrowded, and ill-equipped schools. Poor children, in short, are more likely than more advantaged children to have health problems, to not do well in school, to be in trouble with the law, and, as adults to be unemployed and on welfare. Progressives argue that unless society intervenes with meaningful programs and adequate support, many of these poor children will fail. But the current plan sponsored by Republicans and supported for the most part by Democrats is to reduce programs aimed at helping the poor and their children. This reduction process has been underway since 1981 when President Reagan took office.

Related to poverty is the issue of welfare. We have a social safety net of AFDC, WIC, food stamps, Head Start, subsidized housing, to help those in need. Conservatives see this safety net as the problem since they believe it destroys incentives to work and encourages illegitimacy. They argue that welfare is the problem. Thus, social problems will get worse if we are more generous to the poor; they will get better if less is spent on the poor. This logic is a trifecta for the conservatives — by spending less on the poor, we save money, government is reduced, and the lot of the poor improves. Such a deal!

Using this logic, subsidies for compensatory education programs for poor children such as Head Start for preschool children and Chapter 1 for elementary school children have been slashed. Various welfare programs have been reduced by more than a quarter in the past 20 years. The Contract With America proposes to reduce welfare further by eliminating the earned income tax credit for the near poor, to deny AFDC to unmarried mothers under 18, and reduce Supplemental Security Income, food stamps, subsidized housing and other forms of public assistance to non-citizens; and cap the number of years to receive welfare benefits without providing job training, jobs, decent minimum wage, or child care.

Foremost among the social pathologies that the conservatives are most concerned with is illegitimacy, which they feel is the result of young poor women choosing babies to get welfare benefits. This concern is a bit misplaced, however, because only 8 percent of welfare-dependent households are currently headed by teen mothers.

For Progressives, rising illegitimacy is mostly the result of epochal changes in sexual and family mores, and not the result of AFDC. The esteemed social scientist, Frances Fox Piven points out that out-of-wedlock births are increasing in all strata of society, not just among welfare recipients or potential welfare recipients. Moreover, illegitimate births are increasing in

4. POVERTY AND INEQUALITY

all Western countries. Illegitimacy among the poor is also a consequence of the changing economy. When the mother or the father has a good job, the couple tends to get married. But with jobs leaving the inner cities for the suburbs, leaving for the nonunionized states, and leaving for the low-wage economies overseas during the last two decades, illegitimate births have risen dramatically. This argument runs counter to the conservative one which says that unmarried women are having additional children to increase their AFDC payments. The progressives challenge this prevailing myth with the following empirical facts:

• Since 1972, the value of the average AFDC check has withered by 40 percent, yet the ratio of out-of-wedlock births has risen in the same period by 140 percent.

• States that have lower welfare benefits usually have more out-of-wedlock births than states with higher benefits.

• The teen out-of-wedlock birth rate in the United States is much higher than the rates in countries where welfare benefits are much more generous.

These data surely squelch the argument that generous welfare encourages out-of-wedlock births.

Rising economic inequality. The U. S. has the most unfair distribution of wealth and income in the industrialized world. Moreover, the rate of growth in inequality is faster than in any other industrialized country.

The facts concerning inequality include:

• The richest 1 percent own more wealth ($3.6 trillion in 1992) than the bottom 90 percent ($3.4 trillion).

• Between 1983 and 1989 (the Reagan years) the nation's net worth increased from $13.5 trillion to $20.2 trillion, and 58 percent of that $6.7 trillion increase went to the fortunate top one-half of 1 percent. That works out to a $3.9 million bonanza per wealthy household.

• In 1960, the average CEO earned about as much as 41 factory workers. In 1992 that CEO makes as much as 157 factory workers. In 1995 the average compensation of CEOs (salary, bonus, and stock options) increased by 26.9 percent compared to the 2.8 percent increase in wages for the average worker.

• In a 15-year period ending in 1993, the richest 1 percent almost doubled their income and had their tax rates cut by 23 percent. In sharp contrast, the poorest one-fifth saw their tax rates go up and their incomes go down.

• The real value (adjusted for inflation) of a standard welfare benefit package has declined by some 26 percent since 1972.

• From 1967 to 1979 a full-time, year-round worker paid the minimum wage earned above the official poverty line for a family of three. Now a worker earning the minimum wage of $4.25 earns $8,840 annually, which is $3,427 below the 1995 poverty line for a family of three.

These data point to two related problems — wage decline and wealth stratification. Let's put these problems in historical perspective. After World War II, the United States entered a 30 year period of unprecedented economic growth where each segment of the population, from the top 20 percent to the poorest 20 percent and everyone in between saw their incomes double. But following the Vietnam War, the economy began slowing down. In short order, global competition began heating up and technological innovations changed the workplace forever. In the ensuing years, we have seen accelerated changes involving deindustrialization, capital flight as companies merged, moved their operations to low wage economies over-

seas or to low wage regions within the U. S., corporate downsizing, a decline in union membership and clout, and the replacement of permanent workers by temporaries or independent contractors. With the changing economy, wages stagnated, fewer workers were covered by adequate health care benefits and pensions, and the numbers of the underemployed rose to unprecedented heights.

While the masses lost ground or stagnated because of the economic transformation, the fortunate few have done very well with higher salaries, greater profits, and a rising stock market. A major reduction in taxes under President Reagan increased the gulf between the "haves" and the "have-nots."

This rising inequality gap has enormous consequences.

Economist Lester Thurow is very concerned about this trend:

> These are unchartered waters for American democracy. Since accurate data have been kept, beginning in 1929, America has never experienced falling real wages for a majority of its work force while its per-capita GDP was rising. In effect, we are conducting an enormous social and political experiment — something like putting a pressure cooker on the stove over a full flame and waiting to see how long it takes to explode.

Given the threats to individuals, families, and society that are part of the income/wealth inequality package, what are our policy makers doing to reduce it? Actually, rather than limiting this trend, current policies and proposals are increasing it. I refer to "trickle down" policies such as efforts to cut capital gains taxes and reduce inheritance taxes. Also the various proposals for a flat tax or a consumption tax to replace the income tax clearly will benefit the wealthy at the expense of the less affluent. While the Republicans are more inclined than Democrats toward measures that exacerbate inequality, the leadership in both political parties support policies that lead to a larger gap between the rich and the poor and that increase the constriction of the working class. Rather than fight the conservatives, the Democratic leadership has aimed for some right-of-center compromise. This submissive strategy reminds me of the quip by columnist Mark Shields: "If the Biblical prediction is true that the meek shall inherit the earth, then the Democrats will be land barons."

THE CONSEQUENCES OF DISMANTLING THE WELFARE STATE

What will be the consequences for individuals, families, and society of dismantling the welfare state? The American welfare state, modest in comparison to Canada and the European welfare states, emerged in the 1930s as a reaction to the instability of the Great Depression and capitalism run amuck. Motivated by a fear of radical unrest by the disadvantaged and disaffected and the need to save capitalism from its own self-destructive tendencies (economic instability, rape of the environment, worker exploitation, lack of worker and consumer safety), the creators of the New Deal under Roosevelt and the Great Society under Johnson instituted Social Security, the minimum wage, federal aid to education, health programs, nutrition, subsidized housing, and other services. Although these programs are not as generous as found in the social democracies of Canada and Europe, they have worked. Since Lyndon Johnson's War on Poverty (because it was not an all out war, some have called it Johnson's Skirmish on Poverty) began, for example, the poverty rate has been reduced by 20 percent and the rate of elderly poverty cut by one-half.

22. Dismantling the Welfare State

The prevailing hostility of conservatives to the New Deal legacy of "big government" is based on two fundamental beliefs. First, they believe that the unequal distribution of economic rewards is none of the government's business. They value individualism and a market economy. Inequality is not evil; rather it is good because it motivates people to compete and it weeds out the weak. This laissez-faire approach, consistent with Social Darwinism, guarantees an exaggerated inequality and leads to what economists Robert Frank and Philip Cook have called a "winner-take-all" society.

The second fundamental belief of the conservative creed is that government efforts to reduce poverty and class inequality actually cause the very problems they seek to solve. Welfare dependency, in this view, is the source of poverty, illegitimacy, laziness, crime, unemployment, and other social pathologies. They agree with Charles Murray that only when poor people are confronted with a "sink or swim" world will they ever really develop the will and the skill to stay afloat.

These beliefs lead to the obvious solution — do away with the welfare state and the quicker the better. This leads to the following set of questions: Will dismantling the welfare state be beneficial or will it create chaos? Will reducing or eliminating the safety net to the economically disadvantaged save them or hurt them? Will it make society safer or more dangerous? Are we on the right track when we eliminate all AFDC payments to mothers under age 18? Will poverty be reduced when the pool of unskilled workers is expanded as welfare recipients lose their benefits after a specified period yet there is no effort to create jobs for them or to provide a reasonable minimum wage?

I believe the answers to these questions are self-evident. Society will be worse off rather than better off, as evidenced by a comparison of our society now with the more generous welfare states. The number of people on the economic margins will rise. Homelessness will increase. Family disruption will escalate. Crime rates will swell. Public safety will become much more problematic.

This phenomenon of economic inequality has implications for democracy, crime, and civil unrest. As economist Lester Thurow has asked: "How much inequality can a democracy take? The income gap in America is eroding the social contract. If the promise of a higher standard of living is limited to a few at the top, the rest of the citizenry, as history shows, is likely to grow disaffected, or worse."

Criminologists have shown that poverty, unemployment, and economic inequality are powerful determinants of street crime. Compare, for example, the homicide rate in the United States (9.4 per 100,000) to that of Sweden (1.3), Germany (1.2), and France (1.1). Or contrast the homicide rate in cities of comparable size:

— Chicago with 2.8 million residents had 930 homicides compared to Paris with 2.2 million, which had 88 murders.

— Baltimore with 736,000 inhabitants had 321 murders, compared to Amsterdam with 700,000 residents but only 47 homicides.

Or consider the differences in rape rates, where the number of rapes reported per 100,000 women for the latest period available was 118 for the United States, a rate nearly three times higher than found in Sweden and Denmark.

Another comparison between the United States and the generous welfare states is the incarceration rate. In 1993 the U. S. rate was 519 per 100,000 population, compared to Canada's 116, France's 84, Germany's 80, Sweden's 69, and the Netherlands' 49. Imprisoning so many, as we do, is very expensive, costing about $80,000 to build each cell, and about $25,000 per prisoner per year.

How do the welfare states of Canada and Europe differ from the United States on other dimensions? Most significantly, they are more generous. They each have a comprehensive, universal health care insurance system. They have a much more ample minimum wage. On average, they mandate that workers have a four week paid vacation. These social democracies provide pensions and nursing home care for the elderly. They have paid maternity (and in some cases paternity) leave. Education is free through college. Unemployment benefits are significantly higher than in the United States. These benefits are costly with income, inheritance, and sales taxes considerably higher than in the United States. The trade-off is that poverty is rare, street crime is relatively insignificant, the population feels relatively safe from crime and from the insecurities over income, illness, and old age. Most important, there is large middle and working class with a much stronger feeling of community, of social solidarity than found in the United States.

The United States, in sharp contrast, has the highest poverty rate by far among the industrialized countries, a withering bond among those of different social classes, a growing racial divide, and is moving rapidly toward a two-tiered society. The consequences of an extreme bi-polar society are seen in the following description by journalist James Fallows:

> If you had a million dollars, where would you want to live, Switzerland or the Philippines? Think about all the extra costs, monetary and otherwise, if you chose a vastly unequal country like the Philippines. Maybe you'd pay less in taxes, but you'd wind up shuttling between little fenced-in enclaves. You'd have private security guards. You'd socialize only in private clubs. You'd visit only private parks and beaches. Your kids would go to private schools. They'd study in private libraries.

The United States is not the Philippines but we are already seeing a dramatic rise in private schooling, home schooling, and in the number of walled and gated affluent neighborhood enclaves on the one hand and ever greater segregation of the poor and especially poor racial minorities in segregated and deteriorating neighborhoods and inferior schools, on the other. Personal safety is more and more problematic as the violent crime rates increase among the young. Finally, democracy is on the wane as more and more people opt out of the electoral process, presumably because, among other things, they are alienated and their votes do not count (consistently, the U. S. has the lowest voter turnout among the industrialized nations).

A PROGRESSIVE PLAN TO SOLVE SOCIETY'S SOCIAL PROBLEMS

What would be the planks of a progressive platform? Let me suggest some possibilities.

Eliminating Poverty:

1. Raise the minimum wage to an amount that would keep the worker above the poverty line. This would mean $7.60 for an urban family of four. Index this wage according to the rate of inflation and the methods used to establish the poverty line.

2. Secure the earned income tax credit for the near poor, which would allow them to keep more of their income and thus escape poverty.

4. POVERTY AND INEQUALITY

3. Institute a universal health care insurance plan for all residents.
4. Provide a national, universal child care subsidy.
5. Provide job training and a job. The government can generate jobs in child care, road building and maintenance, building mass transit, waste recycling, parks and national forests maintenance, cleaning up pollution, converting to renewable energy, environmental protection, assisting in schools and hospitals, building and renovating affordable housing, and the like.
6. Furnish compensatory education for preschool and school children from high risk situations.
7. End all redlining and other forms of discrimination by banks and insurance companies that artificially limit the opportunities for the poor and racial minorities.

Family Protection:
1. Paid maternity leave with guaranteed job retention.
2. Access to family planning information and technology, and subsidize birth control.
3. Subsidized child care for working parents, including school and recreation programs for older children to provide adult supervision after school for otherwise latchkey children.
4. Universal health care insurance that provides for prevention of as well as treatment of health problems.
5. Cost-free public education through college.
6. A social security system that provides adequately for retirement and a health care program that supplies the special health needs of the elderly, including home care and nursing home care.

Dealing with Corporate Downsizing: We are in a transition phase where the global economy and the computer chip revolution are causing severe dislocations in businesses, the communities where they are located, and for their workers. Government needs to help ensure the survival and enhance the dignity of downsized workers.
1. Insist that corporations give a sufficient advanced warning to communities and workers affected by downsizing or other corporate moves that have a negative impact.
2. In an economy of shifting employment, benefits (health care, pensions, job training) must be universal.
3. A government-corporate partnership to retrain workers. This happens in Germany, Japan, and Switzerland. Why not here?
4. A full employment policy. The displacement that has become endemic to American industry will be bearable only when other jobs are plentiful. Education and training are lauded as a big part of the solution, but education alone does not produce jobs. Thus, we need a government jobs program that provides work and decent pay while building the physical and social infrastructure of society.

Reduction of Extreme Income/Wealth Disparities. The aim here is not the elimination of wealth inequality but to reduce it. The high taxes in the generous welfare states, for example, still allow for very wealthy individuals and families. What they have done, contrary to our system, is to bring the wealthy down a bit and raise those at the bottom significantly.
1. Restore real progressivity to the tax system by closing loopholes for corporations and the wealthy, and with lower rates for the bottom and higher rates for the top.
2. Increase inheritance taxes.
3. End the discriminatory financing of public schools through private property taxes.

4. Take the power of money out of politics by public financing of local, state, and federal campaigns, free media for debate among candidates, and eliminating all political contributions.

IMPLEMENTATION OF THE PROPOSALS

Three questions remain: (1) Why should we adopt a progressive plan? (2) How do we pay for these programs? (3) Is there any hope to enact progressive solutions to social problems?

Rationale for Adopting a Progressive Plan

Why should we adopt a progressive plan to deal with our social problems? Foremost, these are serious problems and market solutions will not alleviate them. I am convinced that public policy based on abandoning the powerless will exacerbate their problems and societal problems.

A second reason to favor progressive solutions has to do with domestic security. We ignore the problems of poverty and wealth inequality at our own peril. If we continue on the present path of ignoring these problems or reducing or eliminating programs to deal with them, we will be less secure, we will have more problem people that require greater control, and, at an ever greater social and economic cost.

The final argument for a progressive attack on social problems is an ethical one. We need, in my view, to have a moral obligation to others, to our neighbors (broadly defined) and their children, to those unlike us as well as those similar to us, and to future generations. We need to restore a moral commitment to the safety net. As Jonathan Kozol, an expert on poverty, has argued: "There is something ethically embarrassing about resting a national agenda on the basis of sheer greed. It's more important in the long run, more true to the American character at its best, to lodge the argument in terms of simple justice."

Or consider this moral warning from an unlikely source, the very conservative British Chancellor of the Exchequer Kenneth Clarke, who explained his resistance to calls for a minimalist state: "This is a modern state. It is not the fifties, not southeast Asia. I believe in North American free-market economics, but I do not wish to see [here] the dereliction and decay of American cities and the absolute poverty of the American poor."

Financing the Progressive Plan

There are several sources of additional funds. The first is to reduce defense spending. We are the world's mightiest nation by far and there no longer is a Soviet threat, yet we maintain a defense budget of $260 billion that is more than is spent by all of our allies combined. Put another way, we are second in per capita military spending only to the nations of the Mideast. Our per capita defense spending is $1,110, compared to $750 for England, $427 for Canada, and $263 for Japan. We could easily decrease our annual defense spending by $100 billion or so without affecting our safety.

A second source of funds would be to reduce or eliminate corporate welfare and subsidies to the wealthy. At the moment, corporations receive $51 billion in direct subsidies and $53 billion in tax breaks. The wealthiest Americans pay lower tax rates and have more tax loopholes than found in any other modern nation. Annually we have about $400 billion in "tax expenditures" (i. e., money that is legally allowed to escape taxation). The economically advantaged receive most of these tax advantages. Representative George Miller of California has said that if corporations paid today the taxes that they did forty years ago, the deficit would vanish in a year.

Third, we should increase tax revenues. Totalling federal,

state, and local taxes equals 30 percent of the Gross Domestic Product. This is the lowest rate of any industrialized nations. In comparison, the English pay 36 percent, Germans and Canadians pay 37 percent, the French pay 44 percent, and at the high end the Swedes pay 56 percent.

There are also long-term savings that would accrue if a progressive plan were implemented. Research has shown, for example, that for every $1 invested in comprehensive childhood programs, $7 is saved in social costs, unemployment insurance, incarceration, and welfare. Just regarding crime, the costs, including incarceration, amount to $500 billion a year. Consider how much less on a per capita basis the Dutch, Swedes, or Germans pay for crime because of their progressive policies.

The Potential for a New Progressive Era

The crucial question — is there any hope of mounting a successful progressive program? At first the negative side seems overwhelming. Foremost is the fundamental belief in individualism. We celebrate individualism, which is the antithesis of cooperation, social solidarity, and acceptance of the redistribution of resources.

A second barrier to progressive policies is political. To begin, the majorities in the federal and state legislatures are political conservatives, which means they will opt for reducing or eliminating the welfare state rather than expanding these government programs. The political debates in these assemblies are from the right to the political center, with little, if any, voice from the political left. Another political barrier is that the two major parties are financed by big business and wealthy individuals. Also, the two-party system that has evolved in this country makes it structurally difficult for third parties to emerge as viable alternatives.

The social conservatives, who are about 30 percent of those who vote, are very organized. The conservative economic agenda seems assured by the generous contributions to both parties from the business community and wealthy individuals. At present, there are no countervailing pressures from the left on political parties and candidates.

The $5 trillion debt, much of which was generated during the Reagan and Bush presidencies, and the annual deficits are viewed by politicians of both parties as a giant weight on government that makes it difficult to fund existing programs, let alone institute new ones.

One of the necessary ingredients for a generous welfare state is the existence of a heavily unionized workforce. This condition is not present in the United States, as unions have declined significantly in membership since the 1960s.

The provisions of the Contract With America are exactly opposite position from the progressive view. The Republicans, fresh from their victories in the 1994 Congressional elections, where they became the majority in each house of Congress and won a number of governorships, claimed a mandate for their Contract With America. Most political pundits agree that they had such a mandate, but do they? Only 38 percent of the electorate voted in 1994 and only 52 percent of that three-eights or 19.8 percent of the public voted for the Republicans. Those who voted were disproportionately white, relatively affluent, and suburban. What about the poor and the near poor, racial minorities, bluecollar laborers, and city dwellers who chose not to vote? Why didn't they vote? What is the source of their apathy? Are they alienated from the political process because neither political party was speaking to their needs?

These questions lead to the possibility that the progressives might eventually prevail. Although this is unlikely, at least in the near term, there are some plausible arguments for this optimistic view. First, the current crop of Republicans may take what they consider a mandate and go too far with it, thereby alienating substantial portions of the citizenry. We must remind ourselves, however, that there is no guarantee that if the Republicans go too far, that things will only get better. This can occur, but so, too, can the situation steadily worsen.

Second, looking at the lessons from history, exactly 100 years ago the Progressive movement began as a reaction to unchecked capitalism, the robber barons, economic exploitation, and political corruption. Out of the Progressive Era came an activist government that addressed labor problems by instituting workplace safety regulations, prohibiting child labor, mandating the eight-hour workday, and providing disability compensation. The government broke up business monopolies, established a national parks system, and gave women the vote.

If the Republicans and the New Democrats today go too far and the marketplace replaces the welfare state completely, then it may lead to a further unraveling of social solidarity and a less secure society. In short, it may lead to a search for new answers — perhaps a new progressive era, just as it did 100 years ago.

Two necessary conditions for a successful welfare state are a strong union movement and a class-based labor party or Social Democratic party. While such a movement has been moribund for two decades or so in the United States, there is some evidence that unions are undergoing a shift toward more organizing and more ambitious and aggressive approaches against hostile employers and unfriendly laws. If the momentum accelerates, then there is hope for a labor renaissance and an organized push for progressive social policies.

A new progressive era will work only if a class-based political party emerges that addresses the needs of the masses. In particular, progressive leaders need to articulate a vision, a sense of direction, that builds a sense of community. Our future, we must recognize, depends on the welfare of all in the community in which we live, and the society of which we are a part, not just our personal accumulated wealth.

The alternative vision that I have proposed may seem very radical. I do not think so. Most of the suggestions are found in one form or another in each of the Western social democracies except the United States. Can we learn from these progressive societies? Should we learn from them? Can we afford a more generous welfare state? Can we afford not to adopt a more progressive plan?

Article 23

Welfare Fixers

Adam Wolfson

ADAM WOLFSON *is executive editor of the* Public Interest.

IN 1982, the journalist Ken Auletta defined *the* question of the underclass: how do we explain why "violence, arson, hostility, and welfare dependency rose during a time when unemployment dropped, official racial barriers were lowered, and government assistance to the poor escalated"?

Indeed, government spending on welfare increased from about $33 billion in 1964 to over $300 billion in 1992 (both figures in 1992 dollars). During the Reagan and Bush years alone, total welfare spending rose more than 50 percent. But all the while, rates of poverty, illegitimacy, non-work, crime, and family break-up got worse, not better. From 1965 to 1990, the illegitimacy rate for blacks rose from 28 to 65 percent, and for whites from 4 to 21 percent. Meanwhile, work among the poor plummeted, to the point where today only about 11 percent of poor households are headed by a full-time worker. For many, Aid to Families with Dependent Children (AFDC)—what most of us think of when we speak of welfare—has become a permanent condition, with over 50 percent of its recipients remaining on the rolls for over ten years.

One thing, however, has changed. Since 1935, when AFDC was first created, through President Lyndon Johnson's War on Poverty in the 1960's, to Bill Clinton's 1992 promise to "end welfare as we know it," welfare innovation and welfare reform were pretty much a Democratic affair. That is no longer the case. When conservative Republicans gained control of Congress in 1994, they also assumed a major share of responsibility for the nation's welfare system and those trapped in it.

How do they intend to proceed? As it happens, although most conservatives agree on the permanent need to end welfare as a federal entitlement, there have been three different and, to some extent, rival schools of thought about how to reform the system. All three have been incorporated in the Personal Responsibility and Work Opportunity Act, which formed the basis of the Republican welfare bill that President Clinton eventually vetoed this past January, and also in the many state plans now being put into effect by such Republican governors as Tommy Thompson of Wisconsin and John Engler of Michigan. The three approaches therefore bear scrutiny, for it is no exaggeration to say that the well-being of America's welfare population, and indeed of American society, depends upon the conceptual clarity with which we approach this long-festering problem.

THE MOST influential of the three schools is associated preeminently with the name of Charles Murray, and its guiding premise is that humans respond rationally to economic incentives. It is a tribute to the sheer rhetorical force and intellectual brilliance of Murray's extensive writings that, although conservatives often tend to resist mechanistic views of human nature, they have embraced this analysis almost without reservation. The most important parts of the Republican welfare bill, those dealing with "personal responsibility," are in fact based on Murray's logic. I am referring in particular to those sections which attempt to curb the high rates of family disintegration and out-of-wedlock births by the application of negative economic incentives. Under these provisions, states would be permitted (though not required) to deny cash assistance to children born out of wedlock to teenage mothers, and would also be permitted (though again not required) to deny additional cash assistance to mothers on welfare who continue to have more children.

Why, Senator Daniel P. Moynihan asked in connection with this aspect of the conservative reform effort, should children have to pay for the sins of their fathers (and mothers)? The answer is to be found in certain assumptions that were first spelled out by Murray over a decade ago in his now-classic book, *Losing Ground: American Social Policy 1950-1980*. The crucial passage appears midway through the book:

> It is not necessary to invoke the *Zeitgeist* of the 1960's, or changes in the work ethic, or racial differences, or the complexities of post-industrial economies, in order to explain . . . illegitimacy and welfare dependency. All were results that could have been predicted . . . from the changes that social policy made in the rewards and penalties, carrots and sticks, that govern human behavior. All were *rational responses* to changes in the rules of the game of surviving and getting ahead. [Emphasis added]

In other words, according to Murray, the welfare state has provided exactly the wrong incentives to the poor and the underclass by rewarding non-work, family dissolution, and out-of-wedlock births. It follows that if we change the rules of the game, behavior will change with it. Get rid of the economic supports (e.g., AFDC) that enable poor single mothers to support additional children, and they will eventually either abstain from sex, or use birth control, or (one supposes) have abortions.

There is much to Murray's argument. But implementing it might also entail more than the American people and their representatives are willing to swallow. The key to his rationalist approach is "the overriding threat, short-term and tangible." Here is how he describes the threat in a recent article on reducing illegitimacy:

> A major change in the behavior of young women and the adults in their lives will occur only when the prospect of having a child out of wedlock is once again so immediately, tangibly punishing that it overrides everything else. . . . Such a change will take place only when young people have it drummed into their heads from their earliest memories that having a baby without a husband entails awful consequences.

Murray relies heavily on a calculus of pleasure and pain in part because, as a libertarian, he sees no other way. Since government "does not have the right to prescribe how people shall live or to prevent women from having babies," it is left with no options for affecting people's lives other than the tax code. But there is also a deeper reason for Murray's reliance on what he labels "the technology of changing behavior." He thinks it the only effective means of training the human animal. Though he acknowledges the roles of religion and morality in forming people's sensibilities and attitudes, much of the force of these other agencies, he writes, has always been "underwritten by economics."

It is perhaps this oddly materialist version of human volition that has led some conservatives to look beyond Murray for solutions to the welfare problem. What if, they ask, gutting the welfare system does not have the desired effect forthwith? It will take a very resolute legislator indeed to go on applying negative incentives for as long as it takes. And even if we concede that negative incentives have their place in any plan of welfare reform, how can we expect young people to aspire to the roles of motherhood and fatherhood unless we offer a more elevated conception of these roles in their own terms?

Interestingly enough, Murray himself wrote the preface to a recent book, Marvin Olasky's *The Tragedy of American Compassion*, which embodies an alternative to the "technology" of behavior control. The book's legislative impact has thus far been slight, but its influence can be felt in measures that would authorize states to contract out their welfare services to private religious charities and to churches. Its stamp is also to be found on Republican efforts to restore civil society, like Senator Dan Coats's Project for American Renewal. The book has garnered the endorsements of such heavyweights as William J. Bennett and Newt Gingrich, and later this spring a more policy-oriented sequel will be published by the Free Press under the title *Renewing American Compassion*.

Though Olasky (who teaches at the University of Texas at Austin) agrees with Murray that we should scrap the current welfare system, his analysis of how we got where we are is quite different from Murray's and, correctly understood, leads down different paths. In fact, Olasky turns Murray's thesis on its head. Although he acknowledges the impact of economic incentives on people's behavior, in his view the underlying forces are spiritual and, broadly speaking, religious. Thus, according to Olasky, "the key change of the 1960's" was "not so much new benefit programs [Murray's claim] as a change in consciousness concerning established ones, with government officials approving and even advocating not only larger payouts but a war on shame."

To Olasky, American social-welfare policy has always reflected the dominant theology of the day. In the 18th and early 19th centuries, theology emphasized a merciful but just God and a sinful human nature that only God's grace could cure. This produced a hardheaded approach to social policy: aid to the poor was given in kind, but not in cash; charity, understood as "suffering with" the needy, was personal and paternalistic; material aid was considered secondary to, and dependent upon, saving souls; aid was for the "deserving," not the "undeserving," poor.

But this Calvinist theology lost out in the late 19th century to a universalistic, liberalized view that "emphasized God's love but not God's holiness," that jettisoned belief in original sin for a Rousseau-like belief in the natural goodness of man, and that essentially secularized a whole range of Christian beliefs. The effects on social policy were dramatic and devastating—and, in Olasky's opinion, completely predictable. The state took over the care of the poor, crowding out private charity. Shame and the work ethic were supplanted by the attitude that the poor have a constitutional right—that is, an entitlement—to welfare. Emphasis shifted from improving the spiritual conditions of the poor to improving their material conditions. As Owen Lovejoy, president of the National Conference of Social Work, put it in 1920, the goal would no longer be private salvation but rather the creation of "a divine order on earth as it is in heaven."

Olasky's history describes, in short, a descent, a fall from grace. As a nation, he claims sweepingly, we have been making war not on poverty but on God, and "the corruption is general." Therefore, although he too, like Murray, would tear down the welfare state, he does not expect any sudden alteration in behavior. Rather, he sees in the end of the welfare state an opportunity for private charities, and in particular private religious charities, to take over some of the responsibilities of caring for the poor, especially in the (for him) primary arena of their spiritual needs.

After all, writes Olasky, it was the federal government's entry into the welfare arena that "crowded out" private religious charities in the first place. Remove the government, and the charities will come surging back. Yet he is honest enough to admit that the historical record is not entirely clear on this point: which came first, the increasing involvement of professionals and the government in the lives of the poor, or a decline in voluntarism and religiosity? This is a crucial question, for if something in the culture led to a decline in voluntarism prior to the federal government's takeover of welfare, then a simple withdrawal of the latter will not necessarily lead to an increase in the former.

"In the end," predicts Olasky, "not much will be accomplished without a spiritual revival that transforms the everyday advice people give and receive, and the way we lead our lives." If that were really so, it would be reasonable to conclude that public-welfare programs should not be scrapped at all, but rather kept in place until the hoped-for spiritual revival occurs, lest the poor be left without God and without material support at once. Be that as it may, however, there is much else in Olasky's thinking, particularly about the role of private "compassion," that reformers can make use of in the months and years to come.

THIS BRINGS us to the third current. Unlike the first two, both of which see big government as the principal culprit in the welfare mess, this one envisions a role for government in its solution.

Perhaps the principal figure here is Lawrence Mead of New York University. In his book, *The New Politics of Poverty*, Mead argues, against Murray, that the marginal economic disincentives created by welfare do not explain the really staggering extent of non-work and family dissolution in the welfare population. Moreover, having a baby out of wedlock in order to receive a welfare check is not really "rational," in Mead's judgment. Rather, this and other aspects of the behavior of the underclass are the results of a certain personality profile. The non-working poor, says Mead, are defeatist, passive, and psychologically resistant to taking low-skilled jobs. A "culture of poverty" exists that cannot be fully explained by the rationalist model.

What to do? The answer, according to Mead, is workfare, an approach that would require able-bodied recipients of welfare to enter the labor market. By forcing the poor to be like the rest of us, workfare seeks to manage and even (in the words of Congressman Bill Archer) to "transform" them.

The thinking of Mead and others who favor workfare—Mickey Kaus of the *New Republic* is another well-known proponent of such schemes—is evident in the various versions of the Republican welfare-reform bill. All include the basic requirement that for any aid poor people receive from the government, they must work, in the private sphere if possible but in the public sector if not. According to the bill, 50 percent of welfare recipients must be working by 2002; even single mothers with children (over the age of one) should be required to work; and families receiving benefits will be cut off after five years.

Mead argues that workfare represents, in effect, a "new paternalism," a "tutelary regime." And indeed his ideas have alarmed more than a few conservatives, especially those of a libertarian bent. Many believe that any attempt by the government to mold behavior, even that of the poor, marks a break from the American tradition of limited government. Such fears are in Mead's view well-founded. But the appearance of the contemporary underclass itself marks, he believes, a watershed development in our national life, if not "the end . . . of an entire political tradition." That tradition—the tradition of the Founders, and of such classical liberals as Hobbes, Locke, and Montesquieu—"took self-reliance for granted." It assumed that people are, by nature, rational maximizers of their economic interests. But now it appears that many are not; and so a "new tradition," a "new political theory," even a "new political language" is needed.

All this seems somewhat overheated. For some reason, many of those who propose work as a solution to the welfare problem cannot resist militaristic metaphors. (Thus Mickey Kaus, in *The End of Equality*, urges Americans to build a "Work Ethic State.") But we need not really move beyond our own liberal tradition in order to enforce the norm of work. The Founders themselves recognized that humans are frequently irrational, indeed even lazy. And Adam Smith, the classical liberal *par excellence*, was not mincing words when he observed that among the "inferior ranks" of society there was a surfeit of "gross ignorance and stupidity." Rather than positing rational self-interest as a universal human trait, Smith and other classical liberals thought that through persuasion and law, it would be possible to turn men away from their former pursuits of military glory and religious enthusiasm toward "small savings and small gains." A little bit of workfare for those still unmindful of their economic self-interest thus need hardly spell the end of the American political tradition.

WHAT IS especially interesting about the three conservative strands of thought about welfare is that despite the theoretical differences among them, together they provide a coherent guide as to how to fix a broken system. As men are not angels, Charles Murray's negative incentives have their place. But neither are men brutes, and hence something more is needed than a "technology" of behavioral change. As Marvin Olasky reminds us, a rebirth of the spirit of religious charity would change many lives for the better. And as Lawrence Mead reminds us, in a commercial republic such as ours, work is the proper condition for all who are able.

Indeed, the politicians have seen the big picture in a way that is perhaps not so easy for the lone social thinker to do. The Republican welfare-reform bills in Congress, along with the many state plans being put into effect by Republican governors, make use of Murray's incentives, Olasky's religious charities, and Mead's workfare. If there are theoretical and practical difficulties with each of these approaches, it is precisely the combination that may make conservative welfare reform politically palatable and even, in the end, effective.

It's Not Working
Why Many Single Mothers Can't Work Their Way Out of Poverty

Chris Tilly and Randy Albelda

Most current welfare reform proposals assume that all single mothers can simply work their way out of poverty—that it's a matter of will. President Clinton's "two years and out" time limit on benefits before mandatory work, along with the renewed emphasis on job search and short-term training programs, arises from an increasingly common determination to have poor women lift themselves up by their bootstraps.

For many single mothers, this strategy cannot work. In the absence of universal child care, health care, and an abundance of good jobs, welfare plays a crucial role as a safety net. And even if the government were to offset the daunting demands of caring for children, provide health care, and brighten the limited opportunities at the bottom of the labor market, large numbers of single mothers would continue to require public assistance.

It's not for want of trying that single mothers have not been able to make ends meet. They work for pay about as many hours per year, on average, as other mothers: about 1,000 hours a year (a year-round, full-time job logs 2,000 hours). But less than full-time work for most women in this country just doesn't pay enough to feed mouths, make rent payments, and provide care for children while at work.

Not all single mothers are poor—but half of them are (compared to a 5% poverty rate for married couples). For poor single mothers, the labor market usually doesn't provide a ticket off of welfare or out of poverty. That's why AFDC (Aid to Families with Dependent Children, the program known as welfare) works like a revolving door for so many of them.

Heidi Hartmann and Roberta Spalter-Roth of the Institute for Women's Policy Research (IWPR) report that half of single mothers who spend any time on welfare during a two-year period also work for pay. But that work only generates about one-third of their families' incomes. In short, work is not enough; like other mothers, they "package" their income from three sources: work in the labor market, support from men or other family members, and government aid. "Mothers typically need at least two of those sources to survive," says Spalter-Roth.

THE TRIPLE WHAMMY

While all women, especially mothers, face barriers to employment with good wages and benefits, single mothers face a "triple whammy" that sharply limits what they can earn. Three factors—job discrimination against women, the time and money it takes to care for children, and the presence of only one adult—combine to make it nearly impossible for women to move off of welfare through work alone, without sufficient and stable supplemental income supports.

First, the average woman earns about two-thirds as much per hour as her male counterpart. Women who need to rely on AFDC earn even less, since they often have lower skills, less work experience and more physical disabilities than other women. Between 1984 and 1988, IWPR researchers found, welfare mothers who worked for pay averaged a disastrous $4.18 per hour. Welfare mothers with

Chris Tilly teaches public policy at UMass-Lowell. Randy Albelda teaches economics at UMass-Boston. Both are members of the D&S Collective.

jobs received employer-provided health benefits only one-quarter of the time. AFDC mothers are three times as likely as other women to work as maids, cashiers, nursing aides, child care workers, and waitresses — the lowest of the low-paid women's jobs.

Second, these families include kids. Like all mothers, single mothers have to deal with both greater demands on their time and larger financial demands — more "mouths to feed." A 1987 time-budget study found that the average time spent in household work for employed women with two or more children was 51 hours a week. Child care demands limit the time women can put into their jobs, and interrupt them with periodic crises, ranging from a sick child to a school's summer break. This takes its toll on both the amount and the quality of work many mothers can obtain. "There's a sad match between women's needs for a little flexibility and time, and the growth in contingent jobs, part-time jobs, jobs that don't last all year," comments Spalter-Roth. "That's the kind of jobs they're getting."

Finally, and unlike other mothers, single mothers have only one adult in the family to juggle child care and a job. Fewer adults means fewer opportunities for paid work. And while a single mother may receive child support from an absent father, she certainly cannot count on the consistent assistance—be it financial support or help with child care—that a resident father can provide.

No Room at the Bottom?

Suppose Clinton and company make good on their promise to give welfare mothers a quick shove into the labor market. What kind of prospects will they face there? Two-thirds of AFDC recipients hold no more than a high school diploma. The best way to tell how work requirements will work is to look at the women who already have the jobs that welfare recipients would be compelled to seek.

The news is not good. An unforgiving labor market, in recession and recovery alike, has hammered young, less-educated women, according to economists Jared Bernstein and Lawrence Mishel of the Economic Policy Institute, a Washington, D.C. think tank. Between 1979 and 1989, hourly wages plummeted for these women, falling most rapidly for African American women who didn't finish high school. This group's hourly wages, adjusted for inflation, fell 20% in that ten year period. Most young high-school-or-less women continued to lose during 1989-93. At the end of this losing streak, average hourly wages ranged from $5 an hour for younger high school dropouts to $8 an hour for older women with high school diplomas.

Unemployment rates in 1993 for most of these young women are stunning: 42% for black female high school dropouts aged 16-25, and 26% for their Latina counterparts.

But young women don't have a monopoly on labor market distress: workforce-wide hourly wages fell 14% between 1973 and 1993, after controlling for inflation. Given the collapse of wage rates, work simply is not enough to lift many families out of poverty. Two-thirds of all people living in poor families with children — 15 million Americans — lived in families *with a worker* in 1991, report Isaac Shapiro and Robert Greenstein of the Center on Budget and Policy Priorities. And 5.5 million of these people in poverty had a family member who worked *year-round, full-time.*

Reforms That Would Work

The problems of insufficient pay and time to raise children that face single-mother families — and indeed many families — go far beyond the welfare system. So the solution must be much more comprehensive than simply reforming that system. What we need is a set of thorough changes in the relations among work, family, and income. Some of the Clinton administration's proposals actually fit into this larger package, but these positive elements are for the most part buried in get-tough posturing and wishful thinking. Here's what's needed:

• *Provide supports for low-wage workers.* The two most important supports are universal health coverage — going down in flames in Congress at the time of this writing — and a universal child care plan. Two-thirds of welfare recipients leave the rolls within two years, but lack of health insurance and child care drive many of them back: over half of women who leave welfare to work come back to AFDC. A society that expects all able-bodied adults to work — regardless of the age of their children — should also be a society that socializes the costs of going to work, by offering programs to care for children of all ages.

• *Create jobs.* This item seems to have dropped off the national policy agenda. Deficit-phobia has hogtied any attempt at fiscal stimulus, and the Federal Reserve seems bent on stamping out growth in the name of preventing inflation. And yet Clinton and Congress could call for reform at the Fed, use government spending to boost job growth, and even invest in creating public service jobs.

• *Make work pay by changing taxes and government assistance.* Make it pay not only for women working their way off welfare, but for everybody at the low end of the labor market. Clinton's preferred tool for this has been the Earned Income Tax Credit (EITC) — which gives tax credits to low-wage workers with children (this tax provision now outspends AFDC). Although they get the EITC, women on welfare who work suffer a penalty that takes away nearly a dollar of the AFDC grant for every dollar earned. Making work pay would mean reducing or eliminating this penalty.

4. POVERTY AND INEQUALITY

•*Make work pay by shoring up wages and benefits.* To ensure that the private sector does its part, raise the minimum wage. A full-time, year-round minimum wage job pays less than the poverty income threshold for a family of one. Conservatives and the small business lobby will trot out the bogeyman of job destruction, but studies on the last minimum wage increase showed a zero or even positive effect on employment. Hiking the minimum wage does eliminate lousy jobs, but the greater purchasing power created by a higher wage floor generates roughly the same number of *better* jobs. In addition, mandate benefit parity for part-time, temporary, and subcontracted workers. This would close a loophole that a growing number of employers use to dodge fringe benefits.

•*Make a serious commitment to life-long education and training.* Education and training do help welfare recipients and other disadvantaged workers. But significant impacts depend on longer-term, intensive — and expensive — programs. We also need to expand training to a broader constituency, since training targeted only to the worst-off workers helps neither these workers, who get stigmatized in the eyes of employers, nor the remainder of the workforce, who get excluded. In Sweden, half the workforce takes some time off work for education in any given year.

•*Build flexibility into work.* "Increasingly," says Spalter-Roth, "all men and all women are workers *and* nurturers." Some unions have begun to bargain for the ability to move between full-time and part-time work, but in most workplaces changing hours means quitting a job and finding a new one. And though employees now have the right to unpaid family or medical leave, many can't afford to take time off. *Paid* leave would, of course, solve this problem. Failing that, temporary disability insurance (TDI) that is extended beyond disability situations to those facing a wide range of family needs could help. Five states (California, New York, New Jersey, Rhode Island, and Hawaii) currently run TDI systems funded by payroll taxes.

•*Mend the safety net, for times when earnings aren't enough.* Unemployment insurance has important gaps: low-wage earners receive even lower unemployment benefits, the long-term unemployed get cut off, new labor market entrants and re-entrants have no access to benefits, and in many states people seeking part-time work cannot collect. Closing these gaps would help welfare "packagers," as well as others at the low end of the labor market, to make ends meet. But even with all of these policies in place, there will be times when single mothers will either choose or be compelled to set aside paid work, sometimes for extended periods, to care for their families. For the foreseeable future, we still need Aid to Families with Dependent Children as a backstop. But at its current level, AFDC rarely acts as a safety net: Hartmann and Spalter-Roth found that AFDC recipients without significant earnings received incomes worth only two-thirds of the poverty line on average.

So welcome to reality. Most single mothers *cannot* work their way out of poverty — definitely not without supplemental support. There are many possible policy steps that could be taken to help them and other low-wage workers get the most out of an inhospitable labor market. But ultimately, old-fashioned welfare must remain part of the formula.

Resources: Heidi Hartmann and Roberta Spalter-Roth, "The real employment opportunities of women participating in AFDC: What the market can provide" (1993) and "Welfare that works: An assessment of the administration's welfare reform proposal" (1994), Institute for Women's Policy Research; Jared Bernstein and Lawrence Mishel, "Trends in the low-wage labor market and welfare reform: The constraints on making work pay," Economic Policy Institute 1994; Isaac Shapiro and Robert Greenstein, *Making Work Pay: The Unfinished Agenda,* Center on Budget and Policy Priorities 1993; Randy Albelda and Chris Tilly, *Glass Ceilings and Bottomless Pits: Women, Income, and Poverty in Massachusetts,* Women's Statewide Legislative Network (Massachusetts) 1994.

AID TO DEPENDENT CORPORATIONS

EXPOSING FEDERAL HANDOUTS TO THE WEALTHY

CHUCK COLLINS

In 1992 rancher J.R. Simplot of Grandview, Idaho paid the U.S. government $87,000 for grazing rights on federal lands, about one-quarter the rate charged by private landowners. Simplot's implicit subsidy from U.S. taxpayers, $261,000, would have covered the welfare costs of about 60 poor families. With a net worth exceeding $500 million, it's hard to argue that Simplot needed the money.

Since 1987, American Barrick Resources Corporation has pocketed $8.75 billion by extracting gold from a Nevada mine owned by the U.S. government. But Barrick has paid only minimal rent to the Department of the Interior. In 1992 Barrick's founder was rewarded for his business acumen with a $32 million annual salary.

Such discounts are only one form of corporate welfare, dubbed "wealthfare" by some activists, that U.S. taxpayers fund. At a time when Congress is attempting to slash or eliminate the meager benefits received by the poor, we are spending far more to subsidize wealthy corporations and individuals. Wealthfare comes in five main varieties: discounted user fees for public resources; direct grants; corporate tax reductions and loopholes; giveaways of publicly funded research and development (R&D) to private profit-making companies; and tax breaks for wealthy individuals.

Chuck Collins is the Co-Coordinator of the Share the Wealth Project, and works with the Tax Equity Alliance of Massachusetts.

Within the Clinton administration Secretary of Labor Robert Reich and Budget Director Alice Rivlin have attacked "welfare for the rich." Armed with a study from the Progressive Policy Institute, the Democratic Leadership Council's think tank, Reich floated the notion that over $200 billion in corporate welfare could be trimmed over the next five years. In a sign of the problems with our two-party system, Clinton discouraged Reich from taking this campaign further, for fear of alienating big Democratic Party funders.

TAX AVOIDANCE

The largest, yet most invisible, part of wealthfare is tax breaks for corporations and wealthy individuals. The federal Office of Management and Budget (OMB) estimates that these credits, deductions, and exemptions, called "tax expenditures," will cost $440 billion in fiscal 1996. This compares, for example, to the $16 billion annual federal cost of child support programs.

Due both to lower basic tax rates and to myriad loopholes, corporate taxes fell from one-third of total federal revenues in 1953 to less than 10% today (see "Disappearing Corporate Taxes," *Dollars & Sense*, July 1994). Were corporations paying as much tax now as they did in the 1950s, the government would take in another $250 billion a year — more than the entire budget deficit.

The tax code is riddled with tax breaks for the natural resource, construction, corporate agri-business, and financial industries. Some serve legitimate purposes, or did at

4. POVERTY AND INEQUALITY

one time. Others have been distorted to create tax shelters and perpetuate bad business practices. During the 1993 budget battle, New Jersey Senator Bill Bradley attacked the "loophole writing" industry in Washington, where inserting a single sentence into the tax laws can save millions, even billions, in taxes for a corporate client.

Depreciation on equipment and buildings, for example, is a legitimate business expense. But the "accelerated depreciation" rule allows corporations to take this deduction far faster than their assets are wearing out. This simply lets businesses make billions of dollars in untaxed profits. One estimate is that this loophole will cost $164 billion over the next five years.

One particularly generous tax break is the foreign tax credit, which allows U.S.-based multinational corporations to deduct from their U.S. taxes the income taxes they pay to other nations. Donald Barlett and James Steele, authors of *America: Who Really Pays the Taxes*, say that by 1990 this writeoff was worth $25 billion a year.

While in many cases this credit is a valid method of preventing double taxation on profits earned overseas, the oil companies have used it to avoid most of their U.S. tax obligations. Until 1950, Saudi Arabia had no income tax, but charged royalties on all oil taken from their wells. Such royalties are a payment for use of a natural resource. They are a standard business expense, payable *before* a corporation calculates the profits on which it will pay taxes.

These royalties were a major cost to ARAMCO, the oil consortium operating there (consisting of Exxon, Mobil, Chevron, and Texaco). But since royalties are not income taxes, they could not be used to reduce Exxon and friends' tax bills back home.

When King Saud decided to increase the royalty payments, ARAMCO convinced him to institute a corporate income tax and to substitute this for the royalties. The tax was a sham, since it applied only to ARAMCO, not to any other business in Saudi Arabia's relatively primitive economy. The result was that the oil companies avoided hundreds of millions of dollars in their American taxes. Eventually the other oil-producing nations, including Kuwait, Iraq, and Nigeria, followed suit, at huge cost to the U.S. Treasury.

In contrast to the ARAMCO problem, many corporate executive salaries should not be counted as deductible expenses. These salaries and bonuses are often so large today that they constitute disguised profits. Twenty years ago the average top executive made 34 times the wages of the firm's lowest paid workers. Today the ratio is 140 to one. The Hospital Corporation of America, for example, paid its chairman $127 million in 1992—$61,000 on hour! In 1993 the Clinton administration capped the deductibility of salaries at $1 million, but the law has several loopholes that allow for easy evasion.

CHICKEN MCNUGGETS AND OTHER VITAL MATTERS

Taxes are but one form of wealthfare. Subsidized use of public resources, as with J. R. Simplot's grazing and American Barrick's mining, is also widespread. Barrick's profit-making was allowed by the General Mining Law of 1872. Just last year the government finally put a one-year moratorium on this resource-raiding.

In a manner similar to the mining situation, the U.S. Forest Service under-charges timber companies for the logs they take from publicly-owned land. The Forest Service also builds roads and other infrastructure needed by the timber industry, investing $140 million last year.

Many corporations also receive direct payments from the federal government. The libertarian Cato Institute argues that every cabinet department "has become a conduit for government funding of private industry. Within some cabinet agencies, such as the U.S. Department of Agriculture and the Department of Commerce, almost every spending program underwrites private business."

Agriculture subsidies typically flow in greater quantities the larger is the recipient firm. Of the $1.4 billion in annual sugar price supports, for example, 40% of the money goes to the largest 1% of firms, with the largest ones receiving more than $1 million each.

The Agriculture Department also spends $110 million a year to help U.S. companies advertise abroad. In 1992 Sunkist Growers got $10 million, Gallo Wines $4.5 million, M&M/Mars $1.1 million, McDonalds $466,000 to promote Chicken McNuggets, and American Legend Fur Coats $1.2 million.

The Progressive Policy Institute estimates that taxpayers could save $114 billion over five years by eliminating or restricting such direct subsidies. Farm subsidies, for example, could be limited to only small farmers.

The government also pays for scientific research and development, then allows the benefits to be reaped by private firms. This occurs commonly in medical research. One product, the anti-cancer drug Taxol, cost the U.S. government $32 million to develop as part of a joint venture with private industry. But in the end the government gave its share to Bristol-Myers Squibb, which now charges cancer patients almost $1,000 for a three-week supply of the drug.

WHO IS ENTITLED?

Beyond corporate subsidies, the government also spends far more than necessary to help support the lifestyles of wealthy individuals. This largess pertains to several of the most expensive and popular "entitlements" in the federal budget, such as Social Security, Medicare, and the deduct-

25. Aid to Dependent Corporations

ibility of interest on home mortgages. As the current budget-cutting moves in Congress demonstrate, such universal programs have much greater political strength than do programs targetted solely at low-income households.

While this broad appeal is essential to maintain billions of dollars could be saved by restricting the degree to which the wealthy benefit from universal programs. If Social Security and Medicare payments were denied to just the richest 3% of households this would reduce federal spending by $30 to $40 billion a year—more than the total federal cost of food stamps.

Similarly, mortgage interest is currently deductible up to $1 million per home, justifying the term "mansion subsidy" for its use by the rich. The government could continue allowing everyone to use this deduction, but limit it to $250,000 per home. This would affect only the wealthiest 5% of Americans, but would save taxpayers $10 billion a year.

Progressive organizations have mounted a renewed focus on the myriad handouts to the corporate and individual rich. One effort is the Green Scissors coalition, an unusual alliance of environmental groups such as Friends of the Earth, and conservative taxcutters, such as the National Taxpayers Union. Last January Green Scissors proposed cutting $33 billion over the next ten years in subsidies that they contend are wasteful and environmentally damaging. These include boondoggle water projects, public land subsidies, highways, foreign aid projects, and agricultural programs.

Another new organization, Share the Wealth, is a coalition of labor, religious, and economic justice organizations. It recently launched the "Campaign for Wealth-Fare Reform," whose initial proposal targets over $35 billion in annual subsidies that benefit the wealthiest 3% of the population. The campaign rejects the term "corporate welfare" because it reinforces punitive anti-welfare sentiments. Welfare is something a humane society guarantees to people facing poverty, unemployment, low wages, and racism. "Wealthfare," in contrast, is the fees and subsidies extracted from the public by the wealthy and powerful—those who are least in need.

Today's Congress is not sympathetic to such arguments. But the blatant anti-poor, pro-corporate bias of the Republicans has already begun to awaken a dormant public consciousness. This will leave more openings, not less, for progressives to engage in public education around the true nature of government waste.

Resources: Green Scissors Report: Cutting Wasteful and Environmentally Harmful Spending and Subsidies, Friends of the Earth, 1995; *Killing the Sacred Cows,* Anne Crittendon, Penguin Books, 1993; *Aid to Dependent Corporations,* Janice C. Shields, Essential Information, 1994; *Cut-and-Invest to Compete and Win: A Budget Strategy for American Growth,* Robert Shapiro, Progressive Policy Institute, 1994; *America: Who Really Pays the Taxes,* Donald Barlett and James Steele, Simon and Schuster, 1994.

LET US COUNT THE WAYS

A few of the many subsidies received by the wealthy are:

- *The Mansion Subsidy.* Home mortgage interest is deductible up to $1 million per year. Reducing the limit to $250,000 would save the government $10 billion a year.

- *The Accelerated Depreciation Subsidy.* Companies get to depreciate their equipment much faster than it wears out. The cost: $32 billion a year.

- *The Advertising Subsidy.* Corporations fully deduct the cost of their advertising. If only one-fifth of advertising expenses were considered a capital cost of building brand name recognition, and so deductible gradually over time, taxpayers would save $3.5 billion a year.

- *McSubsidies.* $110 million a year goes directly to companies that advertise their products abroad. Beneficiaries include Sunkist, McDonalds, and M&M/Mars.

- *Wealthfare for Mining Companies.* The U.S. lets big mining companies pay peanuts for the use of federally-owned lands — our lands. An 8% royalty would earn $200 million a year.

- *Corporate Agri-Business Subsidy.* The federal government gives $200 million a year to corporate farms that each have incomes over $5 million a year.

Cultural Pluralism and Affirmative Action

America has been referred to as the melting pot of the world because individuals from radically different cultural and ethnic backgrounds have been melded into "Americans." This means that they have largely abandoned their histories, unique cultural heritage, and languages as they have acquired a common identity. Thus we have Irish Americans, Italian Americans, African Americans, and so on, with the focus being on their common identity as Americans, which helped minimize cultural pluralism.

Some argue that this lack of cultural pluralism promoted unity and is what made America strong. As a result, their argument goes, people were not restricted to specific geographic localities by race or ethnicity. Even the Civil War was not fought over cultural factors but over economic ones. The consequences of true cultural pluralism can clearly be seen in such locations as the former Soviet Union, South Africa, and the Balkans.

Other individuals bemoan the fact that Americans expect new migrants to become both assimilated and acculturated. Newly arrived immigrants, they argue, should be able to retain most of their ethnic differences without becoming second-class citizens. Diversity, they claim, is the spice of life and ethnic tolerance the sign of social maturity. But not all Americans have been able to realize the "American dream" equally. To level the playing field—so to speak—specific programs were instituted to reverse the consequences of blatant discrimination. These programs, to clarify their real intent, were called "affirmative action" and applied to those areas that were seen as major avenues to social equality, specifically education, employment, and government service. These programs were very effective in opening doors, keeping them open, and assuring that blatant discrimination was stopped. The major problem now is determining if the effects of historical discrimination have been not only addressed but corrected. If they have been corrected, should these programs be terminated? If they have not been corrected, where and how long should affirmative action continue to be employed?

In "Reclaiming the Vision: What Should We Do after Affirmative Action?" Constance Horner argues that "examining the source of discontent with affirmative action policies and practices will help not only to design sound replacement policies, but also to create and sustain an expectation of good faith in the deliberations about what to do next."

Shea Cunningham and Betsy Reed, in "Balancing Budgets on Women's Backs: The World Bank and the 104th U.S. Congress," document the fact that when nations are faced with serious economic problems, the impact of efforts designed to address these problems is not equally felt by all people. By examining economic data from many nations, they discovered that the most vulnerable members of society, women and children, are forced to bear the bulk of the economic burden associated with balancing budgets. According to Dinesh D'Souza, the most significant problem facing blacks in America today is not white racism but a culture that embraces violence, celebrates ignorance, and lacks a strong, vibrant leadership. These unpleasant facts represent what he labels "Black America's Moment of Truth," and they must be faced head on.

In "A Twofer's Lament," Yolanda Cruz reports on what it is like to be an unintended beneficiary of affirmative action. Cruz thought that she had been admitted to graduate school because of her academic credentials, only

UNIT 5

to be shocked to discover she was a "twofer," which means that she had been admitted because she fulfilled two affirmative action criteria—she was a woman in a male-dominated field, and she was also not white.

The "Crisis of Community: Make America Work for Americans," as William Raspberry calls it, has emerged because of the preoccupation of various groups with correcting wrongs and/or gaining advantage. Communities are being ripped apart because competing groups fail to distinguish between "enemies" (other competing groups) and the "problem." Thus, he says, we engage in a divisive struggle for advantage rather than attacking the problem jointly.

Looking Ahead: Challenge Questions

Just what is meant by the concept "cultural pluralism"? Is cultural pluralism a potentially divisive philosophy, a unifying situation, or an enriching phenomena? Explain your answer.

To what degree are attempts to "balance the budget" real? To what degree is debt a gender-sensitive problem?

Are black leaders focusing on the real problems facing their constituency? Why or why not?

To what degree is white racism the primary problem facing black Americans today?

Are affirmative action programs still needed today? Why or why not? Under what conditions should affirmative action programs be eliminated?

What are the conflicts in rights, values, obligations, and harms that underlie each of the issues covered in this unit?

Reclaiming the Vision

What should we do after affirmative action?

Constance Horner

Constance Horner is a guest scholar in the Brookings Governmental Studies program. A former head of the U.S. Office of Personnel Management, she is a member of the U.S. Civil Rights Commission.

The powerful moral vision that generated America's civil rights movement is on the brink of disintegration. Unless that vision—of a racially integrated society aspiring to equal justice and equal opportunity for black Americans—is reclaimed, the United States will, on the cusp of a new millennium, fail at a crucial political task it has assigned itself since the Civil War. It will also diminish its signal historical standing as creator of exemplary solutions to the deepest dilemmas of civic life.

That is why the current debate over affirmative action matters greatly. How America deals with the challenges posed by this issue and how we explain what we are doing will define our strongest commitments and ideals for our time, just as our words and deeds over the 45 years of the Cold War defined our commitment to political and economic liberty in that time. For there can be no mistaking that civic harmony among racial and ethnic groups is among the most salient global challenges of the next half-century.

With the irony that colors so much of human affairs, the 30-year series of public policy and judicial decisions we know as affirmative action has come to threaten the vision of a just and integrated society that gave birth to it. Yet in the minds of many black Americans, affirmative action is identified with a national commitment to their advancement.

There is little question that affirmative action will be modified, phased out, or even, under a cascade of Supreme Court decisions, state initiatives, and federal legislative action, abruptly terminated. Therefore it is vital to understand the full range of reasons for its rejection. Black Americans should not come to believe that the decision against it constitutes a rejection of the vision that brought it into being, and the country needs an understanding of the elements that would comprise an effective replacement for it.

What Next?

A democratic polity can change the means by which it achieves its ends, provided that it operates in good faith and gives those most vulnerable to democratic decisions, the minority, reason to believe in the majority's good faith. Examining the sources of discontent with affirmative action policies and practices will help not only to design sound replacement policies, but also to create and sustain an expectation of good faith in the deliberations about what to do next.

One of the dilemmas confronting people in public life critical of affirmative action has been the concern that calling it into question would be viewed by black Americans as, at best, indifference to their historic plight or, at worst, a contributor to resurgent racism. Some proponents of affirmative action have, over several decades, taken advantage of this concern to enforce a politically correct silence that has precluded the incremental correction of a public policy gone awry that is preservative of peaceful democratic change. Indeed, some of the explosive force of the current critique results from the unleashed resentment over this intimidation.

Supporters of affirmative action have put forth various economic, political, and psychological explanations for the burgeoning opposition to it. Opposition results, they say, from a generalized hostility ("white male rage") stemming from low wage growth and job loss, exacerbated by partisan Republicans inflaming a "wedge" issue for the next election, or from the flaring up of a permanent or "institutional" racism that can never be fully suppressed, only contained through political *force majeure*.

At root these arguments are premised on an expectation of bad faith and a presumption of economic determinism. As such they deny the strength and persistence in American culture of the premises on which the civil rights movement was based and flourished—a sense of fairness, a belief in racial integration, and a presumption that a civically activist polity will voluntarily (if slowly) make positive social change. Therefore, whatever their truth, these explanations are questionable and limited guides to constructing a sustainable next generation of efforts to increase equality of opportunity.

Moreover, they fail as full explanations for why the broad national revulsion toward the practices of affirmative action (including that felt by women and, to a lesser extent, blacks, its intended beneficiaries) is being expressed at this time and with such force.

Polls defining affirmative action as racial preferences show overwhelming white rejection (in the neighborhood of 75 percent) and an almost equal split for and against among blacks (46 percent for, 52 percent against in an April *Washington Post*–ABC national poll). An entirely separate set of polls taken in the same time frame suggests that an additional or different—and far more benign and hopeful—interpretation of the antipathy to affirmative action is available than the angry class- and race-based explanations advanced to date.

Antipathy toward "Big Government"

Almost 70 percent responding to one national poll indicated a belief that "the federal government controls too much of our daily lives." Another poll had two-thirds of Americans choosing "big government" as the country's gravest peril. In the light of these data and much more confirming their findings, it is hard to escape the observation that antipathy to big government and to affirmative action have emerged together as two of the most powerful political sentiments achieving national expression at this time. Although simultaneity does not demonstrate connection, it is at least suggestive of it.

The 30-year growth of affirmative action's regime of federal statute, regulation, judicial decision, and administrative practice, burgeoning well beyond its straightforward original purposes of nondiscrimination and equality of opportunity, not outcomes, has not occurred in a vacuum. It has developed simultaneously with and as part of the federal government's regulatory curtailment of private, discretionary, and voluntary action in many areas of American life—a curtailment compounded by federal support for social programs embodying and projecting values most Americans reject. The huge Republican victory in November likely reflected a rejection on both counts—a rejection of the degree of regulatory intrusion and a rejection of some of the values embodied in social welfare programs. It is very likely that rejection of affirmative action has been greatly intensified by its association with both, as well as by its provenance in the Democratic party—the "mommy party" for the "nanny state." If this is so, Americans may be viewing affirmative action as a regulatory structure to be dismantled more than a moral vision to be fulfilled. The moral vision was of a nondiscriminatory, integrated society; the regulatory structure, intending to integrate, now separates. The moral vision was democratically and openly implemented through legislation to affirm a commitment to equal opportunity and equal justice before the law; the regulatory structures are developed by fiat-oriented bureaucracies and by judges disdainful of the context of competing values and of the enormous vitality and variety of a culture bursting the bonds of regulatory structures and reverting to earlier, more clearly defined values in many areas.

26. Reclaiming the Vision

Examining the sources of discontent with affirmative action policies and practices will help not only to design sound replacement policies, but also to create and sustain an expectation of good faith in the deliberations about what to do next.

OMB Directive 15

To see these conflicts in play, one need only look at what is happening to a little-known but powerful directive governing the federal government's collection of racial statistics. Promulgated in 1977, Office of Management and Budget Directive 15, "Race and Ethnic Standards for Federal Statistics and Administrative Reporting," governs the categories the Bureau of the Census may use in assessing the country's racial composition. The racial statistics developed through its categories serve as the basis for enforcement of voting and other civil rights by the Department of Justice, the Equal Employment Opportunity Commission, and the civil rights offices of other federal agencies. Billions of dollars of federal spending are targeted for women and racial and ethnic minorities on the basis of these statistics. (Even Small Business Administration set-aside programs directed to "socially and economically disadvantaged" individuals and institutions are *de facto* allocated largely by race and ethnicity because administrative practice and law "presume" that certain racial and ethnic groups are "disadvantaged.") Allocation of not inconsiderable political power, through the drawing of congressional district lines, is based on their collection. Determinations of "adverse impact" in private-sector hiring practices rest on a racial count of an area's labor pool, and lawsuits may follow.

Virtually every arena of activity is affected by these categories. According to a Congressional Research Service report, "targeted funding, in various forms, and minority or disadvantaged set-asides or preferences have been included in major authorization or appropriation measures for agriculture, communications,

5. CULTURAL PLURALISM AND AFFIRMATIVE ACTION

defense, education, public works, transportation, foreign relations, energy and water development, banking, scientific research and space exploration, and other purposes." The distribution of a great deal of public and private money and a considerable amount of political power relies on the racial and ethnic numbers produced under Directive 15.

Currently, people being counted by the census under the categories of OMB Directive 15 are asked to identify themselves racially as white, black, Asian-Pacific Islander, American Indian–Alaskan native, or "other." If "other" is selected, the census taker is expected to "reclassify" that person into one of the four groups on the basis of appearance. (People are also asked an "ethnic" question as to whether they are Hispanic.) These tidy boxes constructed by the federal government bear, as is obvious, very little relationship to the racial and ethnic variety of the country. The OMB, as a result, is agonizing over whether to add a new category, "multiracial." So far it has been unable to do so (except in limited testing). Ironically, and much to the point of the country's discontent with government's affirmative action structures, the use of racial categorization is imposing a powerful set of incentives for those receiving benefits to remain distinctly within their categories.

Thus a governmental structure whose original intent was to overturn a regime of segregation has been transformed into an apparatus supporting its return. Policies growing out of a national commitment to

---

The static regulatory vision of racial America is empirically outmoded. It is also antithetical to the long-term interests of black Americans.

---

racial integration and designed to facilitate integration instead entitle and empower on the basis of separation. To do so, moreover, these policies must deny the reality of an increasingly racially mixed society. Indeed, the government's difficulty in changing OMB Directive 15 must call into question the degree of confidence the public may repose in its commitment to the ideal of racial integration. It surely must at least raise a suspicion that a tolerance (if not a preference) for separatism has infiltrated and is delegitimizing a significant underpinning of affirmative action.

Counterproductive Racial Pigeonholes

Meanwhile, American social vitality defies and discredits the official racial demarcations. In spite of the stresses and strains of historic antagonisms, racial and ethnic groups continue to integrate, even in the most intimate realms of marriage and family. Although the last state anti-miscegenation laws were struck down as recently as 1967, according to the latest census data more than 4 percent of blacks, and 6 percent of black men, are married to nonblacks. Ten percent of black men aged 25–34 have entered interracial marriages, mostly with white women. Thirty percent of Hispanics are married to non-Hispanics. Transracial adoption, strongly discouraged by most state and municipal governments, has nonetheless continued to grow. A provision in the Republican "Contract with America" denies federal funds to agencies that discriminate on the basis of race in child placement. The U.S. Department of Health and Human Services, long an opponent of transracial adoption, has recently, and grudgingly, yielded to public anger over adoptable black children languishing in foster care and issued guidelines that no longer actively discourage such adoptions.

Rita Simon, an academic sociologist who studied transracial adoption for several decades, was quoted in *USA Today* as believing that "Where we come down on transracial adoption should tell us what we really think about integration and separation." She reports polling data indicating that 70 percent of blacks and whites support such adoption. Syndicated columnist Ellen Goodman writes of children like multiracial golf star Tiger Woods, who is of black, Thai, Chinese, and American Indian origins, as affording America a racial "demilitarized zone" and a bridge among the races.

But interracial marriage and transracial adoption are small indicators of the extent to which government's racial and ethnic categories fail to comport with reality, compared with the consequences of the great wave of immigration since those categories were devised. Immigration now accounts for 37 percent of national population growth. More than one million legal and illegal immigrants enter the United States every year—in absolute numbers an historic peak—with the number of countries sending immigrants rising from 21 in 1970 to 27 in 1980 to 41 in 1990. More than 150 languages are spoken in the United States, and Americans claim almost 300 racial and ethnic groups. There is even an intellectual attack on the existence of race as a scientifically reliable concept among geneticists and physical anthropologists. Stanford genetics professor Luigi Cavalli-Sforza believes that classifying by race is a "futile exercise." In a February 13 *Newsweek* poll, one-third of American blacks said that blacks should not be considered a single race.

It would be naive to view these trends as suggesting that the country has, only 30 years after the end of legally sanctioned racial segregation, reached the point where there is no further need for the interventions of government to forestall or punish continuing acts of racial discrimination and to help expand opportunity for advancement. But it is also naive to ignore the belief of opponents of affirmative action that the federal government has far exceeded the bounds of common sense in the structures it has devised to ensure equal opportunity and that indeed many of those structures, like OMB Directive 15, now impede racial integration and advancement by their overreach and by their disconnection from a changing social reality.

Betraying the Vision of an Integrated Society

The static regulatory vision of racial America is empirically outmoded. It is also antithetical to the long-term interests of blacks. Drawing voting district lines on the basis of race and other mechanisms such as those proposed by Lani Guinier, for example, furthers separation by race on so crucial an act of citizenship as voting; it trades a short-term electoral reward for the creation of the idea of permanent, irreducible, separate interests based on race, hardly a vision of an integrated society. Maintaining important permanent structures based on race posits permanent separation and therefore economically, politically, and intellectually counterproductive isolation.

It is simply not possible to reduce the salience of race by enhancing its salience. The widespread application of lower standards for academic admissions, for example, has drawn ill-prepared students in over their heads and created, by this artificially contrived mismatch of student and school, the impression of black intellectual inferiority, a reactivation of the racial stereotype most dangerous to black advancement. These admission policies have allowed white-governed institutions to "feel better about themselves" at the expense of the full development of black intellectual potential. They have produced a college dropout rate for blacks of almost two-thirds. They have led to the spectacle of a university president, however unintentionally, questioning the "genetic, hereditary" capacities of black students. They have led to speculation by the chair of the U.S. Commission on Civil Rights in congressional testimony that if tests determined college admissions and entry-level jobs, "Asians and Jewish Americans would hold the best jobs everywhere and populate almost entirely the best colleges and universities." Similar policies have led municipal police and fire departments to hire, in the 1980s, underqualified recruits who were not able to achieve promotion in the '90s, thereby creating for blacks the appearance of racist promotion policies and for whites the appearance of black incapacity. When public and private institutions have denied or obscured the practices of these admissions and hiring policies, from a concern that the beneficiaries would be embarrassed or a fear that the policies would fail of public support in the sunshine, they have created a cynical distrust about their fairness or good sense that has contributed mightily to the strength of the opposition to affirmative action.

A New Beginning

Now, uphill, good faith must be reclaimed. It would be a tragic failure of American civic genius, and a great cruelty, not to take this opportunity to think the issue of equality of opportunity anew. Along with other 30- to 50-year-old structures of government, affirmative action is crumbling under the pressure of change. But if the regulatory structure on which affirmative action relies is, at best, counterproductive to the accomplishment of the purposes informing the civil rights movement and the antidiscrimination legislation that grew out of it, what then is to be done?

New approaches should align minority interests with ascendant and longstanding American values, so they will be both powerful and sustainable. What might such approaches look like?

First, public policy should promote racial integration, not only at work and at school, but also in the home. Anything weaker will allow separatists and racists, white and black, the wedge they need to indulge the fantasy of a more "comfortable" life, which is socially, if not legally, separate. The moral vision that galvanized national support for the civil rights movement was not "separate but equal."

Second, intentional racial discrimination of the sort the Civil Rights Act of 1964 had in mind should be powerfully stigmatized and punished. The law is a teaching instrument. The moral fuzziness of affirmative action has confused this teaching and undercut it.

Third, where intellectual attainment is properly the basis for decisions, as in some but not all entry-level hiring, promotions, and academic admissions, standards should be applied nonracially with no *de facto* "race norming." That implies the probability of some near-term decline in numbers of blacks in exchange for strengthened confidence in those who, in time, achieve in equal numbers. At the same time, there must be a revitalized commitment to dramatic improvement in the quality of education. Since, for the foreseeable future, almost all students will be in the public education system, its reform must be the focus of attention. There is general agreement on much of what needs to be done. Schools should run year-round. Teachers should be hired competitively and paid accordingly. Black mayors and city councils should stop using schools as jobs programs and be willing to hire the best teachers nationally, regardless of race. Curricula should be basic, tough, challenging, and fad-free. Parents with vouchers might accomplish these and other goals.

For 20 years, public agencies have devoted extraordinary resources to assembling racially representative workforces. The next generation of policies should be clearly non-racial and announced as such. Practical ways to bring race-based hiring to an end should be developed and implemented in ways that assure continuing public confidence in public processes.

Fourth, federal regulation—fiat—to assure outcomes should generally be replaced by greater room for discretion to offer opportunity. That will transform resentful but minimal "compliance" that is taken out of the hides of "beneficiaries" in other ways, into voluntary moral acts of good citizenship and common sense. It's not naive—it's human nature. People resent losing the opportunity to do the right thing freely.

Fifth, tortured attempts to substitute additional economic for racial entitlements should be dropped; as thinly disguised generic redistributionist policies, they are entirely contrary to the thinking of the times, administratively nightmarish, and therefore not at all likely to work.

5. CULTURAL PLURALISM AND AFFIRMATIVE ACTION

Sixth, the bourgeois practices that help poor people improve their circumstances—study, work, saving, marriage, child-bearing, in that order—should be preached, not dismissed or ridiculed as they have been since the 1960s, and rewarded in the design of public policies.

However Long It Takes

Finally, American leaders in every sector and of all races should make the same kind of commitment to racial integration they made after World War II to fighting and winning the Cold War. Affirmative action arose from an impatience that the consequences of several hundred years of slavery and discrimination could not be overcome quickly. For many, those consequences have been largely overcome. For many others, they have not. How long the effort takes is less important than that it be headed in the right direction.

New directions, embodied in values and practices that have worked for America before, may help solve the problems of race in ways that can be sustained. If sustained, they will provide a model for other societies facing worse divisions and internal conflict.

BALANCING BUDGETS ON WOMEN'S BACKS

THE WORLD BANK AND THE 104TH U.S. CONGRESS

SHEA CUNNINGHAM AND BETSY REED

Debt has become the last word in U.S. political debate. Like a magic wand, its very mention conjures up images of welfare moms reclining, their feet propped up in front of TV talk shows, unkempt babies howling. Meanwhile, the real budget-busters — items such as nuclear submarines, corporate subsidies and tons of political "pork" — have escaped the Republicans' fiscal scalpel.

This dangerous phenomenon, blaming national debts on society's most obviously vulnerable members, can be spotted also in economically developing countries, where it was imported by Western financial interests. In the name of reducing debt, "Structural Adjustment" programs (SAPs) administered by the World Bank and International Monetary Fund strip away important social services. Like the Republicans' "Contract with America" in the United States, SAPs target those groups least able to absorb the austerity necessary to make a dent in national debts. Both the Contract and SAPs have an especially severe impact on women, particularly mothers, who often work double-days but earn less than men. Alternatives, such as state-funded welfare in the United States, are disappearing.

SAPs usher women into industries mostly devoted to producing exports, which typically offer long hours, low wages, and trying conditions. The Contract, key provisions of which have won President Clinton's blessing, would have American women flipping hamburgers or taking care of other people's children for poverty-level wages, rather than going to school or looking after their own kids. Because both SAPs and U.S. welfare reform aim to rein in today's social spending, neither supports any investment in a different future.

SAPS AND WOMEN

The World Bank, in its publications, stresses the importance of investing in women to ensure that they benefit from development. From World Bank literature one might conclude that it is a leader in poverty alleviation and equitable, sustainable development. The Bank collects excellent data on many facets of poverty and gender inequality. In fact, the rhetoric in the Banks' slick publications mirrors that of non-governmental organizations working to reform or abolish the Bank and the International Monetary Fund (IMF).

While the Bank claims to be a champion of poor women, it denies that there are clear links between SAPs and the explosion of women's poverty. It ignores that the social costs of structural adjustment continue to be absorbed primarily by females. Since the budget-slashing, pro-business SAPs hurt the poor, and, according to the 1995 UN Development Report, 70% of the world's 1.3 billion people living in poverty are female, women have suffered disproportionately from their austere measures.

Yet few World Bank projects explicitly consider women's issues. According to an unpublished 1994 internal Bank document, only 615 of the Bank's 5,000 projects included "gender components." Of the Bank's 7,200 staff, only 257 have received formal training on gender issues.

Shea Cunningham is a research associate and administrative manager at Focus on the Global South, a new Bangkok-based organization, and co-author of Dark Victory: The United States, Structural Adjustment and Global Poverty *(1994: Food First Books, Oakland CA). Betsy Reed is a D&S editor.*

5. CULTURAL PLURALISM AND AFFIRMATIVE ACTION

The Bank's new strategy, "the gender and development approach," looks better on its face than it turns out to be; since it incorporates gender issues into other areas for staff development, women are no longer treated as a target group deserving of separate attention and assistance.

In practice, SAPs discriminate against women largely because they fail to take women's unpaid labor into account. When SAP policies foster export industries that depend on women working overtime for low wages, the women's work at home must still get done. In the end, many women wind up working double days, or taking their daughters out of school to help them at home. Eighteen-hour work days have become increasingly common for women in much of the developing world. And, according to several studies, when SAP-induced social spending cuts result in recessions, women work more intensively than men to supplement their incomes.

SAPs also disregard the fact that women often take up the slack when the government cuts social services. Reducing social spending often means eroding mechanisms to ease women's burdens, such as state-subsidized child support and care, health care, and educational programs for women and children. In Tanzania, annual per capita health care spending fell from $7 in 1980 to $2 in 1990 after the introduction of a SAP. Often, such cuts not only make women work harder to care for their families, but women also suffer a disproportionate share of the health consequences themselves. The World Health Organization reports that maternal mortality rates are increasing across East, Central and West Africa. In Zimbabwe, after the World Bank introduced medical "user fees," maternal mortality rose from 90 per 100,000 live births in 1990 to 168 per 100,000 in 1993!

In addition, SAPs impoverish women with their bias against small-scale agriculture and businesses, sectors in which women are concentrated, in favor of export-oriented manufacturing and agriculture and trade in services. SAP-induced increases in the prices of farming equipment and other such agricultural inputs have forced many women farmers, who are often unable to get credit because of their gender, out of business. Such women either turn to work in export industries, or they enter the informal sector, hawking products on the street where hours are long, earnings low and job security almost non-existent.

Women who work in the public sector are especially vulnerable to SAP spending cuts. When the government makes cuts in its own ranks to deal with its debts,

UNDERSTANDING STRUCTURAL ADJUSTMENT

The U.S.-led World Bank and International Monetary Fund (IMF), both founded in 1944, were set up to reconstruct and regulate the world economy in the wake of World War II. Today these institutions lend money to economically developing countries with the supposed goals of reducing their debts and poverty and stimulating local economies.

Structural Adjustment is a euphemism for an economic program of wrenching change in economically developing countries, a program to which the countries must submit in order to receive badly-needed loans. The 1982 world debt crisis marked the initiation of SAPs. Strapped with unmanageable debts to private western banks, economically developing countries were facing the cutoff of external funds that they needed to service their debts. With no alternatives, these countries had to accept structural adjustment loans with conditions that would, according to the World Bank and IMF, promote growth, stabilize external accounts, and reduce poverty.

Yet SAP packages have failed to reduce either debt or poverty (see "Reign of Error: The World Bank's Wrongs," *D&S*, September/October 1994). Reminiscent of Reagan-era "trickle-down" economic policies that slash taxes on the rich and social spending in order to promote growth, SAPs generally demand cutbacks in government spending; cutbacks in or containment of wages; privatization of state enterprises and deregulation of the economy. They also focus on developing export industries, with the intent of raising foreign exchange earnings and paying off debt. But by the late 1980s both the IMF and the International Bank for Reconstruction and Development became net recipients of financial resources from sub-Saharan Africa; that is, these countries gave more to the industrialized world than they received.

Overall, the debt burden of economically developing countries escalated from $785 billion in 1982 to over $1.7 trillion in 1995.

women tend to be fired first. According to the U.S. State Department's 1995 Human Rights Survey, women account for 60% of those dismissed from state enterprises, mainly because those enterprises seek to shed their maternity and child-care costs. In Zambia, for example, when a SAP implemented in 1990 required that the government fire 72,000 workers, mostly women lost their jobs.

Among children, it is females again who bear a disproportionate share of the burden. As women must extend their work day, all children are often neglected, but school attendance among girls in particular tends to drop as mothers depend on their daughters to ease their own time constraints. According to a 1990 UN report, 33.6% of the world female population is illiterate compared to 19.4% of the male population.

> IN ZIMBABWE, AFTER THE WORLD BANK INTRODUCED MEDICAL "USER FEES," MATERNAL MORTALITY ROSE FROM 90 PER 100,000 LIVE BIRTHS IN 1990 TO 168 PER 100,000 IN 1993.

When girls are educated, the positive social effects are tremendous: birth rates drop and children live longer, healthier lives. Studies have also linked the education of women to higher rates of economic growth. But in order for girls in poor countries to become equally educated, primary and secondary education must be free for both sexes. Simultaneously, women need to have their work burdens eased so they do not depend on their daughters to help them with household maintenance.

THE DEMANDS OF EXPORT-ORIENTED INDUSTRIALIZATION

SAPs have failed miserably to achieve two of their stated objectives—reducing poverty and debt in developing countries (see "Understanding Structural Adjustment"). But they have had much success in another—making developing economies more export-oriented. Over the past few decades, multinational corporations have relocated much manufacturing from industrialized to developing countries. And the integration of developing countries into the global market has had a profound impact on women's labor force participation. On average, the female labor force has been growing twice as fast as the male labor force. There has been a particularly rapid increase in women working in the export-oriented manufacturing sector, where the rate of growth of global industrial output has been the fastest.

The World Bank unequivocally views this rise as a success of export-oriented industrialization, and hence a success of structural adjustment. But is it beneficial for women? Studies indicate that these labor markets reinforce women's inequality. As a worldwide average, women are paid nearly 40% less than men for the very same work. Employers in the export sector routinely admit that they view women workers as docile, inexpensive, and easily replaceable. The female labor force is highly sensitive to economic fluctuation. When economies grow at fast rates, more women are employed, but when economies grow slowly, less women *vis-a-vis* men are employed. So, whether SAPs create new employment for women at low wages — as they have in Asia and some parts of Latin America, where female employment is rapidly rising — or destroy jobs, as they have in Africa, women continue to be worse off than men.

U.S. WELFARE WOES

While SAPs condemn women to bleak jobs that corporations want to fill abroad, both President Clinton's welfare plan and the Contract with America can be seen as thinly veiled attempts to cater to a rising demand among U.S. businesses for low-wage service workers. Since the early 1970s, America has been undergoing wrenching labor market restructuring. So far, this has meant the loss of good manufacturing jobs — mainly to economically developing countries — and the continued increase in demand for a flexible, temporary, poorly paid service workforce. At this low end of the labor market, the ravaging of the social safety net with little concern about education and training assures a ready supply of candidates for these jobs, candidates with no other choice. As welfare checks shrink, employers can offer ever-lower wages and still have a steady stream of applicants.

While Clinton, backed up by Labor Secretary Robert Reich, has emphasized his support for education and training, after a 2-year grace period the administration has also favored "workfare" for welfare recipients, which forces them to work for their checks and prevents them from obtaining the skills to get good jobs. Following his lead, several Republican governors, including Massachusetts'

5. CULTURAL PLURALISM AND AFFIRMATIVE ACTION

Bill Weld and Wisconsin's Tommy Thompson, have already put such plans in place.

Increasingly, the fate of poor women and children hinges on the whims of governors, as the federal government is withdrawing from the business of insulating families from the free market altogether. In September of this year, Clinton gave a nod of approval to the Senate's welfare-ending plan, which eliminates the status of welfare as a universal "entitlement" funded according to need. The Senate plan would also require half of all welfare recipients to take government-provided jobs in exchange for their benefits. Governors interested in more draconian measures would have room to experiment, but liberal state leaders who want to expand assistance will be effectively hamstrung by the drop in federal funding. Trends in U.S. welfare reform reflect some of the same assumptions underlying SAPs: that child-rearing isn't work, and that women who do it should also take a paid job.

Even the Earned Income Tax Credit (EITC), a mechanism to "make work pay" for welfare recipients and the working poor, is on the chopping block. Clinton has stood by the EITC in the past, as it is the cornerstone of his welfare philosophy: any work is better than welfare. The EITC indeed provides relief to the working poor at a critical time, but it is a compromise measure, quintessentially Clinton: Instead of pushing for a hike in the minimum wage,* which would make *business* pay for the cost of supporting its work force, the EITC asks *government* to subsidize employers' use of low wage labor. The fact that the Republicans want to eliminate this measure at the same time that they shred the safety net, pushing women with children into jobs that can't possibly support a family, is nearly unfathomable. They justify it, again, in terms of debt. House Republicans put the EITC cut into a bill devoted to controlling the deficit by closing "tax loopholes" and cutting spending. Though this bill includes some overdue cuts in "corporate welfare," it saves $23.3 billion by eviscerating the EITC, out of a $38.8 billion total.

These so-called "tough" and simplistic approaches to a complex social issue are inherently flawed. Welfare

[*Editor's Note: In August 1996, Clinton signed a law to increase the minimum wage from $4.25 to $5.15 in two installments—50 cents on October 1, 1996, and 40 cents on September 1, 1997.]

> U.S. WELFARE REFORM CAN BE SEEN AS A THINLY VEILED ATTEMPT TO MEET A RISING DEMAND AMONG U.S. BUSINESSES FOR LOW-WAGE SERVICE WORKERS.

reform advocates recognize the system's glaring failures. But without addressing the root causes of women on welfare—racial and gender discrimination, a substandard educational system, lack of economic alternatives and non-existent childcare—the problem will only get worse.

... AND IT DOESN'T EVEN DEAL WITH DEBT

The type of policymaking behind the "Contract with America" and "Structural Adjustment" places debt reduction in front of basic social needs, especially women's needs. Based on belief in a gender-neutral free market, both of these austere economic packages increase women's hardships relative to men's. An unfortunate but inevitable result of debt servicing? Hardly. In the United States and abroad, debt relief is nowhere in sight. After years of human sacrifice, the U.S. debt now has a ceiling of $4.9 trillion. The bulk of this accumulated during the Reagan years, under an economic program remarkably similar to that of the Republicans today—fat contracts for military contractors, tax cuts for big corporate campaign donors and prosperous individuals. Abroad, the debts of most "structurally adjusted" countries are at least twice as large as they were before the World Bank stepped in with its grand plans for paying them off. And so, to carry their countries further down this fruitless and worn-out path, millions of women continue to shoulder the bulk of the burden. Unlike these myopic policies, debt is gender-sensitive.

Black America's Moment of Truth

The problem facing black America today is not white racism or GOP budget cuts. It's a culture that embraces violence and celebrates ignorance—and a leadership that doesn't have the courage to say so.

Dinesh D'Souza

Dinesh D'Souza, the John M. Olin fellow at the American Enterprise Institute, is the author of The End of Racism *(Free Press), from which this article is adapted.*

The last few decades have witnessed nothing less than a breakdown of civilization within the African-American community. Vital institutions such as the small business, the church, and the family are now greatly weakened; in some areas, they are on the verge of collapsing altogether. And the symptoms of systemic decline are both numerous and ominous—extremely high rates of criminal activity, the normalization of illegitimacy, a preponderance of single-parent families, high levels of drug and alcohol addiction, a parasitic reliance on government provision, hostility to academic achievement, and a scarcity of independent enterprises. The next generation of young blacks is especially vulnerable. "We are in danger of becoming superfluous people in this society," says African-American scholar Anthony Walton. "We are not essential or even integral to the economy." Marian Wright Edelman of the Children's Defense Fund puts it more bluntly: "We have a black child crisis worse than any since slavery."

This crisis did not exist a generation ago. In 1960, 78 percent of all black families were headed by married couples; today that figure is less than 40 percent. In the 1950s black crime rates, while higher than those for whites, were vastly lower than they are today. These figures suggest that the dire circumstances of the black community are not the result of genes or racism. The black gene pool has not changed substantially since mid-century, and racism then was far worse. The main problem facing African Americans is that they have developed a culture that represents an adaptation to past circumstances, but one that is now, in crucial respects, dysfunctional and pathological.

Yet, ever since Daniel Patrick Moynihan issued his infamous 1965 report on the black family, mainstream black scholars and civil rights activists have been quick to attribute public notice of African-American pathology to white racism. When blacks have publicly engaged in criticism of those pathologies, they have often as not found themselves castigated and reviled.

This has been especially true for black conservatives. Michael Williams, an African American who promoted race-neutral scholarships as a senior official in the Bush administration, was denounced by Spike Lee as an Uncle Tom who deserved to be "dragged into the alley and beaten with a Louisville Slugger." Commenting on the writings of Thomas Sowell, columnist Carl Rowan of the *Washington Post* observed that "Sowell is giving aid and comfort to those who . . . are taking food out of the mouths of black children. Vidkun Quisling in his collaboration with the Nazis surely did not do as much damage." Historian John Henrik Clarke goes even further. "Black conservatives," he says, "are really frustrated slaves crawling back to the plantation."

The vehemence and ferocity of this rhetoric is unrivaled in any other area of contemporary debate. For many scholars and activists, whites who criticize black pathologies are racist, and blacks who do so are mouthpieces for white bigots, craven apostates and black impersonators who deserve ostracism, if not physical punishment. "The leadership," remarks Robert Woodson, "has successfully imposed a gag rule on the black community."

Activists denounce criticism of black pathologies as a callous form of "blaming the victim." Yet if the problems endured by African Americans today are substantially the result of cultural pathologies on the part of blacks, these individuals would not be victims but perpetrators. Since the civil rights movement led us to the view that all races are equal in their natural endowments, inequalities between groups have been considered to be the unnatural product of historical and continuing oppression. In this view blacks are portrayed as living largely involuntary lives, wholly manipulated by the structures of visible and invisible racism. It thus becomes unfair to criticize the culture of the failing

5. CULTURAL PLURALISM AND AFFIRMATIVE ACTION

group, because to do so is to avert one's gaze from the conduct of the oppressor. Moreover, such criticism is said to reinforce the oppressor's self-serving racist perception of the victim group as somehow inferior. Black self-criticism becomes a form of self-hate.

Of course, no one is to blame for being a victim. But if, as a reaction to being victimized, a group develops dysfunctional or destructive patterns of behavior that perpetuate a vicious cycle of poverty, dependency, and violence, then continuing to inveigh against the oppressor cannot offer the victim much relief. As African-American scholar Milton Morris puts it, "When black people slaughter other black people on the street, they all come back to: look what the white people made us do." In dealing with pathologies such as black-on-black crime, it may be the victim who is in the best position to address the problem, even though the victim was not primarily responsible for causing it.

For all their regional and economic diversity, blacks in America do share a culture, one that emerged out of the crucible of racism and oppression. Blacks who came from diverse tribes, speaking different languages and practicing varied ways of life, forged a distinct African-American identity in the new country—an identity that was solidified under segregation, when blacks were involuntarily united by the one-drop rule. Although black culture contains elements of Southern rural culture, modern urban culture, and lower-class culture, it has fused these elements into a distinct amalgam.

Some scholars view black culture as largely, if not wholly, imposed from the outside. These scholars are structural determinists: Like Marxists, they tend to view culture as the epiphenomenal product of external forces such as poverty and oppression. William Julius Wilson argues that "cultural values grow out of specific circumstances and life changes and reflect one's position in the class structure." Wilson is right to see ghetto pathologies partly as a response to lack of opportunity. Yet he goes too far in asserting that "cultural values do not determine behavior or success." Culture is not simply an expression of external circumstances; it is also a powerful instrument in shaping those circumstances. Even under the extreme deprivations of slavery, segregation, and racism, black culture was never entirely controlled from without; it was also generated from within.

Blacks in America seem to have developed what some scholars term an "oppositional culture" which is based on a comprehensive rejection of the white man's worldview. African-American psychiatrist Alvin Poussaint articulates a view popular among African-American leaders: "For blacks to mindlessly strive to be like the white middle class in a white racist, capitalist, exploitative society is without question detrimental to the cause of black people." It is not hard to see how this oppositional culture would develop in the black community, which has historically suffered so much at the hands of whites. Until recently, historian Eugene Genovese writes, "there was hardly any room for blacks at the top or in the middle who tried to play by the rules of bourgeois society." Under slavery it made sense to do as little work as possible, because slaves received no share in the master's profits. In some cases even extreme forms of violence constituted a reasonable adaptation, because they were the only way to preserve dignity. But, as Genovese says, "what constituted strategies for survival under one set of circumstances have now emerged, in an entirely different context, as celebrations of self-indulgence and irresponsibility."

Today, that oppositional culture has become an obstacle that prevents blacks from taking advantage of rights and opportunities that have multiplied in a new social environment. Even many middle-class and upper-middle-class blacks routinely permit blackness to be defined by the underclass. Middle-class behavior by African Americans is seen as inauthentic, while lower-class behavior is seen as genuinely black—a reversal of what W.E.B. Du Bois had in mind with his Talented Tenth. While most immigrant groups tend to look to their most successful citizens for emulation and self-definition, on many issues the moral tone in the black community is set from below. The underclass exercises such a powerful influence in shaping black cultural norms because this group is perceived as the most oppositional of all: It is remote from, and organized in resistance to, white middle-class standards.

Booker T. Washington warned almost a century ago of the existence of black pathologies that both strengthened white racism and inhibited black development. Those pathologies have existed in the black community since slavery, but they have been restricted and contained both by white-imposed discipline and black-imposed norms enforced by churches and local community institutions. But those institutions have been greatly weakened since the 1960s, and in a new environment of social permissiveness and government subsidy, black pathologies have proliferated. Today they pose a serious threat to the survival of blacks as a group, as well as to the safety and integrity of the larger society.

The vicious, self-defeating, and repellent nature of these pathologies can no longer be rationalized or ignored. Yet, by dressing these pathologies in sociological cant, complete with the familiar vocabulary of disadvantage and holding society to account, this is precisely what leading black scholars and activists do. Still, the bottom line remains: Society must do its part, and blacks must do theirs. But first, the specific nature of the civilizational crisis facing the black community must be recognized. This crisis points to deficiencies not of biology but of culture; and they are deficiencies that need to be corrected.

Racism as an Excuse
The first dysfunctional aspect of black culture is racial paranoia—a reflexive tendency to blame racism for every failure, even those that are intensely personal. For society as

a whole, promiscuous charges of racism are dangerous because they undermine the credibility of the charge and make it more difficult to identify actual racists. For blacks, the risk of exaggerated charges is that they divert attention from the possibilities of the present and the future. Excessive charges of racism set up a battle with an adversary that sometimes does not exist. Consequently, blacks are in a struggle that they always lose. Racism itself becomes a scapegoat: It is blamed for problems it has little or nothing to do with, such as blacks performing poorly on math tests.

As Eugene Genovese says, "What constituted strategies for survival under one set of circumstances have now emerged, in an entirely different context, as celebrations of self-indulgence and irresponsibility."

The displacement of personal or group problems onto the bugaboo of racism inspires a frenetic assault on society at large. But since racism is not the problem, the assault proves futile. The problem endures, and the frustration mounts. Racism remains the culprit, now accused of having taken an even subtler and more insidious shape. In May 1994 Louis Farrakhan asserted that white racism was responsible for black people killing other blacks. The reason that whites do not stop black-on-black violence, he added, was that whites need the organ donations. "When you're killing each other, they can't wait for you to die," Farrakhan told an audience in Toledo, Ohio. "You've become good for parts."

Farrakhan's intimations of a homicidal white plot are not unique. A 1990 poll of black Americans found that 60 percent thought it was true or possibly true that the government was deliberately encouraging drug use among African Americans, and 29 percent suspected that scientists deliberately created AIDS in order to decimate the black population. "Many of us are convinced there is a conspiracy to anesthetize and ultimately do away with as many blacks in American society as possible," the Rev. Cecil Williams of San Francisco's Glide United Methodist Church told *Newsweek*. "This is genocide, 1990s style."

According to the recent book *I Heard It Through the Grapevine*, blacks are remarkably susceptible to rumors that there is a racist plot to destroy them. Among the false reports that have received widespread circulation and serious attention in the African-American community:

• The Ku Klux Klan owns Church's Fried Chicken, which uses ingredients that make black men sterile.

• The FBI was responsible for the killing of twenty-eight black children in Atlanta as part of an experiment in the genocidal elimination of all blacks in America.

• Popular fruit juices such as Tropical Fantasy are secretly manufactured by white supremacist groups, once again to reduce black sexual potency.

Nothing elicits more cries of racism, however, than the criminal justice system. The Rodney King beating, according to black scholar William Mandel, means that "Negroes are lynched in America." A columnist for the *Los Angeles Sentinel*, the largest African-American newspaper on the West Coast, offered the King verdict as further proof that "the United States is on the verge of becoming a police state, if it is not there already." The author noted "many similarities between present-day USA and early stages of the Nazi Third Reich in prewar Germany."

Although African Americans routinely receive racial preferences in universities and the work force, this does not seem to have inhibited imaginative accusations of racism where none seems evident. When black poet Amiri Baraka was denied tenure at Rutgers University in 1990, he accused fellow professors and administrators of being Nazis and Ku Klux Klansmen in disguise. "We must unmask these powerful Klansmen," Baraka said. "These enemies of people's democracy and Pan-American culture must not be allowed to prevail. Their intellectual presence makes a stink across the campus like the corpses of rotting Nazis."

In *Racism: American Style*, Dempsey Travis accuses heads of major companies of being "Klansmen in pin-striped suits." He compares various job rejection letters that one of his friends received, and is struck by the ominous similarity of their tone and language. Travis writes:

> The letters of rejection are so similar in tone and content that they raise the specter of corporate officers from across the country gathered in back rooms deciding on a kinder and gentler language for rejecting black applicants. Such gatherings could be called a conspiracy and it is undeniable that such schemes are in place.

The Rage of the Privileged Class

A second dysfunctional aspect of black culture, a feature mainly of middle-class African-American life, is a rage that threatens to erupt in an orgy of destruction or self-destruction. When middle-class blacks found their opportunities severely restricted under segregation, black rage was either submerged or nonexistent. But now that middle-class blacks find themselves on the receiving end of racial preferences and government set-asides, many are beside themselves with anger. "I do not believe that we can restore and expand the freedoms that our lives require," June Jordan writes, "until we embrace the justice of our rage."

This rage is not so difficult to comprehend. It represents post-affirmative action angst, the frustration of pursuing unearned privileges and then bristling when they do not bring something that has to be earned—the respect of one's peers. Moreover, black rage arises out of the recognition that, even

5. CULTURAL PLURALISM AND AFFIRMATIVE ACTION

as middle-class activists deplore the pathologies of the underclass, by moving out of inner-city neighborhoods they have sometimes contributed to those pathologies, and are dependent on their continuation for the race-based privileges they enjoy.

The black cultural orientation toward government solutions is understandable. While many whites have traditionally viewed the state as a potential threat to personal liberty and opportunity, government has been the leading protector and guarantor of black rights and freedom.

The effect of this molten frustration is what William Grier and Price Cobbs, in their book *Black Rage,* term "a posture close to paranoid thinking and mental disorder." In *Living With Racism,* Joe Feagin and Melvin Sikes document the state of mind that seems to paralyze many middle-class blacks. The book is intended to spotlight white racism; instead, it provides illuminating insight into some blacks' precarious grip on reality:

> I come in here and scream! I talk to my friends. I come in here and talk to my assistant. She's even seen me cry because I'm so angry until I am to the point of violence. But I know that I have to really, really be cognizant of what I'm doing, because why go to jail for nothing?

> On a scale of one to ten, my level of anger is a ten. Mine has had time to grow over the years more and more until I now feel that my grasp on handling myself is tenuous. I think that now I would strike out to the point of killing, and not think about it. I really wouldn't care.

> One step from suicide! The psychological warfare games that we have to play every day just to survive. It's a mental health problem. It's a wonder we haven't all gone out and killed somebody or killed ourselves.

These are the observations of relatively well-placed men and women: an executive, a government worker, and a college professor. Their rage is utterly disproportionate to any injustices that they claim to have suffered. Indeed, even the sympathetic listener begins to suspect that we are dealing with cases of people who live in a world of make-believe, in mental prisons of their own construction. Antiracist militancy is carried to the point of virtual mental instability.

It is hard to imagine whites feeling secure working with such persons: Surely such inflamed ethnic sensitivities are not what companies have in mind when they extol the diversity of work environments. Yet if these individuals are cranks, they are in respectable company. Leading African-American writers and scholars seem to share their persecution complex and its attendant rage. Here is a fairly typical incident, described by legal scholar Patricia Williams:

> A man with whom I used to work once told me that I made too much of my race. "After all," he said, "I don't even think of you as black." Yet sometime later, when another black woman became engaged in an ultimately unsuccessful tenure battle, he confided to me that he wished the school could find more blacks like me.

It is not hard to envision the scene: A white colleague, probably eager to please, tells Williams that he tries to judge her by her ability rather than her skin color. Apparently confronted on another occasion with an affirmative-action candidate seeking a tenured appointment, the colleague expresses to Williams the hope that the university will find more qualified African-American scholars like herself. Yet here is how Williams reacts:

> I felt myself slip in and out of a shadow, as I become non-black for the purposes of inclusion and black for the purposes of exclusion. I felt the boundaries of my very body manipulated, casually inscribed by definitional demarcations that did not refer to me.

Williams is not finished. Reflecting on similar cases in a kind of stream-of-consciousness rhetoric, she eventually explodes:

> I am afraid of being alien and suspect.... My rage feels dangerous, full of physical violence, like something that will get me arrested. All this impermissible danger floats around in me, boiling, exhausting. I can't kill and I can't teach everyone. So I protect myself. I don't deal with other people if I can help it. I don't risk exposing myself to the rage that will get me arrested.

We have to remind ourselves that this is a highly paid professor at a distinguished university, and one of the leading black female scholars in the nation. Yet her sentiments are shared by others. "I'm a volunteer slave," announces Jill Nelson, an African American, upon being hired as a reporter for the *Washington Post.* "My price? A house, a Volvo, and the illusion of disposable income." This seems to be the kind of servitude that most Americans of any background would settle for. Nelson, however, calls herself a "race woman": She seems both uncomfortable and angry about being a successful African American working in a mainstream profession. She eventually resigned from the *Post* after being suspended for forging an editor's initials on one of her expense reports.

In his autobiography *Parallel Time, New York Times* writer Brent Staples describes how, as a graduate student, he spent his time stalking white people in order to scare them—relishing a game that he, to his great amusement, calls "Scatter the Pigeons":

28. Black America's Moment of Truth

> I became expert in the language of fear. Couples locked arms or reached for each other's hand when they saw me. Some crossed to the other side of the street. People who were carrying on conversations went mute and stared straight ahead, as though avoiding my eyes would save them. A few steps beyond them I stopped and howled with laughter... I felt a surge of power: these people were mine: I could do with them as I wished. If I'd been younger, with less to lose, I'd have robbed them, and it would have been easy.

Recalling these incidents many years later, Staples might be expected to have some regrets about his bizarre youthful behavior, perhaps to offer critical analysis of what was wrong with the situation and his response to it. Yet his perspective on his stalking days seems exactly the same today as it was then. Despite his training as a student of psychology, Staples's recollection is devoid of analytical self-consciousness. He is blind to the immaturity and strangeness of his conduct. The conclusion is hard to avoid: One of America's leading black intellectuals gets kicks out of playing hoodlum, and then proceeds to accuse whites of racism for engaging in what, under the circumstances, can only be termed a rational avoidance of blacks like him.

Government as the Big House
Another destructive stance that seems deeply ingrained in African Americans, especially those in the middle class, is a heavy dependence on government, accompanied by the belief that public programs are the way to solve virtually all social problems.

Today a large fraction of middle-class blacks work for the government. Although blacks make up 10 percent of the civilian work force, about 24 percent of blacks (compared with 14 percent of whites) are employed by the federal, state, and local governments. State and local agencies that service a poor and predominantly black clientele, such as housing and welfare, employ substantial proportions of African Americans: 38 and 23 percent respectively. According to sociologist Bart Landry, about half of black professional males and two-thirds of black professional females work for some arm of the state. In addition, 50 percent of blacks (more than 15 million people) live in households that receive some form of welfare, compared with 19 percent of whites. Blacks are between two and five times more likely than whites to receive means-tested cash assistance and food stamps, and to live in subsidized or public housing. In short, much of the black community is parasitic on government for its basic livelihood.

According to a recent survey, 67 percent of blacks believe that the government, rather than private business or individuals, has the greatest responsibility for creating jobs. Poor and middle-class blacks alike shared this belief. By contrast, whites were far less likely to look to government for jobs; even among the poorest whites, less than 50 percent cited the government's responsibility. Sixty-eight percent of blacks, compared with 36 percent of whites, insisted that increases in government spending were the way to invigorate the economy and provide employment. A typical comment came from Della Simmons, who told the *Wall Street Journal:* "Employment is the big issue the government needs to be dealing with."

Meanwhile, blacks have done very poorly in the one area that is a rapid source of jobs and social mobility for other groups: small business. According to U.S. government data, in 1987 blacks, who make up 12 percent of the population, owned about 420,000 businesses in America, with receipts of $19 billion. Asians, who make up 3 percent of the population, owned 350,000 businesses with receipts of $33 billion. African Americans currently start and run less than 3 percent of the nation's businesses, and take in less than 1 percent of the nation's gross receipts. Black enterprise is so fragile that a large portion of its receipts come from the government; many black businesses would collapse instantly if they were taken off government contracting preferences and set-asides. Only about 4 percent of African-American spending each year goes to black-owned enterprises.

The black cultural orientation toward government solutions is understandable; while many whites have traditionally viewed the state as a potential threat to personal liberty and opportunity, government has been the leading protector and guarantor of black rights and freedom. It was federal intervention in the 1860s that freed the slaves. The basic right of blacks to equal protection of the laws was recognized in the Fourteenth Amendment to the Constitution, which was somewhat incongruous with the rest of the document in that it was passed to expand federal power. In the twentieth century, black leaders found themselves compelled to turn to the federal government to combat state-sanctioned segregation and private discrimination. Government was recruited as an ally out of necessity, not preference.

Today many black scholars affirm their faith in political struggle and state provision. James Farmer of the Congress for Racial Equality wrote, "We might think of the demonstration as a rite of initiation in which the black man is mustered into the sacred order of freedom." Civil rights activist Julian Bond says that blacks are much more likely than whites to view government as "a positive helpful force," for the simple reason that they "have seen government make an enormous difference in their lives. Government eliminated discrimination in the ballot box, government ended segregation."

But the contemporary problems of blacks are less susceptible to a federal solution. It is difficult to envision how the government can keep black families intact, or force mothers to monitor their children's study habits, or counteract the strong peer appeal of juvenile gangs. None of these problems can be addressed by marching to Selma or signing more legislation. Black reliance on government, once justified, may now have become a liability. While blacks detour

5. CULTURAL PLURALISM AND AFFIRMATIVE ACTION

themselves into the racism industry, other ethnic groups are selling cars and computer software, and moving to the suburbs.

Economist John Kasarda argues that many Asians establish small businesses in order to overcome the obstacles that newcomers to this country face: problems of language and access to credit. Kasarda points out that economic solidarity helps Koreans to keep capital circulating within their community. "This is what used to happen in the black community during segregation," Kasarda says. "A single dollar would turn over five or six times, because it would be spent on goods and services provided by other blacks." Ironically this points to the fact that, used correctly, ethnocentrism can be a business asset. People from a homogeneous group who trust each other can avoid the legal and bureaucratic structures that are necessary to adjudicate the transactions of strangers. Groups that succeed through the private sector don't need many civil rights leaders, only entrepreneurs.

Yet, although blacks as a group have an annual income that exceeds the gross national product of many industrialized nations, very little of this money is spent on products sold by black-owned business. Despite a few modestly successful "buy black" campaigns, African Americans famous for their political solidarity have to date shown no comparable sense of economic solidarity. Despite the opportunities provided by entrepreneurship, and the limitations of government in solving social problems, many black leaders continue to counsel resistance as a means to secure greater transfers of wealth from the public treasury to the African-American community. Political scientist Manning Marable defines "the contemporary challenge" as calling for "a renaissance of black militancy" in order to generate "new organizations for collective resistance." Congressman John Lewis offers advice that is more moderate but just as futile: "We have an obligation to organize and mobilize the African-American community like never before. We must continue to push, continue to agitate, continue to get the government to say yes when it may have a desire to say no."

Government, of course, has a legitimate role in providing needed services and a safety net. But the degree of black reliance on government is dysfunctional because it prevents many African Americans from being "spurred by necessity," as John Kasarda puts it, into the risks and rewards of entrepreneurship. That protection from cradle to grave comes at the cost of self-reliance and private initiatives that offer better long-term rewards and, perhaps more important, an abiding sense of personal freedom.

It's a White Thing—You Wouldn't Understand
A further dysfunctional feature of black culture is its repudiation of standard English and academic achievement as forms of "acting white." Resistance to "acting white" is widespread among the African-American underclass, but it also seems to exist for many middle-class blacks. The problem must be viewed in the context of the educational crisis facing black America: A large proportion of the African-American population is illiterate, many adults cannot read beyond the fourth-grade level, the performance of blacks in school lags behind that of whites on every measure of performance, and only about one-third of blacks who go to college graduate within six years.

In the face of this crisis, one might expect a vigorous campaign to improve vocabulary skills, clarity of expression, and educational standards among African-Americans. Instead many students seem to have adopted, with the approval of their elders and civil rights leaders, a hostile stance toward the values of the white world—including the values of scholarship and study. Among some blacks, "getting ignorant" is considered a virtue and a source of self-esteem. Several studies have shown, contrary to popular wisdom, that the self-esteem of young black males is in fact higher than that of any other group.

Skepticism toward the values of whites may be a cultural stance that served blacks well in the past: It was a technique for refusing to be defined by the categories of the oppressor. White values have been historically identified with an assertion of Western cultural superiority, which provided a justification for racism. Blacks in the past who did try to assimilate white values often found themselves rebuffed and scorned. Now an attitude of rejection seems to have set in, and values identified as white are spurned by many blacks.

Proviso West is a racially mixed school outside of Chicago at which the grades and test scores of black students fall consistently behind those of white students. A reporter for the *New York Times* found an atmosphere of ignorance: Herbert Hoover was identified as "the vacuum guy," and Lloyd George was confused with "Boy George." The caliber of writing submitted by African-American students is conveyed by the following excerpts: "The pepel from wen Martin Luther King Jr. Lived and did not wrly get along." "It was three boys with some mast up hair coming into my house. I cute their hair, because they need it." Black students at the school, the *Times* reported, avoid the company of whites and accuse blacks who are on honors tracks of being "nerds" and "sellouts." They are frequently taunted, "Why are you in a class with all the white kids?" and "Why are you using a white man's book?" Laura Banks, a senior, complains, "We're not accepted by the white people because they think we're not smart enough. We're not accepted by black people because they think we're too smart."

Nomathombi Martin, a 19-year-old student at the University of California at Berkeley, recently recalled the intense pressures he encountered in high school to avoid speech and behavior identified as white. "I got a lot of criticism about speaking proper speech," he says. "I don't speak street talk and I never did. When I try I sound really weird. I don't pronounce words wrong like *wif, birfday, bafroom,* and things like that. They would say: Why do you talk so proper? Why do you talk like you're white?"

Many teachers acknowledge the pressures that African-American students fall under—peer pressure not to succeed, but to fail. Marc Elrich, who teaches fourth grade in suburban Maryland, writes that black students treat fellow students of color who try and succeed as white "wannabes." Such wannabes invite amusement and contempt, according to Elrich, "because they pretend to be what they can't be."

Perhaps one of the clearest ways for black students to distance themselves from whites is to communicate with each other in the distinctive lingo of Black English. Scholars estimate that around 80 percent of African Americans use some form of black dialect. There is little doubt, as linguists point out, that the ghetto idiom does possess, like most forms of communication, its own coherent form and structure. Yet it seems equally clear that if African-American students wish to succeed in American society, they cannot do it with Black English.

In her intriguing study *Twice as Less,* Eleanor Wilson Orr argues that, because of its use of unorthodox terms and tenses, Black English hampers African-American students in their logical and mathematical reasoning, because relationships of time, place, and number become garbled. Yet June Jordan exults, "If it's wrong in standard English, it's probably right in Black English." And a recent article in a black studies journal declared, "Whites should not become reference points for how black children are to speak and behave. Black children's encounter with the white world should be filtered through a black frame of reference, which includes the use of Black English."

Among some blacks, "getting ignorant" is considered a virtue and a source of self-esteem. Several studies have shown, contrary to popular wisdom, that the self-esteem of young black males is in fact higher than that of any other group.

Cult of the "Bad Nigger"
Violence has now become a tragic defining feature of life in the black underclass. African Americans in this group seem divided into two factions: perpetrators and potential victims. The violence unleashed by blacks seems to have reached a point where it threatens the future of the African-American community and the stability of society as a whole. Black males are about twice as likely as white males to be victims of robbery, theft, and aggravated assault, and seven times more likely to be victims of murder. The life expectancy of black men in central Harlem is shorter than that of men in Bangladesh. In the District of Columbia, black residents are more likely to be killed than are people in war-torn regions such as Northern Ireland and the Middle East. Partly as a consequence of black crime, American crime rates are the highest in the industrialized world.

One of the main sources for this violence is the African-American cultural archetype of the "bad nigger." This outlaw figure has been revered since the period of slavery. In the view of many blacks, his very badness becomes a symbol of heroic resistance to white oppression. As Eugene Genovese writes, "Oppressed peoples cannot avoid admiring their own nihilists, who are the ones dramatically saying No! and reminding others that there are worse things than death." Harnessed constructively, the oppositional stance of the "bad nigger" can be converted into revolutionary or reformist zeal, as the examples of Frederick Douglass, W.E.B. Du Bois, and Malcolm X suggest. But the "bad nigger" has always had a gross and homicidal underside. Harvard sociologist Orlando Patterson remarks:

> There was a distinct underclass of slaves. They were the incorrigible blacks of whom the slaveowner class was forever complaining. They ran away. They were idle. They were compulsive liars. They seemed immune to punishment. We can trace the underclass, as a persisting social phenomenon, back to this group.

Today, if the statistics are any indication, the culture of the "bad nigger" flourishes in a permissive social climate. Yet most African-American scholars refuse to acknowledge the pathology of violence in the black underclass, apparently convinced that black criminals as well as their targets are both victims: the real culprit is societal racism. Activists recommend federal jobs programs and recruitment into the private sector. Yet it seems unrealistic, bordering on the surreal, to imagine underclass blacks, with their gold chains, limping walk, obscene language, and arsenal of weapons, doing nine-to-five jobs at Procter and Gamble or the State Department. Many of these young men are lacking in the most basic skills required for steady employment: punctuality, dependability, willingness to perform routine tasks, and acceptance of authority. Moreover, there is some evidence that even when jobs are available, many young blacks refuse them, apparently on the grounds that the jobs don't pay enough or that crime is more profitable.

Black scholars who write about underclass violence typically do so in a tone of moral neutrality. Yet statistics and scholarly description do not convey what the culture of the "bad nigger" means on the street level. For this one has to turn to the real experts: members and former members of the black underclass. In his recent book *Monster,* Sanyika Shakur, a former Los Angeles gang member, offers candid reflections on the life of the black underclass, "armed and dangerous," as he describes it, "prowling the concrete jungle."

He articulates an unconventional approach to the work ethic: "Work does not always constitute shooting someone, though this is the ultimate. Anything from spitting on someone to fighting—it's all work. And I was a hard worker."

5. CULTURAL PLURALISM AND AFFIRMATIVE ACTION

Monster relays several stark accounts of his comrades-in-arms boasting about their shootings:

> "So what's up with them niggas across the way? Y'all been droppin' bodies or what?"

Through their struggle over two centuries, blacks have helped to make the principles of the American founding a legal reality not just for themselves but also for other groups. As W.E.B. Du Bois put it, "There are no truer exponents of the pure human spirit of the Declaration of Independence than the American Negroes."

> "Aw, nigga, I thought you knew!" said Li'l Crazy De. "Tell him Joker."
>
> "Monsta, we caught this fool the other night in the hood, writin' on the wall. Cuz, in the 'hood. Can you believe that s---? Anyway, we roll up on boy and ask him: 'Yo, what the f--- you doin'?' Boy breaks and runs and . . ."
>
> "I cut his a-- down wit' a thirty-oh-six wit' a infrared scope!" interrupted Li'l De. "Aw, Monsta, I f---ed cuz up! He was all squirmin' and s---, sufferin' and stuff, so . . ."
>
> "I put this," Joker said, pulling out a Colt .45 from his waistband. "And Kaboom! To the brain, you know. Couldn't stand to see the b-----made m----f---- sufferin' and s---."
>
> "Who was he?" I said.
>
> "S---, we ain't heard yet."

The life of hoodlums like "Monster" Kody, as he was formerly known, is vividly described by journalist Leon Bing in her book *Do or Die*. Bing offers further insight into the pathological behavior and rationalizing that have become commonplace in the African-American underclass, and provides vignettes of underclass black culture by simply quoting its members' descriptions of their lifestyle:

> When I was younger I was a straight killer. They'd have me killin'— everythin'. And when you 11 years old and you get you a gun, you got to be a little shook up. Then you get used to it, no problem. You got to prove a lot when you first start, see. You gotta prove that you down. Then, like I said, you get used to it—it ain't no thing.
>
> My homeboys be doin' rapes. They'll just rape a girl, any girl, if she look good and she don't wanna kick in. Hey, if they want it bad enough, they gonna take it. All of them together. And beat on her too if she try to hold back.
>
> See, when you shoot someone, like with a .45 automatic, one that goes, pow-pow-pow, like that—br-r-r-r-r. Make a big hole. It look real nasty. Take they brain right out. Afterward, you might feel like you want to throw up. Like: Hey, I did this? Just sittin' there, watchin' somebody's brains come out. I saw it. Looked like oatmeal.

Virtually without exception, these morally anaesthetized muggers, drug pushers, and rapists refuse to consider themselves responsible for their behavior, placing the blame instead on whites and the government. Gang member G-Roc says:

> The big enemy is the system, this government. They say they want to help you, they say they are helping you, but then, really, they ain't doin' nothing but killin' you off with words. The government don't want any of us to get no kinda good jobs. And that's based on the color of our skins. So you feel as though, well, I need money and I need it now. So you will go out and try to hurt whoever you can. The government plays a big part in why we kill our own kind.

As with mainstream black scholars, leading African-American politicians and civil rights activists have endorsed the activities of criminals and gang members. The alliance between gangsters and mainstream civil rights activists came into public view after the Los Angeles riots, when Rep. Maxine Waters brought gang leaders before the Congressional Black Caucus to denounce white racism and demand public funding. Waters introduced three young men as "friends, brothers, who are leaders in my community." One of the three said he had been in prison for three years, another for ten years. The third, who did not reveal his prison record, told a news photographer that together they had probably killed about thirty people. "We ask for jobs," the trio complained, "yet when they bring us jobs, it's only 10 jobs, but there are 500 gang members."

Former NAACP chairman Benjamin Chavis offered similar credibility to gang leaders when he invited them to a "summit" where they declared "truces" and signed "peace treaties" like diplomats. Although the treaties turned out to be short-lived, Chavis announced at the time, "This summit has demystified who gang members are. It has shattered the stereotype that gang members are social deviants. They are some of the best members of society, who just need a chance and some encouragement."

Unmarried With Children

Perhaps the most serious of African-American pathologies—no less serious than violence—is the normalization of illegitimacy as a way of life. Nearly 70 percent of young black children born in the United States today are illegitimate, compared with 22 percent of white children. Almost 95 percent of black teen mothers are unmarried, compared with 55 percent of their white peers. Illegitimacy and single-parent households are not exclusive characteristics of the black underclass: College-educated African-American women have children outside of marriage at a rate about seven to eight times higher than college-educated white women. The result is what sociologist Andrew Cherlin terms an "almost complete separation of marriage and childbearing among African Americans."

Whites, too, have a problem with illegitimacy, but the white norm still remains the two-parent household. It is dangerous to euphemize the problem of broken families, because it is connected with a range of other social pathologies. According to the Centers for Disease Control, the AIDS rate among African Americans is about three times higher than that of the U.S. population overall, and more than 50 percent of children with AIDS are black. Part of the reason for this is the common practice of poor black women exchanging sex for drugs. African-American scholar Robert Hampton writes that "sexual abuse, physical child abuse, and family violence are arguably among the most serious social problems in the black community." Richard Majors reports that wife abuse is four times more common among blacks than among whites, and that black men kill their wives at a higher rate than any other ethnic group.

These interrelated pathologies are evident in the Robert Taylor Homes and Cabrini Green, the largest and second largest housing projects in Chicago. William Julius Wilson reports that at Robert Taylor, with a virtually all-black population of twenty thousand, 90 percent of the households are headed by women, and 81 percent receive welfare. Although less than 1 percent of Chicago's population lives in the project, residents commit about 10 percent of all the assaults, rapes, and murders in the city. In Cabrini Green, with a population of fourteen thousand, the single-parent families number about 90 percent, and 70 percent are supported by welfare. Within a few weeks, Wilson counted ten murders and thirty-five woundings by gunshot in the project.

Yet high black illegitimacy rates cannot be blamed on slavery or segregation. For much of the twentieth century, sociologist Sidney Kronus reports, "The personal standards of conduct and behavior of the black middle class were modeled after the prevailing norms of the middle-class white community which stressed responsibility and the leading of the respectable life." After remaining relatively stable for the first half of this century, the black illegitimacy rate seems to have reached a critical mass during the 1960s and has exploded since then.

In most societies young men cannot readily impregnate and leave young women, because these women are protected by their fathers and male relatives. In the black underclass, however, fathers are scarce, and male relatives are either in prison or themselves responsible for producing illegitimate offspring with one or more women. The distinguished African-American urban anthropologist Elijah Anderson points out that unmarried fathers typically do not provide for their children but frequently use government child support to provide for themselves. The welfare economy, he argues, supplements the underground economy as a source of income for young black males. Some youths even speak of "mother's day," a reminder to visit their girlfriends when the welfare check has arrived.

Anderson argues that the women, too, are using the men, although in a less blatant and irresponsible manner. The typical young woman does not seek children outside of marriage—on the contrary, "she wants desperately to believe that if she becomes pregnant he will marry her or at least be more obligated to her." Many young black women frequently offer sex "as a gift in bargaining for the attentions of a young man." Yet once they become pregnant, they find that having illegitimate children "becomes a rite of passage to adulthood." Suddenly these sad young women have something that they can call their own, and "the teenage mother derives status from her baby." They become part of what Anderson calls the "baby club": Many women in the same situation welcome the newcomer into their fold. "Becoming a mother," Anderson writes, "can seem to be a means to authority, maturity and respect."

Welfare is not the reason for getting pregnant, Anderson insists. But it does afford "a limited but steady income" that becomes the means for teenagers to escape the often painful circumstances of their own homes, establish their own independent single-family household, "and at times attract other men who need money." Welfare even encourages young women to endorse the denial of their male sexual partner who does not want to claim paternity. "This incentive is the prospect of a check from the welfare office," Anderson writes, "which is much more dependable than the irregular support payments of a sporadically employed youth." Because of the relative security and emancipation that illegitimate children can bring, Anderson argues that bastardy has—despite the condemnation of some churches—lost most of its earlier stigma and now become "socially acceptable."

Yet despite the scandalous pain generated by illegitimacy in the underclass, leading African-American intellectuals abstain from criticizing, and go so far as to revel in, what they describe as another alternative lifestyle. Radio commentator Julianne Malveaux is appalled that anyone would question the lifestyle of welfare mothers who live off taxpayers or seek to impose work requirements on them; she insists that "being on welfare is hard work." And Nobel Prize-winner Toni Morrison proposes an abandonment of societal restraints and a return to the elemental urges of nature. "The little nuclear family is a paradigm that just doesn't work," she writes. "Why we are hanging onto it, I don't know. I don't think a female running a house is a problem. . . . What is this business that you have to finish school at 18? The body is ready to have babies. Nature wants it done then."

The Big Lie
While mainstream black leaders refuse to criticize African-American pathologies or to seek internal reform, the scholars and activists of the civil rights establishment

5. CULTURAL PLURALISM AND AFFIRMATIVE ACTION

offer only what may be termed excuse theory—an extensive literature offering literally hundreds of reasons why external constraints make it impossible for blacks to succeed. For William Julius Wilson, black pathologies are the product, not cause, of a lack of jobs: In Wilson's view, blacks just happened to arrive in the cities around the time that unskilled jobs were leaving. Douglas Massey and Nancy Denton argue that residential separation accounts for the formation and persistence of the underclass. Bernard Boxill even argues that black pathologies should be classed as a cultural injury which entitles African Americans to further subsidies from the government.

One problem with this approach is that by emphasizing how poverty and deprivation cause crime and other pathologies, excuse theorists cannot explain why the majority of poor people remain law abiding. Another problem is that by focusing almost entirely on the cause of pathologies, excuse theorists offer no coherent vision of what to do about them. Since external factors are always to blame, these activists cannot do better than propose societal remedies such as redoubled federal funding, more social engineering, or a renewed campaign to root out white racism. In some cases government programs may help, yet none of them show any prospect of reducing the pathologies themselves. The reason for the moral paralysis of mainstream black intellectuals is that their relativist framework makes it impossible to identify black pathology without placing the onus for it on society at large.

Blacks as a group stand at a historic junction. Very few people in the civil rights leadership recognize this: Convinced that racism of a hundred varieties stands between African Americans and success, most of the activists are ready to battle once again with this seemingly elusive and invincible foe. Yet the agenda of securing legal rights for blacks has now been accomplished, and there is no reason for blacks to increase the temperature of accusations of racism. Historically whites have used racism to serve powerful entrenched interests, but what interests does racism serve now? Most whites have no economic stake in the ghetto. They have absolutely nothing to gain from oppressing poor blacks. Indeed the only concern that whites seem to have about the underclass is its potential for crime and its reliance on the public purse.

By contrast, the civil rights industry now has a vested interest in the persistence of the ghetto, because the miseries of poor blacks are the best advertisement for continuing programs of racial preferences and set-asides. No one is more committed to the status quo, and more likely to resist its demise, than those professional blacks whose livelihoods depend on maintaining a large and resentful African-American coalition. Publicly inconsolable about the fact that racism continues, these activists seem privately terrified that it has abated. Formerly a beacon of moral argument and social responsibility, the civil rights leadership has lost much of its moral credibility, and has a fair representation of charlatans who exploit the sufferings of the underclass to collect research grants, minority scholarships, racial preferences, and other subsidies for themselves.

The supreme challenge faced by African Americans is the one that Booker T. Washington outlined almost a century ago: the mission of building the civilizational resources of a people whose culture is frequently unsuited to the requirements of the modern world. Sadly, the habits that were needed to resist racist oppression or secure legal rights are not the ones needed to exercise personal freedom or achieve success today. As urged by black reformers, both conservative and liberal, the task ahead is one of rebuilding broken families, developing educational and job skills, fostering black entrepreneurship, and curbing the epidemic of violence in the inner cities.

Government can help, mainly by doing no harm. Undoubtedly the state has the responsibility of providing police protection in the inner city, and of ceasing to subsidize illegitimacy and provide incentives for socially destructive behavior. Other policies, such as vouchers for school choice and enterprise zones to boost urban investment, may also prove beneficial. Yet in a free society, the government necessarily has a limited role in shaping the private lives of citizens. Consequently the government is not in a good position to reform the socialization practices of African Americans, and the primary responsibility for cultural restoration must lie with the black community itself. "When we finally achieve the right of full participation in American life," Ralph Ellison wrote, "what we make of it will depend upon our sense of cultural values, and our creative use of freedom, not upon our racial identification."

Nobel Prize–winner Toni Morrison proposes an abandonment of societal restraints and a return to the elemental urges of nature. "The little nuclear family is a paradigm that just doesn't work," she writes. "Why we are hanging onto it, I don't know. I don't think a female running a house is a problem."

We can sympathize with the magnitude of the project facing African Americans. In order to succeed, they must rid themselves of aspects of their past that have become aspects of themselves. The most telling refutation of racism, as Frederick Douglass once said about slavery, "is the presence of an industrious, enterprising, thrifty and intelligent free black population." For many black scholars and activists, such proposals are anathema because they seem to involve ideological sellout to the white man and thus are viewed as not authentically black.

Frantz Fanon, a leading black anticolonialist writer, did not agree. What is needed after the revolution, Fanon wrote, is "the liberation of the man of color from himself. However painful it may be for me to accept this conclusion, I am obliged to state it: for the black man, there is only one destiny, and it is white." In this Fanon is right: for generations, blacks have attempted to straighten their hair, lighten their skin, and pass for white. But what blacks need to do is "act white"—that is to say, abandon idiotic Back-to-Africa schemes and embrace mainstream cultural norms, so that they can effectively compete with other groups.

There is no self-esteem to be found in Africa or even in dubious ideologies of blackness. "Let the sun be proud of its achievement," Frederick Douglass said. Instead, African Americans should take genuine pride in their collective moral achievement in this country's history. Blacks as a group have made a vital contribution to the expansion of the franchise of liberty and opportunity in America. Through their struggle over two centuries, blacks have helped to make the principles of the American founding a legal reality not just for themselves but also for other groups. As W.E.B. Du Bois put it, "There are no truer exponents of the pure human spirit of the Declaration of Independence than the American Negroes."

Yet rejection in this country produced what Du Bois termed a "double consciousness," so that blacks experience a kind of schizophrenia between their racial and American identities. Only now, for the first time in history, is it possible for African Americans to transcend this inner polarization and become truly modern, unhyphenated Americans. Black success and social acceptance now are both tied to rebuilding the African-American community. If blacks can achieve such a cultural renaissance, they will forever dispel rumors of inferiority and bring America to the promised land that has eluded the nation for so long—a destination that we can call the end of racism.

A Twofer's Lament

YOLANDA CRUZ

Yolanda Cruz is an associate professor of biology at Oberlin College.

This piece was adapted from a 1993 speech delivered at Oberlin College.

I grew up and graduated college in the Philippines; I've spent the last twenty years in the United States. I see a tremendous difference between the perception of education there versus here, then versus now—of whether securing an education is viewed as an opportunity or as a privilege.

I received a bachelor of science degree from the University of the Philippines. I was an agricultural science major, but I had just as many courses in engineering as in philosophy, in language as in math, in literature as in physics, in physical education as in the arts. The five-year curriculum was extremely strict: inflexible in terms of course choices, not only rigorous but quite brutal. There was no entrance exam; your freshman year *was* the entrance exam, and it was trial by fire. Anyone who survived the thirty-six-credit requirement was permitted to continue. We took those painful but marvelously edifying years one at a time, savoring and suffering every midterm exam, sweating every horrific term paper, including the dreaded senior thesis. Every student wrote one based on original research—that is, every student who made it to senior year. Many didn't.

Courses were taught in English (in a country whose citizens speak approximately 100 non-English languages and dialects) by a faculty that was 40 percent women. These women were not merely technicians or teaching assistants but professors, deans, lecturers and research scientists, with Ph.D.s from American, Canadian, Australian and European universities, just like the men. At the time, our student body was also about 40 percent women, although in recent years, I'm told, this figure has grown to about 50 percent. We enjoyed no financial aid or student loans; we went to university the old-fashioned way—on full scholarship, paying full tuition or working. There was only one criterion for admission: academic excellence. The occasional congressman's son or niece got in, but the brutal freshman year was a great equalizer. It didn't matter that your grandfather had graduated fifty years before either, because that didn't guarantee whether or not you would do well in your courses. It didn't matter how tall you were, what ethnic group you represented, what sport you played or what sex you were; or that you came from a finishing school in Switzerland or a public school in the boonies. The only criterion for admission—and for success—was that you could do your stuff and do it well.

It struck me as extraordinary, therefore, that when I matriculated at the University of California at Berkeley, I had to identify myself by sex, ethnicity and other criteria such as financial need. I considered my Graduate Record Examination and Test of English as a Foreign Language scores as relevant; after all, I was to be a graduate student in an English-speaking country. But sex? Ethnicity? I wasn't even sure what "ethnicity" meant. (Even today, I'm not sure whether I'm Asian, Filipino or Pacific Islander. I usually end up checking the box marked "other.") Financial need? That was my concern; I intended to work my way through graduate school. I wasn't asking for privileges, only opportunities.

Imagine my shock, then, when one of the second-year grad students came up to me, shook my hand and said that he had been looking forward to meeting the "twofer" who had been accepted that year. I discovered later that "twofer" meant I was a double whammy; not only was I a woman in a male-dominated field, but I was also not white. Little did that second-year student know that I was transferring from another department and had been accepted into his department because I had aced all the courses there. I remember feeling diminished by his remark; it was as if I had somehow been accepted because my sex and skin color made up for my lack of smarts. Years later I had a similar jolt. In 1986, shortly after I took my present teaching job, I asked one of my colleagues if my sex and ethnicity had anything to do with my getting hired. He said yes: it was affirmative action. And there I was, assuming I had gotten the job because I was good.

Until recently I thought nothing of this. I figured it came with the territory of living a foreigner's life in an alien country. Then a talk with one of my research students, a Hispanic-American woman, brought back a bit of the pain. Last year this student was accepted into Ph.D. programs in molecular biology at Harvard, Cal-Tech and the University of California at San Francisco. After recounting for me the back and forth of her interviews, she asked a poignant question: Did I think she'd been admitted to these universities because she is a twofer? At that moment, I realized my experience at Berkeley had nothing to do with my being foreign. It had to do with the American perception of education as a privilege, deserved or undeserved. My student did not want an undeserved privilege. Like me, all she wanted was an opportunity. How cruel that a person so young, so bright, is made to feel that she is being given a handout, not a hand.

More recently I encountered, in an exchange with my daughter, Elsa, the confusion that seems to accompany the delineation between opportunity and privilege. After Elsa came home with a perfect eighth-grade report card, she regaled

29. Twofer's Lament

me with tales of her classmates who, after earning high marks, had received from their parents gifts, allowance increases, shopping sprees, and spring breaks in the Caribbean. "Why can't I get $20 for every A I bring home, Mama?" Elsa asked.

Smart kid. She knew she had me cornered. I searched my mind for a fitting response. Without losing my cool, I said, "My dear, I love you very much, but in this household you do not get paid for A's. Instead, you will have to pay me for every grade of B or lower that you bring home." Elsa realized that an A was simply an opportunity to move farther in her coursework; it did not entitle her to an automatic privilege.

Being awarded a privilege and given an opportunity are similar in that the odds are stacked in the recipient's favor. With privilege, however, the odds are handed to you; with opportunity, you stack the odds in your own favor. It is hard not to see the dignity in the latter enterprise—the sublime feeling of self-worth, self-respect and pride that it engenders.

Crisis of Community
MAKE AMERICA WORK FOR AMERICANS

WILLIAM RASPBERRY, *Columnist for* Washington Post

Delivered at the Landon Lecture Series at Kansas State University, Manhattan, Kansas, April 13, 1995

I'VE been writing a good deal of late about the violence in our streets, the apathy in our schools, and the hopelessness among our young people — the crisis in our community. But this morning, I want to talk about a deeper, more pervasive and ultimately more serious crisis. Let me call it our crisis of community.

America has a crisis of community that is as deep and wide as it is unnoticed. And it threatens to destroy our solidarity as a nation, in much the same fashion as a similar crisis in community has ripped apart the former Soviet Union and what used to be Yugoslavia.

I refer, of course, to the gender wars newly resurrected by the latest battles in the Clarence Thomas/Anita Hill holy wars; to the ethnic battles over university canons and multi-culturalism, to the political warfare that makes party advantage more important than the success of the nation, and to the racial animosities and suspicions fueled by everything from the rantings of Khalid Abdul Muhammed to the O.J. Simpson trial to Charles Murray's pseudo-intellectual call for racial abandonment.

But when I express my fear that we are coming unglued, I'm thinking about far more than these things.

I'm talking about more even than the normal give and take among the various sectors and ideologies of the society. I am talking about our growing inability to act — even to *think* — in the interest of the nation.

It's almost as though there IS no national interest, apart from the aggregate interests of the various components. The whole society seems to be disintegrating into special interests.

And not just in politics. College campuses are being ripped apart by the insistence of one group after another on proving their victimization at the hands of white males, and therefore their right to special exemptions and privileges.

One example of what I'm talking about: A few years ago, the Federal Aviation Administration adopted a rule that would bar emergency exit row seating to passengers who are blind, deaf, obese, frail or otherwise likely to inhibit movement during an emergency evacuation. Common sense? Only if you think of the common interests of all the passengers.

Surely it is reasonable to have those emergency seats occupied by people who can hear the instructions of the crew, read the directions for operating the emergency doors and assist other passengers in their escape.

But some organizations representing the deaf, blind and otherwise disabled reacted to the regulation only as a form of discrimination against their clients who, they insist, have a "right" to the emergency seats.

It is true that the majority must never be allowed to run roughshod over the rights of minorities. That is one of the tenets of the American system. But the notion of fairness to particular groups as an element of fairness to the whole has been perverted into a wholesale jockeying for group advantage.

Mutual fairness, with regard to both rights and responsibilities, can be the glue that bonds this polyglot society into a nation. Single-minded pursuit of group advantage threatens to rip us apart at the seams. The struggle for group advantage has us so preoccupied with one another's ethnicity that we are losing our ability to deal with each other as fellow humans.

What are we to make of this dismaying evidence that the relationships among us are getting worse — even among our college students? I believe two things are happening, and that they reinforce one another. The first is the racism and bigotry that never went away, even though it was relatively quiet for a time.

The second is what has been called the politics of difference. There is a pattern I have seen repeated on campuses across America. A black group, perhaps motivated by some combination of discomfort and rejection, goes looking (always successfully) for demonstrable evidence of racism.

I used to marvel at this search. Of course there was racism on campus, but what was the point of PROSPECTING for it, as though panning for gold?

I mean, where was the assay office to which one took these nuggets of racism and traded them in for something of value?

Well, it turns out that there IS such an assay office. It's called the Administration Building. Turn in enough nuggets and you get your reward: a Black Student Union, a special course offering, an African American wing in a preferred dormitory — whatever. All it takes is proof that you are a victim.

But despite the reports one hears these days, college students aren't exactly stupid. They are bright enough to see that there are rewards in the politics of difference, in demonstrated victimism. So the victories won by black students become models for similar prizes for gay students or Hispanic students or female students, all of whom gather up their nuggets of victimism and take them to the administration building for redemption.

Cornell University, one of the finest institutions in America, has a dormitory called Ujamaa College, a residence for black students; Akwe:kon, a dorm for Native Americans, and also the Latin Living Center.

That's the trend when the accent is on difference. And finally, it turns out that everybody gets something out of the politics of difference except white males, who start to feel sorry for themselves.

And if they can't find anyone to reward them for their sense of being slighted, they may turn to behavior that was once unthinkable — the "acting out" that manifests itself in incivility, reactionary politics, open bigotry and, on occasion, violence.

Every gain by minority groups justifies the sense of victimism on the part of white males, and every repugnant act of white males becomes a new nugget for a minority to take to the assay office.

Two things get lost in this sad ritual. The first is that the administration seldom gives up any of its own power: the gains of one group of students are extracted from other groups of students, who then must play up their own disadvantage to wrest some small advantage from another group. The administration's power remains intact.

The second overlooked aspect is that the process turns the campus into warring factions — each, no doubt, imagining itself as the moral successor to the heroes of the Civil Rights Movement. There's a difference, though. Dr. King's constantly repeated goal was not special advantage but unity. His dream was not of a time when blacks would finally overcome whites; his dream was that we should overcome, black and white together.

His hope was not that we should celebrate our differences but that we should recognize the relative unimportance of these differences. The differences do not *seem* unimportant, of course. Sometimes we seem to notice ONLY our differences.

That's why I find it helpful to look at what used to be the Soviet Union and what used to be Yugoslavia. From this distance, it seems clear that the similarities between the Serbs and Croats and other ethnic neighbors in Bosnia-Herzegovinia should outweigh their differences.

They share the history of a place and indeed many were intermarried. But now that Yugoslavia has broken up, even the marriages have been ripped apart.

I find myself wishing these erstwhile Yugoslavs could see for themselves what distance makes clear to us. And I wish we could learn to see ourselves as from a distance. Maybe we'd learn to appreciate how great are our similarities and how trivial our differences, and get OUR act together.

A "Star Trek" episode of some years ago makes my point. Capt. Kirk and his crew rescue a humanoid who, on his left side, is completely black. His right side, it turns out, is altogether white.

They are in the process of trying to learn the origins of this stranger — Lokai, he is called — when they are confronted by a similar humanoid named Bele — this one black on his right side and white on the left. The Enterprise crew, of course, can hardly tell them apart. But the humanoids can see themselves only as complete opposites — which, of course in one sense they are. And not just opposites. Though they are from the same planet, they are also sworn enemies.

I won't try to tell you the whole episode, but let me recall this much. Lokai is thought to be a political traitor, and Bele, an official of their home planet's Commission on Political Traitors, has been chasing him throughout the galaxy for a thousand years.

Lokai tries to convince the Enterprise crew that Bele and his kind are murderous oppressors. Bele counters that Lokai and his kind are ungrateful savages. The Enterprise crew decides to travel near the strangers' planet.

When they come within sensor range they are surprised to learn there is no sapient life there. The cities are intact, vegetation and lower animals abound, but the people are dead. They have annihilated each other. These two have survived only because they happened to be in the business of chasing each other down.

And what do they do when they learn what has happened to their planet? They lunge at each other in furious battle. Though the Enterprise crew is appalled, Kirk is unable to convince the two enemies of the futility of their war.

"To expect sense from two mentalities of such extreme viewpoints is not logical," says Spock. "They are playing out the drama of which they have become the captives, just as their compatriots did."

"But their people are dead," Sulu says slowly. "How can it matter to them now which one is right?"

"It does to them," says Spock. "And at the same time, in a sense, it doesn't. A thousand years of hating and running have become all of life."

We don't learn from this "Star Trek" episode the nature of the original problem between these warring humanoids, though we can be certain each felt fully justified in continuing the war. They had made a mistake that too many of us make in real life: They had forgotten the difference between problems and enemies.

And so have we. Virtually every issue that strikes us as urgent or important is made more intractable by our insistence on seeing it as a matter of us against them.

Give us a problem, and we'll find an enemy. Is the U.S. economy in trouble? Make the Japanese the enemy. Are we concerned about the discouraged and dangerous underclass? Blame white racists.

Members of my own profession seem unable to tell a story, no matter how significant, unless they can transform it into a case of one person, or one group, against another — unless they can make it a matter of enemies.

It is not so much that the enemies we identify are innocent as that identifying and pursuing them takes time and attention away from the search for solutions.

It was no trouble at all to come up with evidence that the Japanese were hurting the American economy through predatory pricing, product dumping and nonreciprocity, and certainly all these things merited attention.

But the U.S. auto industry improved its position relative to Japan's auto industry not when we all became expert at bashing our Japanese enemy but when Detroit started making better cars.

And that's the point. The failure to distinguish between the enemy and the problem has us looking balefully at one another instead of jointly attacking the problem which, in most cases, is as much a problem for us as for those we attack.

Take the current fight over affirmative action, for instance. Politicians who lack the imagination to address the *problem* settle for giving us each other to attack. White men — particularly those with a high school education or less — are not imag-

5. CULTURAL PLURALISM AND AFFIRMATIVE ACTION

ining things when they feel less secure economically than their fathers were. But they make a mistake when they suppose that their jobs have somehow been handed over to black people in the name of affirmative action. More likely those jobs are in Taiwan or Singapore or have gone up in the smoke of corporate mergers and downsizing. We've got a problem, and we waste our time assaulting enemies.

Honest communication about the problem might lead us to look for ways to restore our industrial base, expand our economy, improve the quality of our products and put our people to work. Focusing on enemies produces stirring speeches and little else.

You've heard the speeches. You've watched as communities have been ripped apart by those who deliver these speeches. There's how Teresa Heinz, widow of the late Pennsylvania senator, described them in a recent speech:

> ". . . critical of everything, impossible to please, indifferent to nuance, incapable of compromise. They laud perfection but oddly never see it in anybody but themselves. They are right all the time, eager to say I told you so, and relentlessly unforgiving. They occasionally may mean well, but the effect of even their good intentions is to destroy. They corrode self-confidence and good will; they cultivate guilt; they rule by fear and ridicule.
>
> "They are creatures of opportunity as much as of principles, extremists of the left and the right who feed on our fear and promote it, who dress up their opponents in ugly costumes, who drive a bitter wedge between us and the Other, the one not like us, the one who sees the world just a shade differently. . . . They demonize us by our parts and tear our country into pieces."

My own formulation is less eloquent; they focus on enemies rather than on problems. They forget that, at the end of the day, when we've all taken our unfair shots at one another, this simple truth remains: The *problem* is the problem.

Our politicians and our factional leaders never miss an opportunity to list the atrocities the *enemy* has committed against us. But nothing changes.

Sometimes we're not even sure what we want to change, or what we want the people we call enemies to do. We say we want things to get better, when sometimes I think we only want to score points.

We say we want a society in which all of us can live together as brothers and sisters, and the whole time we are saying it we are busy creating another group of barriers to place between us.

It's a strange sort of progress we have made since the death of Dr. King. We have "progressed" to the point where we are embarrassed to speak of brotherhood, of black and white together, of our shared status as Americans.

That's not an accusation; it's a confession. All of us are capable of getting so caught up in the distance that remains to be run that we forget to give ourselves full credit for the distance we've come.

Yet, every now and then, we manage to overcome our embarrassment and see things as from a distance. In that spirit, I'd like to share something I wrote a while back — something I still believe but something I may have trouble saying again.

Here it is: The immigration applications, the legal and illegal dodges for getting into this country, the longings you hear in virtually every other part of the world all attest to two astounding facts.

The first, widely accepted though not always with good grace, is that "everybody" wants to be an American. The second, of which we take almost no notice, is that virtually anybody can *become* an American.

To see just how extraordinary a fact that is, imagine hearing anyone — black, white or Asian — saying he wants to "become Japanese." It sounds like a joke. One can *live* in Japan (or Ghana or Sweden or Mexico) — can live there permanently, and prosper. But it's essentially impossible to imagine anyone born anywhere else becoming anything else — except American.

It's a thought that crosses my mind whenever I hear demands that the government protect the ethnic or language heritage of particular groups: when African Americans demand that the *public* schools adopt an Afrocentric curriculum, for instance, or when immigrants from Latin America are sworn in as American citizens — in Spanish.

It crossed my mind again when I came across Jim Sleeper's essay, "In Defense of Civic Culture."

I won't try to characterize Sleeper's piece or to summarize its recommendations. [the Washington-based Progressive Foundation] I won't even tell you I agree with everything Sleeper has to say on the subject of race and ethnicity.

But he says some things that echo my own feelings, especially when I ponder the extraordinary possibility of becoming American.

He acknowledges the obvious: that the America that counted my great-great-grandfather as only three-fifths of a human being has never been free of ethnic and racial bigotry, and that that bigotry has sometimes achieved the status of law, of philosophy — even of religion.

But he notes something else: that America is one of the few places on the globe where accusation of such bigotry is a serious indictment. Even when America has been at its ugliest in fact — slavery, the slaughter of Native Americans, the internment of the Japanese and the full range of private and public atrocities, "yet always America held out the promise that, as Ralph Waldo Emerson put it, 'in this asylum of all nations, the energy of . . . all the European tribes [and] of the Africans, and of the Polynesians will construct a new race.'"

The civic culture Sleeper writes about includes this notion of Americans as a new and different race, but it also entails what he describes as characteristic American virtues: tolerance, optimism, self-restraint, self-reliance, reason, public-mindedness — virtues that are "taught and caught in the daily life of local institutions and in the examples set by neighbors, co-workers and public leaders."

It is, he suggests, the internalizing of these virtues that defines "becoming American."

But the transformation works both ways. If people from an awesome range of colors, cultures and ethnicities have become Americans, so has America become what it is (and continues to become) by absorbing and embracing these myriad influences.

Some of us are angry, and ought to be, that our academic texts and teachings still disregard or underestimate our part of these influences.

Some of us are disappointed that what we bring to the smorgasbord is often undervalued, even brutally rejected.

But surely the cure is in working for greater inclusion, not cultural isolation. That's what observers as different as Sleeper, Arthur Schlesinger and John Gardner have been saying. That's what Gary Trudeau was saying in that hilarious (and sobering) series of "Doonesbury" strips that ended with black students — already having attained their separate courses and dormitories — demanding, at last, separate drinking fountains. Sleeper's insight is that there is nothing "natural" or automatic about those values and attitudes that used to be called "the American way." Educators must teach them, he says, and also "teach that self-esteem is enhanced not simply through pride in one's own cultural origins but, more importantly, by taking pride in one's mastery of civic virtues and graces that all Americans share and admire in building our society."

Critics of this view will argue that Sleeper's virtuous and graceful American is a figment, that America is a deeply — perhaps irredeemably — racist society.

I prefer to think that Americans are still becoming Americans, just as America is still becoming America.

How can we accelerate that becoming? By recognizing its importance, by understanding that hating and running must not become all of life, and by working to grasp the difference between problems and enemies.

Confront a difficulty as a problem, and you have taken the first steps toward creating the climate for change.

Confront it as the work of enemies and you create the necessity for DEFEATING someone, of intimidating someone, of browbeating someone into doing something against his will.

Enemies have to be sought out, branded and punished. Which, naturally, gives them one more reason to find an opportunity to strike back at us. And the beat goes on.

Problems, on the other hand, admit of cooperative solutions that can help build community.

Searching for enemies is most often a pessimist's game, calculated less to resolve difficulties than to establish that the difficulties are someone else's fault. Identifying *problems* is by its very nature optimistic and healing. The whole point of delineating problems is to fashion solutions.

Maybe that's what President Clinton had in mind when he called on America to bring back "the old spirit of partnership, of optimism, of renewed dedication to common efforts."

"We need," he said, "an array of devoted, visionary, healing leaders throughout this nation, willing to work in their communities to end the long years of denial and neglect and divisiveness and blame, to give the American people their country back."

And that is precisely what we need. America has had enough of the politics of difference, the marketing of disadvantage, the search for enemies. It's about time we started to work on what may be the most important problem we face:

How to heal our crisis of community and make America work — not for blacks or whites or women or gays; not for ethnics; not for Christians, Moslems or Jews — but for Americans.

Cities, Urban Growth, and the Quality of Life

- Cities (Articles 31 and 32)
- Quality of Life (Articles 33 and 34)
- Education (Articles 35 and 36)
- Drugs (Articles 37 and 38)

Until 1965 the major problems facing cities were those associated with growth. People by the millions were abandoning rural areas and migrating to the cities. Due to many factors, this trend has been reversed. Since 1965 many U.S. cities have lost major industries and businesses as well as the people involved. These losses have produced a declining tax base and an aging infrastructure, both of which have spawned a host of related social problems such as unemployment, underemployment, homelessness, crime, and gangs. But what is fascinating is that while many major cities are in trouble, others are thriving. In *Annual Editions: Social Problems 97/98*, we have created four categories for this unit's articles: *Cities, Quality of Life, Education,* and *Drugs*. The last two topics constitute major social problems with unique implications for American society as a whole, hence their own categories.

To stop the deterioration of the cities, we must discover the basic causes. In "Can We Stop the Decline of Our Cities?" Stephen Moore and Dean Stansel contrast cities in decline with those that are not. They reveal that decline or growth is directly dependent on the costs and quality of services provided by the city and on the size of the city's governmental bureaucracy.

In "Can Churches Save America?" it is noted that some politicians believe that they can and cite statistics to support this claim. Regular church attendance is associated with lower teenage drug use, optimism about the future, lower criminal recidivism, lower family disintegration rates, and greater emotional stability. These plus other statistics suggest that the church is potentially a significant force in solving society's problems. But some people maintain that it is totally unrealistic to think that small local churches can even begin to solve society's huge problems.

Rob Gurwitt observes that the practice of concentrating public housing projects in high-rise, high-density areas may provide low-income housing, but it also causes heightened crime, drugs, and other community blights. To correct this problem, Gurwitt argues, in "The Projects Come Down," that high-density projects must be replaced by housing scattered throughout the community. He reports on cities where this has happened and finds the results very encouraging. The article "Time Out" reports that more and more individuals are discovering that life holds much more than the time and stress associated with the acquisition of large salaries, power, or high-status positions. In addition, foreign competition, advanced technologies, and corporate restructuring have forced many people to reevaluate what they want out of life. Increasingly, individuals are leaving corporate America to spend more time with their families, for home-based work, and for a chance to do something more meaningful with their lives.

Are public schools doing their job? A great number of parents and concerned citizens are saying, no! Lower test scores in many academic areas, coupled with a growing need for colleges to require remedial instruction in basic courses, have an ever-increasing cross section of Americans questioning what is happening in the schools. In "Can the Schools Be Saved?" Chester E. Finn Jr., former assistant secretary of education, writes that saving the schools requires a combination of grassroots involvement, reduction in bureaucracy, and viable educational alternatives.

Karen Lehrman's article, "Off Course," examines what is happening in many women's studies courses around the country. Lehrman visited many campuses and discovered that "discussions run from the personal to the political and back again, with mere 'pit stops' at the academic."

"A Society of Suspects: The War on Drugs and Civil Liberties" reflects the paradoxes of trying to balance effective measures against crime with the upholding of civil rights. Law enforcement personnel insist their hands are being tied, negating their chances to win the war on drugs. But those constraints were established by the Founding Fathers to protect the rights of every American citizen to be free from unwarranted searches, forced entry into homes, and arrest. Steven Wisotsky argues that constitutional rights are being systematically limited and redefined because of this "war."

Why don't our welfare, medical, and criminal justice systems work? In "It's Drugs, Alcohol, and Tobacco, Stupid!" Joseph A. Califano Jr. reports that until we develop

UNIT 6

workable policies concerning drugs, the rest of the system cannot function effectively. Drug abuse creates costly medical problems for the users, their families, and society. Drug addiction creates a demand that cannot be realized legally and thus clogs the courts and prisons. It appears that this problem facing society cannot be addressed until we find a solution to the problems created by drugs.

Looking Ahead: Challenge Questions

Describe the possibilities for rescuing our inner cities. In countering urban blight and decline, how effective has the infusion of federal dollars into cities been? What are the implications of tearing down substandard housing and rebuilding high-rise subsidized housing exclusively for the poor?

What role can churches play in the revitalizing of America? Is there any evidence that active involvement in a church has impacted various types of social pathologies?

What should be the primary objective(s) of any university academic program?

Is public education failing in the United States? What can be done to make students in America competitive with comparable students elsewhere? Distinguish between the harmful effects of drugs approved by society (alcohol, tobacco, and caffeine) and those that are not (heroin, cocaine, and marijuana).

In what major ways can raising people's fears affect their rights as citizens? In the attempt to deal with the drug abuse problem, does the end always justify the means? Defend your answer.

Why are so many people rejecting success in the corporate world?

Which of the three sociological theoretical positions do you think most clearly helps to understand the issues covered in this unit?

What are the values, rights, obligations, and harms associated with each of the issues examined in this unit?

Can We Stop the Decline of Our CITIES?

*Unless and until they start putting people first—
by cutting service costs and anti-growth tax rates—no amount of
Federal aid can reverse the trend.*

Stephen Moore and Dean Stansel

The authors are, respectively, director of fiscal policy studies and a research assistant, Cato Institute, Washington, D.C.

FOR MORE THAN a quarter-century, Americans have been voting with their feet against the economic policies and social conditions of the inner cities. Fifteen of the largest 25 U.S. cities have lost 4,000,000 people since 1965, while the total U.S. population has risen by 60,000,000. The exodus no longer is just "white flight"—minorities also are leaving the cities in record numbers.

In recent years, the departure of businesses, jobs, and middle-income families from the old central cities has begun to resemble a stampede. For example, since the late 1970s, more than 50 Fortune 500 company headquarters have fled New York City, representing a loss of over 500,000 jobs. Cleveland, Detroit, Philadelphia, St. Louis, and other major cities also are suffering from severe out-migration of capital and people. Those once-mighty industrial centers are becoming hollow cores of poverty and crime.

Ever since the Los Angeles riots and looting, urban lobbyists—including mayors, public employee unions, urban scholars, and many members of Congress—have been arguing that the inner cities were victims of Federal neglect under Presidents Ronald Reagan and George Bush. "There was, quite literally, a massive Federal disinvestment in the cities in the 1980s," according to Congressional delegate Eleanor Holmes Norton of Washington, D.C. To revive them, the U.S. Conference of Mayors is asking for $35,000,000,000 in new Federal funds—a "Marshall Plan for the cities."

The Federal government already has tried the equivalent of some 25 Marshall Plans to revive the cities. Since 1965, it has spent an estimated 2.5 trillion dollars on the War on Poverty and urban aid. (That figure includes welfare, Medicaid, housing, education, job training, and infrastructure and direct aid to cities.) Economist Walter Williams has calculated that this is enough money to purchase all the assets of the Fortune 500 companies *plus* all of the farmland in the U.S., but it has not spurred urban revival. In 1992 alone, Federal aid to states and cities rose to $150,000,000,000. Adjusted for inflation, that is the largest amount of Federal intergovernmental aid ever extended—hardly a massive disinvestment.

Central cities' budgets on the rise

The budgets of Cleveland, Detroit, Philadelphia, New York, St. Louis, and other large central cities have not been shrinking; they have been rapidly expanding for decades. In constant 1990 dollars, local governments spent, on average, $435 per resident in 1950, $571 in 1965, and $1,004 in 1990. The largest cities saw an even faster budget rise. In real dollars, New York's budget nearly tripled from $13,000,000,000 in 1965 to $37,000,000,000 in 1990. Philadelphia, another nearly bankrupt city, allowed its budget to rise by 125% between 1965 and 1990—from $1,600,000,000 to $3,500,000,000. During the same time period, the city lost 20% of its population. In short, 25 years of doubling and even tripling city budgets have not prevented urban bleeding.

Not all U.S. cities are in decline. Among the nation's largest urban areas, there are dozens—many on the West Coast, in the Sunbelt, and in the Southeast—that have been booming financially and economically for at least the past 20 years. Las Vegas, Nev.; Phoenix, Ariz.; Arlington and Austin, Tex.; Sacramento and San Diego, Calif.; Raleigh and Charlotte, N.C.; and Jacksonville, Fla., all have rapidly rising incomes, populations, and employment and low poverty and crime rates.

What do growth cities—Phoenix, Raleigh, and San Diego, for example—do differently from shrinking cities such as Buffalo, Cleveland, and Detroit? The answer is found, at least partially, in their fiscal policies. Bureau of the Census finance data from 1965, 1980, and 1990 for the 76 largest cities reveal significant and consistent patterns of higher spending and taxes in the low-growth cities than in the high-growth ones.

• For every dollar of per capita expenditures (excluding those spent on anti-poverty programs, education, and health care) in the highest-growth cities, the shrinking cities spend $1.71.

• In 1990, a typical family of four living in one of the shrinking cities paid $1,100 per year more in taxes than it would if it lived in one of the highest-growth cities.

• Shrinking cities' bureaucracies are twice as large as those of growth cities. In 1990, the latter had, on average, 99 city employees per 10,000 residents; the former, 235.

• Cities with high spending and taxes lost

31. Stop the Decline of Our Cities?

population in the 1980s; those with low spending and taxes gained. High spending and taxes are a cause, not just a consequence, of urban decline.

Expenditures are high and rising in large central cities primarily because their governments generally have above-average unit costs for educating children, collecting garbage, building roads, policing neighborhoods, and providing other basic services. In 1988, for example, the shrinking cities spent roughly $4,950 per pupil on education, whereas the high-growth cities spent $3,600. The $1,350 cost differential can not be explained by better schools in places such as Detroit and Newark.

The influence of municipal employee unions also contributes to higher costs in declining central cities. Compensation for unionized local employees tends to be roughly 30% above wages for comparably skilled private-sector workers. In New York, the average school janitor is paid $57,000 a year. In Philadelphia, the average municipal employee receives more than $50,000 a year in salary and benefits. According to the Census Bureau, cities with populations over 500,000 pay their mostly unionized workers more than 50% more than those with populations under 75,000, whose workforces are less likely to be unionized. In short, thriving cities are places where costs are lower, bureaucracies are smaller, and services are better.

Some city officials are beginning to recognize economic reality. Philadelphia Mayor Edward Rendell is challenging the entrenched municipal unions and other spending constituencies with a budget plan that calls for $1,100,000,000 in savings over five years. He has spurned more Federal aid as the poison that produced Philadelphia's near-insolvency in 1992. Chicago Mayor Richard M. Daley and Indianapolis Mayor Stephen Goldsmith have contracted out dozens of services to private providers and have slowed the growth of massive, bloated budgets to a crawl.

The decline of America's major cities is not inevitable. They can and should be saved. For generations, they have served as the nation's centers, not only of industrial might, but of culture, diversity, and intellect. Through an aggressive agenda of budget control, tax reduction, privatization, and deregulation, America's declining cities can rise again in prominence and prosperity.

The riots in Los Angeles in the spring of 1992 dramatized the social and economic deterioration of many central cities. At a rally held in Washington a month after the riots, then-New York Mayor David Dinkins aptly described the inner cities as places of "only grief and despair." Almost all quality-of-life indicators for many of the fastest shrinking cities confirm that gloomy assessment. Consider these examples:

- The 1990 census data reveal that 48% of all Detroit households are headed by a single female.
- Newark has a lower real per capita income today than it did in 1969.
- Cleveland had nearly 1,000,000 residents in the 1950s; now, it has fewer than 500,000.
- The Chicago area has averaged 10,000 manufacturing job losses annually for the past 15 years. A Feb. 14, 1992, *Chicago Tribune* headline said it all: "Factory Flight Hits Record Pace."
- St. Louis has lost more than two of every five jobs it had in 1965.
- Philadelphia has a four-year $1,000,000,000 budget deficit. Its bond rating sunk so low that it effectively is blocked from municipal capital markets and must borrow through a state oversight agency.
- Washington, D.C., has been dubbed the "murder capital of the world."
- In Baltimore, 56% of black males between the ages of 18 and 35 were in trouble with the law in 1991, according to a study by the National Center on Institutions and Alternatives.

The root cause of almost all the social, economic, and fiscal problems of America's depressed cities is the steady flight of businesses and middle- and upper-income families. In the past quarter-century, 15 major cities combined have lost 3,800,000 people and roughly the same number of jobs. No longer is the issue just white flight. The 1990 census data suggest that middle-income minorities are fleeing to the suburbs in record numbers.

The urban crisis is not shared by all cities. The declines in the most depressed cities are matched by impressive gains in the highest-growth cities. Raleigh, for instance, has seen its population almost double, jobs more than double, and real per capita income grow by better than 40% since the mid 1960s. The wide diversity in the economic performance of central cities suggests that the individual policies of each play an important role in explaining urban growth and decline.

In conventional analyses of the urban crisis, cities are portrayed primarily as victims of national trends, Federal mandates, and other conditions beyond their control. Factors blamed include the recession, Reagan budget cuts, suburbanization, decline in manufacturing, rise of the automobile, immigration, an aging infrastructure, racism, AIDS, homelessness, urban gangs, guns, and drugs. Each of these can place considerable strain on municipal budgets, yet city officials are impotent to combat them.

There is some truth to this. The last recession is painful evidence that no city is immune from the impact of national economic conditions and policies. For instance, from the 1950s through the 1980s,

most California cities enjoyed spectacular rates of growth. The state widely was considered recession proof. Yet, the recession had a devastating impact on California localities—many of which are at the top of the list of growth cities. One-third of all job losses in the past two years has occurred in California. Another example is the impact that wide fluctuations in international oil prices in the 1970s and 1980s had on the economies and budgets of Texas cities.

Unquestionably, there are regional factors at play in determining relative rates of growth. Most of the declining cities are the once-mighty industrial centers in the Northeast and Midwest—the Rust Belt. Most of the growth cities are in the Sun Belt, on the West Coast, and in the Southeast. A related factor that usually correlates with the rate of economic growth is age—older cities tend to experience less economic growth, a phenomenon that often is attributed to their aging infrastructures.

Still, the fact that some cities have been flourishing for long periods of time as others have been deteriorating suggests that self-imposed policies play an important role in determining their economic fates. Regional factors can not explain fully the different growth rates of cities. Although the South has been a high-growth region, three of the fastest declining cities—Birmingham, Louisville, and New Orleans—are in that area.

Even within states, there are significant differences in the economic performance of cities. Why is Oakland declining, but Santa Ana growing? Why is Arlington doing so much better than Fort Worth, or Colorado Springs better than Denver? What explains the fact that the eighth fastest growing city, Lexington, and the ninth fastest declining city, Louisville, both are in Kentucky? An important factor is that their spending and taxing policies are very different. Growth cities have pro-growth fiscal policies; declining cities, anti-growth.

Taxes and spending. There are several potential problems in comparing spending and taxes by city. The major one is the division of responsibilities for program funding among state governments, counties, school districts, and cities differs from state to state. For example, in some, the responsibility for funding welfare assistance is delegated to the counties; in others, it is paid for by the state government; and in still others, the cities bear the cost. Another complication is that some cities are not part of independent counties, and so the city government funds all the activities of the county. By contrast, most cities are part of larger counties, which means that the costs of funding services such as hospitals and courts in the inner cities are spread to the suburbs. Finally, in some cities, school funding is handled by school districts or in

6. CITIES, URBAN GROWTH: Cities

large part by the state, not the city, government.

Another potential difficulty with comparing city growth rates over time is that borders change. Many cities, such as Portland, Ore., aggressively have annexed neighboring suburbs. In the case of Portland, annexation has been a major source of population growth. However, that annexation probably is as much a consequence of the city's economic success as it is an explanation. In many cases, localities have merged with central cities because it has been in their economic interests to do so. By the same token, more and more localities are attempting to secede from declining central cities because the latter's economic policies, such as tax rates and service costs, have grown too burdensome.

Population growth. The best indication of the livability of a city probably is whether people are moving to or out of it. Those that had large population losses from 1965 to 1990 spend, tax, and hire city workers at roughly twice the rates of the cities that had large population gains. For instance, the highest-growth cities (those with population gains of 100% or more) spend six percent of personal income, whereas the lowest-growth cities (those that lost at least 15% of their populations) expend 12% of personal income. The highest-growth cities had 107 employees per 10,000 residents vs. 217 in the lowest-growth cities. Those findings suggest that cities with high taxes and service costs are driving people away.

Job growth. Just as population changes are a measure of how people are voting with their feet, so are employment patterns. Firms and capital that create jobs tend to migrate to areas with pro-growth and pro-business climates. Per capita spending is roughly 50% higher in the lowest-growth cities than in those with the greatest increase in employment. The cities with the lowest growth in employment spend almost $1.40 for every $1.00 of municipal expenditures in the highest-growth cities. High taxes and spending are driving businesses and jobs away from shrinking cities.

Female-headed households. One indication of poverty and family disintegration in an area is the percentage of female-headed households. Fatherless homes are about eight times more likely to be poor than are intact families. Moreover, children who grow up in fatherless households are much more likely to commit crimes and engage in other socially unacceptable behavior of the type that plagues inner cities.

Between 1970 and 1990, the percentage of female-headed households rose at an alarming pace nationwide. The largest growth occurred in the inner cities, where Census Bureau data indicate an increase from about 15 to 35% in fatherless homes. Cities with the largest rise in female-headed households have spending and taxes at least 50% higher than those with small increases.

Per capita income growth is one of the best measures of the economic growth rate of a city and the standard of living of its residents. The cities with the lowest growth in income (less than 10%) are characterized by per capita spending and taxes that are about 10% higher than those of cities with the highest growth in income (more than 25%). As a share of personal income, spending and taxes are about 25% higher in cities with the lowest growth.

Where the money goes

The urban lobby and scholars invariably argue that low-growth cities spend more because their needs are greater. Poor cities have to meet increased demands for anti-poverty spending, subsidized child care, homeless assistance, drug abatement and rehabilitation, crime control, job training, etc. Because declining cities have less wealth and fewer workers and businesses to pay the cost of those programs, they have to impose higher tax burdens on their residents and businesses to raise the same amount of revenue wealthy and prospering cities do.

Education is a major budget item that should not cost much more on a per student basis in a low-growth city. If per student education costs were uniform across cities, one would expect to find higher per pupil expenditures in growing and affluent areas if only because the residents have more money to devote to the schools. Indeed, the education lobby successfully has argued before several state supreme courts that school financing is inequitable because wealthy areas spend more on schools than do poor areas.

The data show that, on average, per pupil expenditures on schools in the lowest job growth cities are approximately $1,800 higher than they are in the highest job growth cities. Yet, spending on schools and student performance appear to be wholly unrelated. For example, Washington, D.C., which spent almost $6,000 per student in 1988, has among the worst inner-city schools in the nation. Conversely, San Diego spent about $3,500 per child in 1988 for schools that are considered above average for cities its size.

Low-growth cities appear to provide education, one of the major items in local budgets, much less cost-efficiently than do high-growth cities. If declining central cities simply could lower their education costs to the national average for large cities, they could reduce their per family tax burden by hundreds of dollars.

City taxes and economic growth. A city's economic performance is influenced not only by its over-all tax burden, but by the composition of its taxes. In particular, income taxes have a consistently strong negative effect on city growth rates. Only one high-growth city—Lexington, Ky.—imposes a city income tax, whereas the low-growth cities have average per capita income taxes of approximately $100-200. The evidence suggests that imposition of a city income tax is a recipe for economic decline.

City property taxes also have a negative impact on economic growth, but not to the extent that income taxes do. Cities with high population and job growth, as well as those with low poverty rates, have substantially lower property taxes than cities in decline.

City sales taxes have no apparent positive or negative impact on economic growth. If anything, high-growth cities tend to rely heavily on sales taxes for revenue; declining cities, on income taxes.

Impact of state taxes. Workers and businesses are affected not only by local tax burdens, but by state taxes as well. High-growth cities tend to be located in states that have low combined city-state tax burdens. State and city taxes are about $360 per person higher in cities with population losses of 15% or more since 1965 than in those with population gains of 100% or more.

Even more dramatic is the destructive impact of high combined state and city income taxes on the economic performance of cities. Growth cities tend to have state and local income tax burdens that are, on average, about 60% of those of shrinking cities.

Federal aid. Cities have a multitude of problems, but too little money is not one of them. Real per capita spending escalated from $435 in 1950 to $571 in 1965 to $1,004 in 1990. At the start of that spending binge, America's largest cities were at the peak of prosperity—indeed, they had higher per capita incomes than the suburbs. Now, incomes are 50% lower than in the suburbs. A nearly 2.5-fold increase in outlays has not prevented urban bleeding; if anything, it has accelerated it.

Those figures understate the true extent of the budget buildup in cities. In the 1950s and early 1960s, cities had primary responsibility for funding welfare programs and indigent health care, whereas today the burden of funding anti-poverty programs is borne mostly by the Federal government and the states. In fact, most cities spend a much smaller share of their budgets on health and welfare than they did 40 years ago.

The cities that are least underfunded are not the smaller, more affluent communities, but the largest ones (*i.e.*, those that would be the beneficiaries of more Federal aid). Cities with populations over 500,000 spend roughly $1,200 per resident (excluding health, education, and welfare), whereas communities with populations under 75,000 spend about $550. The primary reason the expenditures of large central cities are so excessive is not that they have more responsibilities; it is that they are increasingly inefficient in providing basic services. It costs New York, Chicago, Los Angeles, Philadelphia, and other big cities twice as much to educate a child, collect garbage, build a road, police a neighborhood, and provide other basic services as it does smaller communities.

From 1980 to 1990, direct Federal aid to cities was reduced by about 50%. That reduction was made up in various ways. A 1990 study in the *American Economic Review* reports that, for every dollar the cities lost in Federal aid, they received an additional 80 cents in state aid. Moreover, while direct Federal aid to cities was cut in the Reagan years, aid to poor people living in cities increased. Federal social welfare spending—on education, training, social services, employment, low-income assistance, community development, and transportation—rose from $255,000,000,000 to $285,000,000,000 from 1980 to 1992. Those figures exclude Social Security, Medicare, and Medicaid—programs that have mushroomed in cost and significantly benefit inner-city residents as well.

Since 1989, domestic spending across the board, including outlays on urban aid, has exploded. Federal domestic spending under George Bush rose by almost 25%, the fastest rate of growth of the domestic budget under any president in 30 years. In real terms, cities and states received more Federal money in 1992 than in any previous year.

Since the late 1960s, the Federal government has spent 2.5 trillion dollars on urban renewal and the War on Poverty, or the equivalent of 25 Marshall Plans. The reason the money has not caused an urban revival is that the programs, particularly those that were abolished in the 1980s, did not work. For example, Urban Development Action Grants, which finally were abolished in 1987, subsidized the construction of major chain hotels, such as Hyatts, and luxury housing developments with rooftop tennis courts, health spas, and indoor tennis courts in Detroit.

Despite Federal grants totaling more than $50,000,000,000 for urban transit since the mid 1960s, total ridership has declined. The Federal government spent over $2,000,000,000 to build Miami's Metrorail, which the local population calls Metrofail; today, it has less than 20% of predicted ridership and its operating subsidies are in the hundreds of millions of dollars a year. A Congressional Budget Office audit of Federal wastewater treatment grants to cities found that construction costs were 30% higher when plants were built with Federal funding than when local taxpayers footed the bill.

Even cities that have received huge infusions of direct Federal aid have not been able to leverage those funds to resuscitate their economies. For instance, Gary, Ind., got more than $150,000,000 from 1968 to 1972 for urban renewal—or about $1,000 per resident—yet the city's deterioration continued.

Cuts in urban aid can not account for the Los Angeles riots, the exodus from New York City, sky-high poverty in Detroit, and other woes of the central cities. The period of catastrophic decline of population, incomes, and jobs was the 1970s, when urban aid exploded. By every meaningful measure of social and economic progress, the 1970s were the worst decade for cities since at least the 1930s.

By contrast, the Reagan years were a period of economic gains for many big cities. "The 1980s was the best decade in this century for the old central business districts," maintains *Washington Post* reporter Joel Garreau in his 1992 book on urban America, *Edge City*. "From Boston to Philadelphia to Washington to Los Angeles to San Francisco to Seattle to Houston to Dallas to Atlanta, the business districts of the downtowns thrived." Adjusted for inflation, the tax base of America's inner cities expanded by 50% in the 1980s, compared to 20% in the 1970s. The entrepreneurial explosion unleashed by Reaganomics had a very positive effect on the finances of big cities.

Medium-sized areas fared even better. More than 150 thriving new suburban cities have grown up during the past decade, such as Fairfax, Va.; Mesa, Ariz.; and Irvine, Calif. They quickly have become centers of enterprise and job creation. Until the recent recession, the problem confronting those booming cities was too much development and business investment, not too little.

An agenda for urban renaissance

Reviving America's depressed cities will require implementation of a growth-oriented agenda on the part of the Federal government, states, and cities. The overrid-

31. Stop the Decline of Our Cities?

ing goal of such a strategy must be to provide incentives for people, businesses, and capital to return to the inner cities.

Federal role. A proper adherence to the constitutional principles of federalism would dictate that the Federal government have almost no direct relationship with cities. All Federal programs that give direct aid to local governments and Federal regulations that mandate local spending should be abolished. Federal aid, to the extent that it continues, should be provided to the states. If cities and other jurisdictions of the states are in need of financial aid, it should be provided by the state legislatures.

If Congress feels compelled to assist areas that have deep pockets of poverty, money should be given directly to poor people, not city bureaucracies or service providers. There seems to be a bipartisan consensus emerging on such a strategy. A 1982 Brookings Institution report on urban decline emphasized that policymakers "should consider switching more federal aid from *empowering governments* to deliver services, to *empowering individuals and households* to purchase services or provide their own."

There are practical as well as philosophical objections to Federal grants to cities. As with Federal aid to foreign countries, little of the money ever gets filtered through the city bureaucracies. For example, according to the Wisconsin Policy Research Institute, Milwaukee spends about $1,100,000,000 in Federal, state, and local money annually on poverty abatement. That money flows through 68 programs that spend almost $30,000 per poor family, but only about 35 cents of every dollar ever gets to the poor. Most of the Federal money funds a massive welfare industry.

Probably the only effective Federal agenda for aiding the cities is the promotion of national economic growth. The lesson of the past three decades is that central cities' fiscal fortunes often turn with the national economy. In the slow-growth, high-inflation 1970s, cities rapidly deteriorated; in the prosperous 1980s, they partially revived; but in the recessionary 1990s, cities again are financially strapped. Reducing Federal deficit spending through expenditure control, growth-oriented reductions in payroll and capital gains taxes, a noninflationary monetary policy, and regulatory relief will have a very positive effect on cities. It would be best to provide the stimulus of tax cuts and regulatory relief to all areas nationwide, but, if politics precludes doing so, the enterprise zone concept is viable, as long as it means tax reduction and deregulation, not a new subsidy program.

States' role. State governments substantially increased their aid to cities in the 1980s, but during the 1990s, with budget problems in statehouses, aid to localities

6. CITIES, URBAN GROWTH: Cities

has declined. In general, such reduction is appropriate. States should not be in the business of paying for locally provided services or acting as the cities' tax collectors. Whenever possible, local services should be paid for with local taxes. State aid to localities is defensible only when it distributes funds exclusively to lower-income jurisdictions or pays for services that provide a direct benefit for the entire state.

The principal way state governments can promote the economic growth of their cities is to reduce the over-all state tax burden on individuals and businesses. States without an income tax—such as Florida and Texas—tend to have healthy and growing cities. Northeastern cities clearly have been harmed by the high state taxes in that region. For example, New York City has estimated that a $900,000,000 tax increase proposed by the state would cost the city some 300,000 jobs.

Cutting service costs

Cities' role. The key to restoring economic vitality and capital investment to declining cities is to reduce the costs of providing municipal services and then slash the heavy tax burdens that are required to pay for them. For no justifiable reason, unit service costs are substantially higher in large cities than in small ones—whether for education, public transportation, street cleaning, park maintenance, garbage collection, or police protection.

Labor costs appear to explain much of the inefficiency. Salaries of government workers in suburbs average $2,150 a month, compared to $2,700 a month in cities with populations over 500,000. If benefits are added, the disparities are even wider. Large cities also pay their employees substantially more than comparably skilled private-sector workers receive, and the gap is growing larger.

If cities had the political will to cut service costs and taxes, they could do so without sacrificing vital services. One way to begin is through competitive contracts. Smaller cities routinely contract out municipal services; large unionized cities seldom do. Indeed, some, such as La Mirada, Calif., contract out almost all their services and thus have tiny city bureaucracies. Several dozen studies verify that unit costs are reduced 20-50% by contracting out to the private sector. Moreover, the quality of contracted-out services is rated higher than that of services offered in-house.

However, public employee unions are so powerful in some large cities that not only is contracting out effectively prohibited, but any private-sector competition with the government monopoly service provider is forbidden by regulation. For instance, in New York City, private van and jitney services are providing fast, reliable transit for Manhattan commuters, yet the city transit agency has acted to shut such operators down. The action is contrary to the interests of residents and area workers. City provisions that prohibit private competition with the government should be ended.

Education is one of the largest items in city budgets. The declining cities tend to have much higher per-student costs, despite generally lower-quality public schools. In most large cities today, private schools provide a better education for half the cost of inner-city public schools. Central cities can cut costs significantly by recognizing that, when parents send their children to private schools, there are huge savings for the public school system. Even accounting for the fact that some school costs are fixed, if cities were to provide inner-city parents with incomes under $30,000 a voucher of, say, $2,500 per year to send each child to a private school, the public school systems significantly could reduce their operating costs and educational opportunities for children of low-income families would be improved.

There are diseconomies of scale in municipal services. Some cities, like New York, have such powerful special-interest lobbies and entrenched bureaucracies that it has become politically impossible to cut service costs, even in times of severe crisis. Large cities could reduce costs by splitting service responsibilities and conferring taxing authority on city districts, villages, or even homeowner associations. If the provision of services and the levying of service taxes were closer to homeowners and businesses, taxpayers would have greater influence on decisions about which services they need and which they do not, and they could place greater pressure on the government to reduce costs by diluting the influence of special-interest groups.

An even more radical idea is for cities to acknowledge that they have become unmanageable at their current size and to split up into separate smaller jurisdictions. It would make sense, for instance, for each borough of New York City to become a separate city, as Staten Island is attempting to do. Service responsibilities that can not be divided conveniently—poverty programs perhaps—then could be borne by the county government.

Finally, cities can lure businesses and people back by changing the composition of taxes. If all cities with income taxes were to replace them and other levies on industry with sales taxes, those cities substantially could improve the business climate within their borders. City income taxes are defended by local officials as fair because they fall primarily on upper-income individuals and big business. In practice, those taxes are paid by the working poor, the middle class, and small business owners, because the wealthy have fled most cities with income taxes or have received legislative exemptions from high taxes.

Urban advocates are right when they say that America's inner cities have been victims of destructive government policies, but not those of the Federal government. The wounds of the central cities largely are self-inflicted. A 1988 private audit of nearly bankrupt Scranton, Pa., stated that "the city government appears to exist for the benefit of its employees instead of the people." Those words could describe the operating principles and skewed priorities of too many ailing cities. Unless and until America's central cities start putting people first, cutting service costs and anti-growth tax rates, no amount of Federal aid can reverse the decline of urban America.

CAN CHURCHES SAVE AMERICA?

*Some politicians say faith-based programs cure social ills.
But the needs are huge and a backlash grows*

In the beginning came the governor's challenge: If each of Mississippi's churches would help just one poor family back on its feet, welfare could end. "God, not government, will be the savior of welfare recipients," Gov. Kirk Fordice likes to say. That's how Frances White, a jobless mother facing eviction from a crumbling house, suddenly found herself adopted by a suburban church. There were good deeds (church members paid White's bills), communion (they took her kids shopping) and finally redemption (through one congregant, White found a job as a hospital records clerk and now is partly off welfare).

Not long ago, no governor or politician could get away with asking churches to do the antipoverty work that is normally the responsibility of government. And it would have been the height of hokeyness to suggest that churches, synagogues and mosques could cure the drug addicted, feed the hungry, house the homeless, rehabilitate the criminal and lift the poor better than government. Yet with President Clinton and Congress agreeing to end welfare as we know it, there is talk of a second welfare revolution: Let churches and charities, not government, provide more of the social safety net.

Republican presidential nominee Bob Dole is a convert. Dole—who relied on private hometown charity when he returned from war severely injured—has endorsed a plan to shift tens of billions of social welfare dollars to direct tax credits. The money would reimburse taxpayers who donate up to $500 to poverty-fighting charities. And while it is largely conservatives and Republicans who want faith programs to do more, there are Democratic and liberal believers, too, like Joseph Califano, an architect of Lyndon Johnson's Great Society and cabinet secretary to Jimmy Carter. Califano, now head of Columbia University's Center on Addiction and Substance Abuse, says he was surprised when, on a tour of center programs, nearly every ex-drug addict he met cited religious belief as a key to rehabilitation. "I don't see anything wrong with public funding for a drug-treatment program that provides for spiritual needs," says Califano, "if that's what an individual needs to shake cocaine, to shake alcohol, to shake heroin."

Supporters of turning over social programs to churches and charities say their plan will reawaken American compassion. They claim neighbors no longer help neighbors, as they did before the rise of big government, because they assume their taxes pay for assistance programs. Further, advocates say charitable programs are more effective and efficient than government. Religion is often crucial to turning lives around, they contend.

Detractors see a dangerous trend that will distribute money unfairly, further cutting scarce resources to the poor. Worse, they say, counting on charities ignores history: The government social safety net grew because churches and volunteers could no longer deal with the entrenched poverty, the demands of a mobile society and the runaway health care costs of the late 20th century. Further, charities are already overburdened trying to respond to existing cuts in government spending on the poor, disabled and needy. Finally, nonprofit institutions are not a substitute for government, critics maintain, because government already provides 37 percent of the funding for charities, which is used to run programs from housing for the elderly to job training for the homeless and unemployed.

Here are the biggest questions shaping the debate:

CAN CHURCHES REPLACE THE SOCIAL SAFETY NET?
Frances White's journey off welfare is the kind of tale that supporters of Mississippi's Faith and Families highlight. But like most such odysseys, it included false starts and setbacks that show both the promise and the limitations of reliance on church-based programs.

It was Christmas 1994 when White, a divorced mother of three, got laid off from her third temporary job. ("Oh, God, please, my mother is out of a job again. Don't let this be happening," wrote White's 9-year-old daughter in her diary.) Governor Fordice had proposed Faith and Families just months before, and critics charged it was political gimmickry.

But for White, the program was a lifeline. In a state where about 25 percent of residents get some public assistance, For-

> Americans gave $143.9 billion to charities last year, up 7.8 percent from the previous year, the biggest increase in a decade.

6. CITIES, URBAN GROWTH: Cities

dice (who in 1992 controversially declared America a "Christian nation") challenged churches to assist some of the 141,000 people on Aid to Families With Dependent Children. "Government screwed it up," says Fordice of welfare. "People, one on one, can fix it." White volunteered to be "adopted" and was taken in by members of the Crossgates Baptist Church in Brandon, a middle-class suburb east of Jackson. "They paid my phone bill, found me a refrigerator, got my car running and some tires for it, caught up my rent, bought food. If it weren't for them, I would've been evicted," says the 40-year-old White.

Yet for White there was something more meaningful than the church's financial help: "The personal contact was better than anything. When you apply for welfare, you are a case number. But the church officials would call me, and we'd just talk."

The suburban church provided a ready-made job network, too. Through a church member, she landed work as a records clerk at a Jackson hospital. Making more than minimum wage, White left her substandard home. Although her rent jumped from $175 a month to $465, she doesn't have to worry, in her new home, about a leaky roof or cold winds whipping through the walls.

Still, White has climbed only a few tenuous rungs up the ladder from poverty. When she signed up with Faith and Families, she was receiving $144 per month in welfare and $360 per month in food stamps. She still gets $247 in food stamps because her salary is low and her new job does not provide health benefits. The Mississippi program promises to keep providing Medicaid a full year after a participant finds a job, but White and others found themselves bumped, because other state health officials were unaware of the pledge. For White, the results were disastrous. She lost her Medicaid just as her daughter was about to enter the hospital for a tonsillectomy. Unable to pay the $4,000 bill, White filed for bankruptcy in June.

At Crossgates Baptist Church, White counts as a success. But it's not one parishioners will try to replicate soon. The church spent $1,400 on her, far more than anticipated, and some members were miffed about aiding someone outside the congregation. "We're holding off on helping others right now," says Ken Box, chairman of the congregation's Benevolence Committee. "We have so many of our own church members who need help. We should take care of our own first."

The biggest disappointment with Faith and Families, Fordice concedes, is that there have been too few Frances Whites. In a year and a half, only 21 participants got off welfare. One problem is that just 98 welfare families—the most motivated—volunteered for the program. Explaining such reluctance, state coordinator Margaret Luckett notes, "The No. 1 question is always, 'Do we have to go to church?'" But White says that although she was often invited to Crossgates, she never felt compelled to attend. Congregations, too, are hesitant. Only 267 of the state's 5,500 signed up, and most of those never followed through on their commitment. Today, only about 15 churches are matched with families. Critics like the Rev. Rims Barber, who dismisses the program as "Faith and Foofoo," say that despite breakthroughs with mothers like White, it stigmatizes welfare seekers. "It's based on myths," says Barber, a Jackson Presbyterian minister. "The main one is that the problem with poor people is that they don't love Jesus."

As much as churches can add to the safety net, it is doubtful they can ever replace more than a fraction of government assistance. In White's case, for example, perhaps the most long-lasting aid she received was state-paid tuition to a local college to seek a bachelor's degree in clinical psychology.

Making matters harder, government is slashing funding for social services. Overall, each of the nation's 258,000 churches, synagogues and mosques would have to donate at least $225,000 a year to make up for congressionally proposed cuts, estimates Father Fred Kammer of Catholic Charities USA. But the average total budget of a congregation today, Kammer notes, is only around $100,000.

Paradoxically, at the exact moment churches are being asked to do more to save the poor, they face a backlash. This November, for example, voters in Colorado could make theirs the first state to tax the property of churches and many charities. In Hartford, Conn., the City Council has imposed a moratorium on the expansion of any new charity-run social services, like soup kitchens and drug clinics. A badly needed health center in one of the most distressed parts of town has been blocked. And although the moratorium and consequent zoning changes seem to hurt the poor most, it was the city's Puerto Rican deputy mayor and a black state representative, herself a former welfare mother, who pushed hardest for the moratorium. They say the impoverished city's many social services drive private businesses out of downtown.

DOES FAITH WORK BEST?

What's the surest guarantee that an African-American urban youth will not fall to drugs or crime? Regular church attendance turns out to be a better predictor than family structure or income, according to a study by Harvard University economist Richard Freeman. Call it the "faith factor." The link between religious participation and avoidance of drug abuse, alcoholism, crime and other social pathologies is grist for some intriguing new research. Says Brookings Institution political scientist John DiIulio, "It's remarkable how much good empirical evidence there is that religious belief can make a positive difference." Policy makers are loath to promote faith because of their intellectual bias, he argues. But in most inner cities, where government, schools and other institutions fail the poor, says DiIulio, it is church programs that are "leveraging 10 times their own weight and solving social problems for us." And they offer personal salvation. A survey by John Gartner of Loyola College of Maryland and David Larson of Duke University Medical Center found over 30 studies that show a correlation between religious participation and avoidance of crime and substance abuse.

New research goes further, exploring how faith can turn around already troubled lives. Federal prisoners who got leadership training from Prison Fellowship, a prison ministry started by Watergate conspirator Charles Colson, were 11 percent less likely to be rearrested after 14 years, according to one survey. A more rigorous follow-up found similar drops in recidivism among inmates who regularly attended prison Bible classes. Similarly, Alcoholics Anonymous invokes a "higher power" to overcome addiction. "Religion is the forgotten factor," says criminologist Byron Johnson of Lamar University in Texas. He studied Prison Fellowship ministries and concluded faith is essential to preventing recidivism. Family support, education, job training and other factors matter, too, he notes. But religion is the one piece of the puzzle that

> The two most reliable predictors of teenage drug avoidance: optimism about the future and regular church attendance.

most troubles government. One counselor at a Kentucky prison, on state orders, left the room when inmates in an alcohol-treatment group ended their daily meetings by standing up, holding hands and reciting the Serenity prayer. Complains Johnson: "We use pet therapy, horticulture therapy, acupuncture in prisons, but if you mention God, there's a problem."

Religion provides a set of values and moral beliefs. That's obvious. And churches provide a supportive community. That can be crucial. But belief seems to work at a more redeeming level. "To overcome addictions, you have to have phenomenal motivation," notes Patrick Fagan, who has reviewed studies on the impact of religion for the Heritage Foundation. Most addicts have so often disappointed loved ones that no one trusts them. So there is powerful blessing, says Fagan, in "a knowledge of God's acceptance, that he accepts the person and their sinfulness, that even as they fail they will not face rejection from God."

That seems clear at a singular drug rehab program in San Antonio. It is midday on the city's blighted West Side and the streets are ominously empty while some nearby buildings house cocaine and heroin dealers. But on one block, buzzing voices fill the street. More than 100 men and women sit in a parched yard, praying and discussing the Bible. At a tiny outdoor chapel, men crowd together in gos-

> **The divorce rate for regular churchgoers is 18 percent; for those who attend services less than once a year, 34 percent.**

pel song, prayer and witness. These are addicts, killers, thieves and gang members. But there are no guards and no security. "There's no fence. I could leave, but I still don't leave," says Joe, a founder of the Mexican Mafia, a brutal Texas prison gang. A tattoo of a snarling panther crawls up his muscular forearm, the better to hide the needle tracks of 32 years of drug addiction. "I can't explain it," says Joe, "but whatever it is, is working." To Freddie García, the ex-heroin addict who started the Victory Fellowship with his wife, Ninfa, the answer is simple: "The human spirit needs God."

It's hard to say what makes the program work for a man like Joe. Belief? "Now, I know a little bit about Jesus and he's nobody to fool around with," he offers reluctantly. Or the fact that Joe finds García's approach more culturally relevant than his too-many-to-count stays at professionalized drug clinics?

Unlike most government programs, for example, an addicted mother can come to Victory Fellowship without placing her kids with relatives or in foster care (even if it means the mother sleeps on the floor). On this day, García has reunited Joe with his skeptical wife, a reunion that has left the hulking ex-con fighting tears.

If faith is the forgotten factor that can turn around lives, García knows government is too often needlessly meddlesome. In the past year, Texas state officials objected to his referring to his program as "rehabilitation," since it did not hire licensed counselors. Another San Antonio faith-based treatment program, Teen Challenge of South Texas, faced a more serious threat last year when Texas revoked its license. Director James Heurich says state officials insisted it use more traditional therapeutic approaches and could not understand the group's view of drug addiction as sin,

---

MARVIN OLASKY'S APPEAL

A golden age of charity

Writer Martin Olasky dressed himself in rags and roamed Washington, D.C.'s streets posing as a homeless man. He got food, pills and shelter. But at a church soup kitchen, when he asked for a Bible, he was told, "I'm sorry, we don't have any Bibles."

Olasky is arguably the nation's most influential thinker on charities, and that moment six years ago sums up everything he thinks is wrong with the state of compassion. Charity fails, he argues, when it asks nothing in return. Recipients must be expected to change their lives fundamentally. Donors must personally get involved beyond signing their names on a check. "For compassion to be effective, it needs to be challenging, personal and spiritual," he says. "Federal government programs have provided entitlement, bureaucracy and an attempt to banish God." Most religious charities, he adds, expect too little. In the soup kitchen, Olasky wasn't even asked to remove his tray after he ate.

Olasky seeks a return to the pre-20th century era when people helped one another. That vision of a golden age of philanthropy, which Olasky examines in his seminal book *The Tragedy of American Compassion,* has inspired some Washington policy makers like House Speaker Newt Gingrich. Olasky would like to see government assistance phased out over time. That would lead to a revival of private charity, Olasky argues.

Good old days? Other students of charity argue that his glorification of the past is misinformed. "Charity wasn't sufficient in Dickens times, it wasn't sufficient in Hoover times and it isn't sufficient now," says Robin Garr, author of *Reinvesting in America,* a survey of community-based anti-poverty programs, including many run by churches. Indeed, University of Pennsylvania historian Michael Katz found that contrary to Olasky's depiction of massive private charity, in 1890s Buffalo, three quarters of public assistance to the poor was funded by government.

Olasky is a searcher. His study of charity resulted from his own reconciliation of being a Christian with service to others. The man who would put God back into charity was born Jewish, bar mitzvahed at 13 and became an atheist by 14 and a card-carrying member of the U.S. Communist party who traveled across the Soviet Union in his early 20s. A religious conversion in 1976 led him to fundamental Christianity. He now teaches journalism at the University of Texas at Austin.— J.P.S.

not disease. After members of the successful program protested at the Alamo, the state backed off.

That raises the tough constitutional question: Should public funding go to proselytizing groups? The separation of church and state, a valued constitutional principle, has allowed religious diversity to flourish. Still, some legal scholars, like Yale University law professor Stephen Carter, author of *The Culture of Disbelief,* say courts interpret this separation too strictly, thereby depriving churches of their proper "involvement in the public square." Indeed, silly bureaucracy can stifle the efficacy of faith, like New York officials who threatened to sue the City Mission in Schenectady because it refused to let the homeless and addicted men who lived there keep pornography.

Yet balancing religion and rights can be very tricky. In Fort Worth, the sheriff created a special living unit in the county jail for religious inmates to hold daily Bible studies and nightly chapel. But only fundamental Christianity is accepted and other mainstream religions, including Unitarianism, the Church of Jesus Christ of Latter-day Saints and Christian Science, are taught to be "cults." Other inmates sued, and a Texas court soon will consider whether the so-called God Pod creates a preferential living setup.

CAN GOVERNMENT HELP CHARITIES FLOURISH?

Betty Christian got job training from a state program—and she got some from an evangelical ministry. "Government training programs are so impersonal," she says. But at the Joy of Jesus, there were daily Bible lessons to talk about diligence, responsibility and other values that make good workers. And there was prayerful discussion of the loneliness and financial woes in her life. "It renewed my faith; it made me feel everything was not lost," she says.

The Detroit ministry was so successful at finding jobs for unemployed workers like Christian that two years ago Michigan officials asked it to accept state funding. There was one catch: The prayers and Bible studies had to go. The result, says Kevin Feldman of the ministry, was disastrous. The program, which placed 60 percent of its students in jobs while it relied on a faith orientation, saw the rate drop to near zero. Classroom sessions were disrupted by noisy, uninterested enrollees. Last December, Joy of Jesus closed its work program and returned the state money. It has been unable to reopen because few donors—other than government—provide money for adult job training.

> **Frequent churchgoers are about 50 percent less likely to report psychological problems and 71 percent less likely to be alcoholics.**

One proposed, but controversial, solution: vouchers. The new welfare reform law would allow states to give vouchers to unemployed workers, who, in turn, could choose a faith-based work program like Joy of Jesus. But letting individuals, not government, direct money to religious groups may still violate the separation between church and state. Ohio last week became the first state to give public money to students to pay for religious schools. But the Wisconsin Supreme Court last year stopped a similar plan in that state. Bob Dole champions school vouchers; President Clinton opposes them.

An idea also in vogue is using tax credits. Sen. Dan Coats would give taxpayers a credit equal to their donations to antipoverty charities—up to $500 for individuals and $1,000 for couples. The Indiana Republican estimates his tax credit would cost the Treasury $25 billion a year, and those funds would come largely from cuts in social programs.

Some 20 state experiments with tax credits show they sometimes infuse charities with needed funds. But tax credits can be costly to government, create unwanted paperwork for charities and, ultimately, distribute money unevenly. In Michigan, groups that aid the homeless get the credit, but only if that is their sole mission. So while other food pantries qualify, the American Red Cross Regional Food Distribution Center in Lansing is excluded. Further, food pantry officials say, it is unclear whether the credit stimulates substantial new giving or simply takes state revenues to reward those who give anyway.

Coats says tax credits let taxpayers, instead of government, decide which charities get money, "creating a free market of charitable giving." Citing a Beacon Hill Institute study, he argues that 67 percent of federal social service dollars go to overhead and public employees, instead of those who need services.

For many of the nation's most successful religious charities, however, whether to take public or private donations is not an either-or question. For them, government is a crucial partner. When blocks of central Newark, N.J., burned down during 1967 riots, William Linder was a parish priest. Today, Monsignor Linder presides over one of the nation's most astonishing urban rebirths. Linder started the New Community Corp., a community development corporation that used public and private money to revitalize the city's economy by building housing, nursing homes, day-care centers, restaurants—even a shopping mall with a supermarket so residents no longer have to leave the neighborhood to shop.

Faith fuels a uniquely successful type of compassion. That truth is also on display at Hope House, a residential facility for 51 abandoned and severely abused children in Nampa, Idaho. The 12 full-time staffers work for free, allowing the ranch to operate at a fraction of the cost of state facilities. "We believe in our heart we are called to serve," says founder Donnalee Velvick. Of the nine residents who will graduate from the facility's Christian academy in June, two are headed for military service and five for college. The remaining two, with mental retardation, will live in the program's adult house. Thousands of such faith-based charities across the country fill holes in the social safety net. The issue is whether they can fill even more.

BY JOSEPH P. SHAPIRO WITH
ANDREA R. WRIGHT IN MISSISSIPPI

THE PROJECTS COME DOWN

America's worst public housing facilities are finally getting the bulldozer they deserve. That isn't a solution to inner-city decay, but it could be the beginning of one.

Rob Gurwitt

Rob Gurwitt can be reached by e-mail at robbg@well.com.

Cotter and Lang Homes are twin housing projects that sit on about 70 bleak acres amid a set of working-class black neighborhoods in Louisville, Kentucky. While they may not quite rank on the same scale of affliction as projects in, say, Newark or Chicago, they are not pleasant places to live. Two-story, barracks-style apartments built in the 1940s with flat, leaky roofs, they are planted hard by what is, not coincidentally, Louisville's most notoriously violent and drug-ridden intersection. Much like a rust-eaten chassis that threatens an otherwise serviceable car, the projects form the corrosive heart of the poorest census tracts in all of Jefferson County. So Louisville has decided to do something about them.

Until recently, that would have meant, at best, making cosmetic changes that might improve the complexes slightly—upgrading the apartments, putting in awnings, doing some landscaping—but would avoid the central problems. Now, however, it means something entirely different. Cotter and Lang are coming down. If all goes according to plan, the Louisville Housing Authority will not only bulldoze them but replace them on the same site with about 1,700 units of single-family homes, townhouses and garden apartments, only a third of which will be reserved for public housing residents. The rest will be subsidized housing for the working poor and market-rate housing for middle-class families in search of value.

Then, the city plans to take the Cotter and Lang residents who don't make it back onto the redeveloped site and scatter them in small, mixed-income multifamily units elsewhere in the city, hoping to use its own construction as a way to begin rebuilding confidence in other shaky neighborhoods. "You can use it to start turning the tide of a situation where all you see is disinvestment," says Barry Alberts, executive director of the Louisville Development Authority.

This is a moment of radical change in public housing and in the history of the ghetto, in Louisville and cities all over the country, large and small. After decades of debate, the projects are starting to come down. It is far from clear just what will ultimately replace them, or what the result will be for those who have been trapped in them for the past generation. It is not clear what the job of a public housing authority will even be five years from now. Inevitably, at the end of the decade, there will still be concentrated housing projects in large cities. Still, there is no doubt that a turning point has been reached.

In much of inner-city America, 1995 will be remembered as the year of demolition. No fewer than 32 housing projects around the country are slated for total or partial destruction before it is over. In the next few years, there will be many more. Most of the 13 cities that have big high-rise projects have made it clear they would be just as happy without them. "Just about everyone wants to get rid of the family high-rise," says Wayne Sherwood, former director of research for the Council of Large Public Housing Authorities.

But if the rush to dismantle actual buildings produces the most drama, the true revolution these days is less visible: It is a new determination to change the basic dynamics of public housing. Instead of trying to administer dense congregations of the poor, housing authorities want to "deconcentrate" their tenants.

As in Louisville, many are turning quietly but decisively to scattered-site housing, whether in blocks of a few dozen units or in single-family homes. They are beefing up their use of Section 8 vouchers, the federally administered program that subsidizes rental housing for poor families. They are experimenting with so-called "residential mobility" programs that move inner-city residents to middle-class neighborhoods. They are trying to piece together funding for mixed-income developments in an effort to transform both the physical and the social environment of what is now the ghetto. And even where they are simply tearing down projects and then rebuilding on the site, they are planning to create far less dense developments geared toward improving tenants' odds of eventually escaping the system altogether.

In Denver, for instance, the local housing authority has torn down 422 units in its Lincoln Park development, and will replace them with 206 townhouses. It plans to create a "family investment center" on the site that will provide day care and educational programs aimed at helping residents find work and become self-sufficient. In Charlotte, North Carolina, the housing authority plans to demolish the 400-unit Earle Village and replace it, in part, with what it is calling "family transition" units and a community services center.

"To participate in the program," says John Kinsey, who is coordinating the Charlotte effort, "tenants will have to identify specific goals over a five-year period of time. They will have to agree to improve their education level or take job training or whatever is required to move

6. CITIES, URBAN GROWTH: Quality of Life

up and out of public housing within five years and be off all public assistance."

At the heart of all these moves is a conviction that by changing the circumstances in which tenants live, public housing will help the roughly 55 percent of residents who are neither elderly nor disabled to change their lives. "One of the reasons we're in this business," says Richard Gentry, executive director of Richmond, Virginia's housing authority, "is the belief that by changing the habitat, the surroundings, you can better their lives. We're trying to create a context within which normal life can occur."

Some skepticism is appropriate. Public housing authorities have talked this way at other times in their 60-year history. They talked this way, a generation ago, of the very projects they are now planning to level. Yet the underclass population they are dealing with today is depressingly vast: There are an estimated 4 million people across the country who live in public housing, 3.3 million of them officially and the rest off the books. As yet, deconcentration strategies have affected only a small minority of them; some of the schemes—most notably the residential mobility programs—still amount at this point to little more than boutique policies.

But none of this challenges the fundamental point. The era of massive housing projects crammed with the poor is coming to an end. "What we as public housing administrators have inherited are clearly ghettos of the very poorest people in projects that isolate them in the least desirable areas," says Andrea Duncan, executive director of the Louisville Housing Authority. "They are simply out of step with the times."

The notion that it might not, in fact, be the best social policy to sequester the poor in housing projects set outside the main currents of civic life is hardly new. Lee Rainwater, a Harvard sociologist, made the point as early as 1970 in his study of Pruitt-Igoe, the notorious St. Louis highrise development that foreshadowed current trends when large portions of it were pulled down a few years later.

In *Behind Ghetto Walls*, Rainwater detailed the design problems that amplified the project's troubles and talked about the ruinous social effects of putting large numbers of families together who, because of welfare and housing eligibility rules, were overwhelmingly poor, disproportionately headed by women, and predominately minority and unemployed.

Rainwater was writing at a time when there was still a deep reservoir of belief in the beneficial effects even of large public housing projects. It was possible in 1970 to read without snickering what Franklin D. Roosevelt had said 35 years earlier as he dedicated Atlanta's Techwood Housing Project and launched the federal public housing program. "Within a very short time," Roosevelt told the crowd, "people who never before could get a decent roof over their heads will live here in reasonable comfort and healthful, worthwhile surroundings." By May of this year, when demolition began on Techwood, the words seemed hopelessly innocent.

At the time Roosevelt made his remarks, of course, public housing was envisioned as a place for working families to get a toehold on upward mobility. Those days are long gone. The median income of families living in public housing now is less than $6,500, compared with over $35,000 nationally. HUD has estimated that some 80 percent of nonelderly public housing residents live below the poverty line, but that only gives a hint of the level of desperation. Most tenant households in large cities report income that is less than a fifth of the local median income, while overall, the portion of public housing residents making less than a tenth of the local median rose from just over 1 percent in 1974 to almost 20 percent in 1991. Public housing has, in short, become the place we put the very poorest of our citizens.

That policy carries a price. In the years since Rainwater published his Pruitt-Igoe study, a rich and extensive literature has grown up detailing the devastating effects of concentrating poverty, whether in public housing or simply in neighborhoods segregated by race and class. What is somewhat less familiar, although equally disturbing, is the effect the worst of the projects can have on the communities around them. A recent study of census tracts in Philadelphia by two academics at the University of Pennsylvania's Wharton School found that public housing gradually dragged surrounding neighborhoods into the web of poverty over the course of a decade. "As the distance to a large public housing project decreases," wrote law professor Michael Schill and real estate professor Susan Wachter, "the likelihood that poverty in a census tract will rise goes up."

That is not to say that public housing need serve its tenants poorly or drag down the surrounding community. New York City's high-rises are often safer and in better shape than their surrounding neighborhoods, in no small part because the city's tight housing market has helped them—until recently, at least—to safeguard their mixed-income character.

Even the worst of large, concentrated housing projects can be rejuvenated. Boston's Commonwealth development, for instance, is "arguably the single greatest success story in the country of turning around a severely distressed develop-

> **Housing authorities are ready to abandon their most distressed projects, soured not just by crime and social despair but by the sheer burden of trying to keep aging buildings going.**

ment," says Lawrence Vale, an associate professor of planning at MIT who has studied it. It took a determined—and $30 million—effort by the housing authority, the project's tenant organization and a private developer to make it happen, but the result, says Vale, is not just a tenant population that has the highest employment rate among public housing residents in the city, it is a housing project where suburbanites actually park their cars before taking public transit downtown.

Still, Commonwealth is an exception. In the vast majority of cities, public housing authorities are ready to abandon their most distressed projects, soured not just by the gangs, drug use, rampant unemployment and social despair but by the sheer burden of trying to keep aging buildings going. No new large-scale developments have been built in this country since the early 1970s, and much of the country's public housing stock

33. Projects Come Down

'We learned a hard lesson from urban renewal 35 years ago,' says one expert. 'There is a risk that viable communities are going to be swept away' in the new wave of bulldozing.

dates from decades earlier. Estimates of the cost to modernize existing public housing range from $14 billion to $29 billion.

Small wonder, then, that any number of cities want to start over. "Our buildings make no sense in 1995," says Zack Germroth, spokesman for the Baltimore Housing Authority, which intends to tear down all of its family high-rises. "In 1948, they probably made some sense, although that's arguable to a lot of people. They just were not designed for poor people to have any amenities; they're vertical warehouses."

Until fairly recently, local authorities did not have the freedom even to consider radical change. "Not only have our residents been isolated by past housing policy and the restrictive nature of where they could live but we as entities have not been allowed to be players in our communities," says Andrea Duncan. "We had no flexibility with money; our only choice was to dump it back into projects that were poorly designed in the first place."

Now, however, those restrictions have been lifted, in large part by the changed outlook of their patron and regulator, the U.S. Department of Housing and Urban Development. For the first time, HUD is willing to countenance public housing options that don't involve simply modernizing existing buildings. In fact, it is promoting them.

"One of the major tenets of what we are trying to do is, when you have bad public housing, tear it down," says Deputy Assistant Secretary Kevin Marchman. "If you build back up on the same site, put it up in a way that will blend into the community and that will allow people pride in where they live, or disperse it into other neighborhoods of the city. And where you don't build or acquire hard units, give people opportunities by having vouchers, so they can find housing in the private market." In fact, HUD has recently made it clear that it intends Section 8 vouchers eventually to take the place of much of the existing federal housing program.

And so, driven chiefly by their own desires and by some slacking of HUD's tight hold on the reins, there is tremendous ferment among housing authorities. In San Antonio, Texas, and Pueblo, Colorado, they are moving out of the business of running traditional housing projects by buying up properties taken over after the savings and loan debacle by the Resolution Trust Corporation. In Richmond and several other cities, they are buying existing housing with an eye toward selling it to tenants within a few years. In Baltimore, the housing authority has joined forces with the state to develop privately managed low-income rental housing scattered around the city.

In Hartford, Connecticut, Mayor Michael Peters pushed hard not just for the demolition of the sprawling Charter Oak Terrace but for redevelopment of the site for commercial uses. Instead, the city and the housing authority have reached an agreement to tear down one part of the project and replace it with about half the number of units, some single-family homes and some townhouses, with the goal of helping tenants eventually buy them. The other part of the project would mix residential and commercial development. "Everything we build from now on will be built not for housing for the poor," says Paul Capra, special assistant to housing authority director John Wardlaw, "it's going to be housing in which the poor might temporarily live, but it will be attractive to homeownership."

And then, of course, there are projects like Louisville's—or a similar proposal for a corner of Chicago's Cabrini-Green—that aim not just to reduce the number of public housing families living on a given site but essentially to develop the area economically and socially by making it attractive to working and middle-income families. No one has had much experience making such projects work, but its backers in Louisville are optimistic. An earlier project in another rundown neighborhood, overseen by the Louisville Development Authority, filled most of its market-rate units before its subsidized units. And a marketing study for the new project produced results that were surprisingly positive. "What we found," says Barry Alberts, "was that for African Americans, and particularly for some who had lived in that area before, there was an interest in returning and living in the inner city—in traditional African-American neighborhoods—if, in fact, the projects were to disappear and there was a good housing product."

There is, of course, another route that housing authorities are pursuing in their efforts to diffuse poverty, and that is to scatter it across an entire city or metropolitan area. Their thinking is that poor people will be more likely to find a way out of poverty and dependency if they can be helped to put the ghetto behind them.

One of the most extensive scattered-site programs in the country, in reach if not in absolute numbers, is Omaha's. After a modest start in 1987 buying houses around the city, housing authority President Robert Armstrong decided to push the program to the next stage by tearing down the city's worst project; then, after much controversy, he persuaded the city council to place the tenants equally throughout the city's seven council districts. In the years since, that notion has actually become overall city policy; eventually, each council district will have an equal number of public housing units spread throughout it. "Our plan," says Armstrong, "is over time to get rid of all our developments."

As Omaha's experience suggests, one clear factor in the success of scattered-site programs is how well housing officials manage to spread them out. That is not just because it would be pointless to reconstitute the ghetto in a new neighborhood; it is also because the politics of public housing become much dicier when you begin mixing public housing residents into working- and middle-class neighborhoods.

Springfield, Illinois, for instance, is just beginning to recover from a bruising fight over the housing authority's plans to build complexes in several neighborhoods. The numbers were not large by inner-city standards—mostly four to eight units—but their concentration was more than enough for the middle-class homeowners in whose neighborhoods they were to be located.

"When you put five or six units on one of our blocks, it's half the block," says

6. CITIES, URBAN GROWTH: Quality of Life

Charles Redpath, an alderman who represents one of the wards involved and a leading opponent of the authority's plans. "What you were doing was making concentrated housing in well-established neighborhoods that frankly didn't want it. The argument was we were basing our opposition on race, but that wasn't it—we had neighborhoods where some of the neighbors were black or Hispanic, and they said, 'We don't want this stuff, we worked hard to get here, we don't want concentrated housing in our neighborhood.'" In the end, the housing authority agreed to scatter the sites more broadly throughout the city.

The tangled politics of race and class have also recently put a dent in the other leading deconcentration strategy, known as residential mobility. Residential mobility programs help public housing tenants move to working-class or middle-income neighborhoods, either within the city or in the suburbs, subsidized by Section 8 vouchers. The notion is modeled after Chicago's Gautreaux program, which has placed about 5,600 public housing families around Chicago and in more than 115 suburban communities since 1976. Administered by the nonprofit Leadership Council for Metropolitan Open Communities, Gautreaux has been, in a sense, geared for success: It screens out families with poor rental records and more than three children; it avoids sending families to communities thought to be near a racial "tipping point"; and it provides its families abundant help in finding housing.

The results from Chicago and a smaller program in Cincinnati have been compelling enough that HUD set up five other programs around the country to study the impact more closely. The effort, known as Moving to Opportunity, was launched last year in Boston, Baltimore, Chicago, New York and Los Angeles. Earlier this year, however, Congress canceled the program's second year of funding after the first steps to move public housing tenants in Baltimore ran into fierce opposition in several suburban blue-collar communities that feared they would be overrun by inner-city blacks. Though the project continues to work with the first year's families in Baltimore, it will not expand as planned.

In the eyes of many housing analysts, MTO's mistake was that it operated blatantly. "Part of the success of Gautreaux and Cincinnati is they're essentially stealth programs," says Paul Fischer, a political scientist at Lake Forest College in Illinois who studied the Cincinnati effort. "In some communities, there may be many families who have moved out of public housing, but there's no heightened identifiability. The problem in Baltimore was that the people running the program decided to have outreach in communities where some of these families were going to be going. It lent itself to a tremendous amount of demagoguery."

Indeed, says MIT's Lawrence Vale, that may be enough to limit the use of such programs. "To me, the results of Gautreaux would suggest that it's very much worth having in one's repertoire," he says, "but I would guess that the level of community resistance is so high that cities that want to do it run the risk of proposing a program that is politically not feasible to carry out." Vale's point becomes even more compelling when you reflect that Gautreaux covers only a relative handful of Chicago's public housing tenants, and the best-functioning ones at that. How it could be meaningfully applied even to the 12,000 inhabitants of Robert Taylor Homes is difficult to imagine.

In fact, all of the strategies that housing authorities are trying these days present formidable problems, though none of them are insurmountable. Scattered-site housing can not only be difficult to find, it is more difficult to administer and manage than concentrated developments. That may not be an impossible burden for well-run housing authorities, but 100 of them around the country are currently on HUD's administratively "troubled" list, and others will no doubt find effective management difficult.

Mixed-income developments have their own limits. They are, to begin with, tricky to piece together, requiring housing authorities to develop the kind of adroitness at assembling funds that nonprofit developers gained during the 1980s. And even then, they are not going to work everywhere. "It's incredibly difficult to do in most areas where there is concentrated poverty," says George Galster, director of housing research at the Urban Institute. "At Cabrini-Green, because of its proximity to the Loop, it's conceivable that non-poor individuals could be induced to live with poor individuals because it's a great location. But Robert Taylor Homes? Uh uh, I don't believe it. It's a strategy with pockets of possibility, but it can't be thought of as generalizable."

As much as we know about the baneful effects of concentrating poverty, we don't actually know much about the impact of trying to deconcentrate it. It is more than a little ironic—bitterly so, in some cases—that renewing urban areas now involves, in part, tearing down the very places built for those who were displaced by the urban renewal of three decades ago. Nothing says that dispersing public housing residents won't carry hidden costs as well.

"We learned a hard lesson from urban renewal 35 years ago about destroying communities because of a rundown physical appearance," says Vale. "There is a risk, at least in some cases, that viable communities are going to be swept away, either intentionally or unintentionally, by deconcentration efforts."

And finally, creating new forms of public housing will undoubtedly require public housing authorities themselves to change. Already, the most forward-looking of them have abandoned their somewhat separate status as federal enclaves and begun to work far more closely than in the past with other local government agencies and service providers. "Our role has changed from being property managers to being owners of property where we must create an atmosphere that's conducive to positive learning, where we can assist, educate, help and employ individuals so they can become self-sufficient," says Omaha's Robert Armstrong. "The only way to do that is to have collaborative relationships with schools, police departments and social agencies that can provide the kinds of opportunities people need."

"What we're evolving into we still don't know," says Louisville's Andrea Duncan. "One possibility is a downsized entity that becomes more of an asset manager protecting the options for low-income people in the community—making sure that in the housing that's developed we're crafting a place for low-income people, because no one else will take that responsibility."

There is, of course, enormous risk for housing authorities as they abandon the past. If they're going to be putting up mixed-income developments, for instance, it's reasonable to ask whether they are the best qualified to take on the tasks involved. And if they are to adopt the goal of revitalizing poor communities, should they be taking the lead, or should they make way for development professionals to do so? Finding a new role may, in fact, turn out to be the greatest challenge housing authorities now face. As Lawrence Vale says, "If they aren't going to provide and manage public housing developments, who are they?"

Time Out

Plagued by stress, a growing number of people say they think time is becoming more precious than money and they're trying to slow down

John Marks

Behind the locked door of the emergency-ward bathroom of Sinai Hospital in Detroit, Dr. Cynthia Shelby-Lane desperately seeks to withdraw. It is past midnight; she has been soaring on adrenalin for hours, dealing with gun wounds and bloodied bodies from car accidents, and now she craves silence, an ounce of space to meditate and unwind. But the staff knows about her refuge. An EKG scan slides under the bathroom door. "What are you doing?" Shelby-Lane sputters. "You know this is my quiet time!"

Her cry defines an era. Five years before the end of what has been called the American Century, Americans say they have become worn down. But the exhaustion represents a paradox. They are extraordinarily stressed out even though they make more money, have more leisure time, spend more on recreation and enjoy more time-saving and efficient technology than adults did a generation or two ago. The reasons underlying this paradox are varied. Many Americans, especially married women, are working longer at their jobs now than they were then—although they have cut back on the amount of work they do around their homes. The anxieties wrought by the increasingly competitive global economy also have put many on edge, not to mention the fact that work-related intrusions via fax, E-mail or cellular phones can take place anywhere, from a living room to the family minivan.

In addition, analysts have found that the proliferation of labor-saving devices in the home makes it easier for folks to fret more about how to spend their free time—an option they largely lacked as recently as 50 years ago. Moreover, some of the stress arises from consumers' burgeoning sense of entitlement and expectations about what life ought to provide them. Finally, there's the thesis of economist Steffan Linder, who argued in a book called *The Harried Leisure Class* that affluence itself engendered "an increasing scarcity of time." His argument: Productivity increases the "value" of time spent at work, and folks who want to maximize their worth then feel they should work more.

Now, though, a growing number of citizens have begun to unplug their lives from a system they feel leaves little or no time to recharge. They have begun to retreat, like Dr. Shelby-Lane, into their private corners and demand at least some quiet time. In a comprehensive new quality-of-life poll conducted by U.S. *News* and the advertising agency Bozell Worldwide Inc., half of all Americans say they have taken steps in the past five years that could simplify their lives—steps as dramatic as moving to communities with a less hectic way of life, cutting back their hours at work, lowering their commitments or expectations and declining promotions. "People say they are pretty satisfied with their jobs but are searching for new simplicity," says U.S. *News* pollster Ed Goeas, who, with Celinda Lake, consulted on the survey. "They are uneasy about the time they're spending on the job, and some are already trying to act on it. For

48%
OF AMERICANS SAY THEY HAVE TAKEN STEPS IN THE PAST FIVE YEARS THAT COULD SIMPLIFY THEIR LIVES

others, it's a looming conflict." Fully 51 percent of those surveyed in the *U.S. News*/Bozell poll say they would rather have more time for themselves, even if it means less money.

Signs of stress and retreat are everywhere. Jeffrey Stiefler, the president of American Express, quit his job to "work a less intense pace and spend more time with my family." William Galston, a key adviser to President Clinton, walked away from his White House job this year after his 10-year-old son wrote him a letter: "Baseball's not fun when there's no one there to applaud you." Anna Quindlen, a high-profile columnist rumored to be the first woman in line for the top editor's job at the *New York Times,* quit her post to be home with her elementary-school-age children and write novels.

And a growing legion of female executives are leaving their corporate jobs. After 15 years climbing the ladder at DuPont, Stephanie Hood, 38, went from full-time to part-time work—30 hours a week. According to a recent study conducted by the corporation, in the last 10 years, 21 percent of DuPont employees have refused overtime or a job with more pressure; 24 percent refused jobs that require increased travel. Before Hood cut back, she passed up a job that would have involved trips to Asia and, inevitably, long hours away from home. "We wanted a better balance in our lives," she says of herself and her husband.

In Seattle, an entire social infrastructure has evolved out of the desire of many to make their lives less complicated. The movement, known as voluntary simplicity, boasts scores of newsletters and a growing number of devotees around the country who exchange ideas about consuming less, slowing down, saying no and getting out. Indeed, 28 percent of

6. CITIES, URBAN GROWTH: Quality of Life

respondents told pollsters for the Merck Family Fund they had voluntarily made changes in their lives that resulted in their making less money in the last five years.

Those who are taking steps to simplify represent a decidedly upscale group. According to the *U.S. News*/Bozell poll, which was done by KRC Research, the higher the income, the greater the likelihood that a respondent would place a premium on time and the more likely the chance that a respondent had moved to a less hectic place or declined a more stressful job. For instance, computer specialist Henry Hill, 48, was offered a post as sales director for a big software firm, but the job entailed his moving from Texas to California and would cut deeply into his time with his children, especially daughter Hayley, 8. "These jobs are man-eaters," he acknowledged and then turned it down.

Yet, for vast swaths of the middle class and working class, any thoughts about finding more free time are secondary to their anxious quest to make ends meet. In Troy, Ohio, Scott and Beth Lavy, both 46, show the stresses that money worries bring into those families. They have three sons, ages 14, 16 and 22, and Scott is a troubleshooter for the *Dayton Daily News*. Last year, in a company downsizing, two district representatives were fired, and he had to pick up the slack. Meanwhile, Beth delivers newspapers at around 6:30 a.m., one of the few moments in the day she gets time by herself. Some mornings, she'll switch on a Christian radio station and listen to people chat. It's not much, but it is solitude. Adds Scott: "The real world isn't simplified at all. If the man went to work, and what he made was sufficient for the family, and the wife would stay at home, that would be an ideal world."

A physical malady. Whatever its source, there is no doubt that a pervasive force in this culture is stress, a derivation of the Latin word that means "to be drawn tight." No corner of society is safe because virtually anything can cause stress. For instance, run-down consumers can buy stress maps, stress audits and diminutive stress cards that show, with the touch of a finger, whether

---

U.S. News/Bozell poll of 1,009 adults conducted by KRC Research Nov. 3–7, 1995, with consulting by *U.S. News* pollsters Celinda Lake of Lake Research and Ed Goeas of the Tarrance Group. Margin of error: plus or minus 3.1 percent.

TIME OR MONEY
Asked to pick between more free time or more money

- Some **51%** say they would rather have more free time even if it means less money.
- And **35%** say they would rather earn more money even if it requires more time.

---

a person is mellow or overwhelmed. Knowledge itself can cause trouble. "Twenty-five years ago, the random doctor in the random town thought he could do a fair amount of good and not much harm," explains John-Henry Pfifferling, who has been studying and treating burned-out physicians for over two decades. "Now, as a doctor, I am more aware not only of lawsuits but all the things in the health care system that can do potent harm."

And that stress literally racks the body. As pressures mount—the physician's knowledge of the harm his medicine might do, long working hours, family needs or guilt over the death of a patient—the body and mind rise to meet the occasion. Energy burns at a higher rate. Respiration, heart rate and blood pressure increase. Body temperature rises. Metabolism shoots up. The immune system braces for an onslaught.

On a short-term basis, this extra energy can work to advantage. If one has control of the various strands of his or her life, stress can be a kind of labor aphrodisiac, says Paul Rosch, professor of medicine at the American Institute of Stress. "It's a question of setting goals that are appropriate," he says, "beyond your grasp but within reach."

But if a person feels overwhelmed rather than challenged, stress becomes a burden. It causes depression and insomnia, strains the heart and nervous system and, over time, poses a danger to the entire body. Acute and chronic stress can lead to migraine headaches, hypertension, chest pains, ulcers and, ultimately, heart disease.

The malady, though hardly specific to the modern moment, is a national one. Seven out of 10 respondents told the *U.S. News*/Bozell survey that they feel stress at some point during a typical weekday—30 percent say they experience a lot of stress; 40 percent say they feel some stress. Forty-three percent of all adults suffer noticeable physical and emotional symptoms from burnout. Somewhere between 75 percent and 90 percent of all visits to the doctor's office stem from it. By one account, the country loses $7,500 per worker per year to stress, either through absenteeism, decreases in productivity or workers' compensation benefits.

No wonder those who suffer its worst effects are ready for a change. Michael Grossman is a case in point. After a bad day at work three years ago, the 55-year-old Hollywood movie executive returned home complaining of chest pains and wound up in an emergency room on the brink of a heart attack. He was treated quickly enough, but his values permanently changed. Within months of

---

THE HAPPINESS INDEX

Americans were asked how close they are to meeting their ideal goals on a scale where 1 means the goal has not been met and 10 means the goal has been reached. Analysts at KRC Research used the answers to develop measures of happiness called "quality quotients." Answers above 8 denote general happiness; below 7 denotes relative unhappiness.

	PERCENT WHO RANK ISSUE AS ONE OF THE TOP THREE PRIORITIES IN LIFE	QUALITY QUOTIENT
1. Family life	68%	8.18
2. Spiritual life	46%	8.25
3. Health	44%	7.68
4. Financial situation	25%	5.98
5. Their jobs	23%	6.82
6. Romantic life	18%	7.71
7. Leisure time	14%	6.14
8. Their homes	11%	8.12

QUALITY-OF-LIFE ENHANCERS

Asked to name the three things that most contribute to their quality of life, Americans cited these factors:

MEN	
1. Job/career satisfaction	32%
2. Relationship with family	28%
3. Money I earn from job/financial independence	18%
4. Good health	12%
5. Where I live (city/state/urban/suburban/rural)	11%
6. Religion/spirituality	11%
7. Relationship with spouse/significant other	10%
8. Relationship with friends	10%
9. Education level	8%
10. My home	7%

WOMEN	
1. Relationship with family	33%
2. Job/career satisfaction	28%
3. Good health	19%
4. Religion/spirituality	18%
5. Money I earn from job	17%
6. Relationship with children	14%
7. Relationship with friends	12%
8. Where I live (city/state/urban/suburban rural)	10%
9. Relationship with spouse/significant other	9%
10. My home	8%

quitting the job in movies, Grossman hired on at a fishing tackle store. "I went from $200,000 a year to $5 an hour," he says. "What I wanted out of life changed. The race was no longer worth the candle."

Hindi Greenberg, founder of an organization called Lawyers in Transition, helps attorneys like Grossman find alternative uses for their degrees. When Greenberg held the group's first meeting in the mid '80s, 25 lawyers showed up. Now, more than 70 regularly attend her seminars, most of them looking for a way out of a profession they feel has become too adversarial and time consuming.

Others throttle back when they feel their spirits getting sick. Fifteen years ago, Nancy Castleman quit her job with a New York City foundation, and her companion and business partner, Marc Eisenson, quit his contracting firm in the suburbs; they then fled for country life in upstate New York. There, they publish a newsletter called *The Pocket Change Investor* which advises readers how to save more. They rent a farmhouse and live on $15,000 a year. "People are much more responsive to our message than they were a decade ago," says Castleman. "Everybody can voluntarily cut back a little. You don't have to give up your day job or grow your own food."

Not everyone can afford to climb out of a former life. Instead, people make the smaller changes they can. Esther Thompson is like many who juggle their schedules and only fleetingly find guilt-free time for themselves. She works the assembly line at the General Motors plant in Flint, Mich. To make her morning shift, from 6 a.m. to 3:30 p.m., Thompson rises at 5 a.m. and gets her daughter, Darias, to company-run day care by 5:30 a.m. In the afternoons and evenings, Darias goes to ballet and gymnastics, and her bedtime is 8 p.m. Her mother's bedtime is 9 p.m.

Respite. It's hard to find gaps in the schedule. But recently, her shift ended early, and she was faced with that unimaginable rarity—two full, unplanned hours. She considered picking Darias up early from day care, then had a change of heart: "In my mind, I said, 'You deserve a couple of hours free.' So I stopped by my favorite Chinese place, went home, jumped into bed for a rest, then picked up Darias at the regular time."

Another kind of simplifying involves managing stuff rather than time. Bette Gall-Vaughn, 61, worked full time until three years ago, when a spinal-related disability began to slow her down. She and her husband, Ric, moved out of a 2,700-square-foot house into a 700-square-foot mobile home on Discovery Bay in the Seattle area. In spring 1994, their belongings went into storage—exercise equipment, bikes, a chaise longue, a washer and dryer and a dining room set with four chairs, among other items—and to this day she says she hasn't missed them. The storage space is only a mile away from her home, but she isn't interested. She now figures she had too much stuff in the first place.

Frugality. Gall-Vaughn is not alone, says Bob Lilienfeld, founder of Use Less Stuff, an organization devoted to the efficient reduction of waste in America. Two years ago, inspiration hit Lilienfeld when he helped a plastics company conduct a recycling campaign that was a big success. He started to publish a

A REGIONAL PROFILE
Here are some of the most fascinating characteristics of the people in each region:

WEST	CENTRAL	SOUTH	NORTHEAST
■ Residents least likely to be happy with their job situation. ■ Residents most likely to spend extra free time pursuing leisure activities. ■ Residents most likely to have lowered their expectations about what they need out of life. ■ Residents most likely to cite crime, violence and gangs as detractors from their quality of life. ■ Residents most likely to cite government as a detractor from their quality of life.	■ Residents most likely to want to spend extra free time with their families. ■ Residents least likely to want to spend extra free time on their own personal pursuits. ■ Residents most likely to cite health as a detractor from their quality of life. ■ Residents most likely to cite finances and bills as detractors from their quality of life. ■ Residents least likely to have reduced their work hours.	■ Residents most likely to say they have a great deal of control over improving the quality of their lives. ■ Residents happiest with meeting their spiritual goals. ■ Residents happiest with meeting their romantic goals. ■ Residents most likely to say their income is falling behind the cost of living. ■ Residents most likely to cite rudeness and selfishness of others as detractors from their quality of life.	■ Residents happiest with meeting their health goals. ■ Residents most likely to say their children's quality of life will be better than theirs. ■ Residents most likely to want to spend their extra free time sleeping. ■ Residents most likely to think of their work as a career rather than a job. ■ Residents most likely to cite relationship with their children as enhancing their quality of life.

6. CITIES, URBAN GROWTH: Quality of Life

newsletter and last month announced the first Use Less Stuff Day to encourage folks to conserve over the holidays. For instance, he says that in New York City, one less grocery bag per person per week would save 5 million pounds of waste per year and $250,000 in disposal costs.

His suggestions echo those of *Simple Living* magazine and the *Tightwad Gazette*, a Maine-based publication that advocates frugality. His holiday decorating recommendations: "Use collected pine cones, dried flowers and other natural materials to decorate wreaths and make centerpieces. Use clove-studded oranges for room fresheners, or make homemade potpourri." This is quintessential 1990s simplicity.

Generation X has its own version. Its flannel-covered simplicity influences every aspect of today's youth culture. It tends to be antimaterialistic, like its older cousin, but contains a bit more anger and irony. Gen X simplicity can be heard in the new passion for acoustic, or "unplugged," music, pioneered by MTV in the early 1990s and now ubiquitous. Simplicity can also be seen in a multitude of sitcoms from "Seinfeld" to the current smash, "Friends." In these shows, young people lead ridiculously uncomplicated existences. Their attentions are focused on minutiae: on watching a video, a trip to the dentist or body odor. It is life at a microcosmic level, simplicity for people not ready to settle down.

Kahane Corn, 33, shares many of these sentiments. A free-lance filmmaker who lives in California, she has spent most of her post-collegiate life with an empty bank account and an idealistic sense of what she wanted out of life. She has struggled to get her own films made and produced her own documentaries, a genre with few lucrative prospects. She has enjoyed the simplicities of the unattached, single life. But as she gets older, Corn says, she thinks increasingly about making money and settling down. "I'm much more willing to make a compromise now," she says.

The evolution of her sensibility raises a problematic aspect of the simplicity trend. Not everyone is convinced that a genuine national movement is gelling. Skeptics say changes in the lives of the baby boomers account for the high numbers of people who said in the *U.S. News*/Bozell poll they had taken steps that could simplify their lives. Any generation at midlife will start focusing less on work and more on family, notes psychologist and stress expert Georgia Witkin. And people undergoing big changes like moving because of economic difficulties or scaling down grandiose career aspirations often rationalize those events. They say they are trying to simplify their lives when in fact other forces have propelled their decisions, notes sociology professor John Robinson, director of the Americans' Use of Time Project at the University of Maryland.

Perceptions. Another problem with making judgments about the relationship between work and stress, say experts like Robinson, has to do with perception. People rarely give an accurate account of the time they spend at work. Usually, they feel they are working far longer hours than is actually the case. People actually have more free time than they realize. How they use that time is another story.

But Fran Rodgers, CEO of Work-Family Directions, a prominent workplace consulting firm, believes something is going on. "There's an incredibly increased desire to have more control over one's life," Rodgers says, "to have more control over what is controllable, to spend more time with family." In the *U.S. News*/Bozell poll, about half the respondents said they'd come pretty close to meeting their goals for controlling the way they spend their time; the other half expressed qualms. In all, Rodgers does not believe many people are actually putting their desire into effect. On the contrary, she says, men have not made many changes at all. If anything, women have begun to leave the workplace to spend more time with children, and men are working harder than ever.

Maybe Dr. Shelby-Lane can teach them a thing or two back in Detroit. Last August, after falling asleep at the wheel of her car, she went to her superiors, and they quickly granted her request for mixed daytime and night shifts. She has learned how to say no to commitments that would bog her down, even if she wants to do them. Most important, she has tapped into her own sense of humor as a source of release—she's a stand-up comedian on the side.

But until leisure becomes a synonym for health rather than guilt, until companies uniformly encourage employees to spend as much time at home as they can, few people will see a need to slow down. Stress, for the time being, remains an American constant; simplification a tantalizing possibility for the truly daring.

Can the Schools Be Saved?

Chester E. Finn, Jr.

CHESTER E. FINN, JR., *a former Assistant Secretary of Education, is John M. Olin fellow at the Hudson Institute in Washington, D.C.*

PUBLIC EDUCATION in the United States is a vast enterprise, involving some 45 million young people, or three-fifths of all Americans under the age of nineteen; 85,000 schools; 5 million employees; and a cost to taxpayers of more than a quarter-trillion dollars annually. Vast as this enterprise is, it is also increasingly precarious. The evidence is by now so familiar as hardly to bear repeating:

- In 1994, according to the National Assessment of Educational Progress (NAEP), six out of seven eighth-graders were not "proficient" in American history, and 57 percent of high-school seniors registered "below basic" in this subject.
- In the same year, three out of four seniors were less than "proficient" in geography; 30 percent lacked even rudimentary understanding.
- Two out of five fourth-graders, including two out of every three black and Hispanic youngsters, can hardly read at all. Among high-school seniors in 1994, only 36 percent were "proficient" in this most basic skill.
- Almost half the entering freshmen in the California state-university system in 1994 required remedial instruction in reading or math or both—the fifth straight year in which this number has increased.

This ever-expanding catalogue of inadequacy and outright failure has led to a development full of significance for the future: a broad cross-section of America is losing faith in public education itself. In one 1995 survey, 72 percent of respondents voiced concern about drugs and violence in local schools, 61 percent complained of low standards, 60 percent of a lack of attention to "basics." In the words of the report summarizing the data, "support for public education is fragile and porous." Even Albert Shanker, president of the American Federation of Teachers, has declared that "time is running out on public education. . . . The dissatisfaction that people feel is very basic." Or, in the recent words of David Mathews, a former secretary of the Department of Health, Education, and Welfare, "Americans today seem to be halfway out the schoolhouse door."

Indeed they do. Hundreds of thousands of families are already paying private firms to tutor their children in skills and branches of knowledge they are not acquiring in school. Others have increasingly begun to opt out altogether. Private-school enrollment, though still below the "market share" it attained in the 60's and early 70's, has risen faster than public enrollment for the past five years; Catholic schools have arrested their long-term slippage; and home schooling is on the rise.

6. CITIES, URBAN GROWTH: Education

Can the institution of public education be saved? Should it be?

However one answers these questions, there can be no doubt that the *principle* of public education runs deep in our society. The American Founders clearly believed that some general provision of education was necessary if democracy was to prosper. Though the word "education" appears nowhere in the Constitution—it was a responsibility reserved for the states—the writings of the Founders are replete with references to the need, as Thomas Jefferson put it, to "enlighten the people generally [so that] tyranny and oppressions of body and mind will vanish...."

By the mid-19th century, the rationale for public education had become very ambitious: universal schooling would not only equip the poor as well as the rich with skills that would enable all to succeed economically, but it would also foster and reinforce American civic ideals, in particular by easing the assimilation of immigrants and thus strengthening national unity. Horace Mann (1796-1859), the foremost educational thinker of his day, conceived of schooling as "the great equalizer of the conditions of men—the balance-wheel of the social machinery."

In fulfillment of these ideas, government gradually but inexorably became involved with the delivery of education. By the 1840's, Massachusetts, under Mann's leadership (and borrowing heavily from the Prussian model), had established the elements of a universal system of government-financed and government-run elementary schools. By the 1850's, New York City had a semblance of public schooling. And as newer territories joined the union, their constitutions all embraced a state obligation to furnish education to the citizenry.

But no sooner was the mid-19th-century model established than it began to change under the pressure of various movements of "reform." The cumulative effect of these waves of reform has much to do with where we are today.

Perhaps the earliest such wave was romantic progressivism, which introduced into the schools Rousseauian notions of "natural" learning. Rather than imposing the judgment of adults about what was worth learning and how it should be taught, classroom activity, it was held, should center on the interests and proclivities of the young. By the 1920's, progressivism was joined and extended by the "mental-hygiene" movement, whose thrust, as the psychiatrist Yale Kramer has written, "was to transform the goals of the public school from education to therapy. The schools were to produce not informed and skilled students but students with healthy personalities."

Progressivism had its undeniable virtues, especially when contrasted with the exclusive reliance on rote memorization of some 19th-century classrooms: children do learn best when the material they study is interesting, the teacher is encouraging, and their minds are engaged. But progressivism was nevertheless in tension with the notion of externally set standards and vigorous, teacher-led instruction. In many schools today it has been carried farther still. There is no punishment, no confining rows of desks, no wearying homework, no prescribed program of study, little respect for authority, and less discipline. In contributing to this state of affairs, progressivism and the various contemporary incarnations of the mental-hygiene movement have been a calamity, depriving children of basic skills while providing an amplitude of activity designed to augment empty concepts like self-esteem.

A different fruit of the progressive movement in American history was civil-service reform. The changes it brought had less impact on what went on in the classroom and more on how schools were governed. Reacting against the widespread practice of reserving educational slots for patronage, reformers lobbied for the creation of nonpartisan structures. In localities across the country, they succeeded in establishing elected or appointed boards consisting of eminent, public-spirited citizens who in turn entrusted their executive duties to certified professionals.

This had the beneficial effect of putting public schools beyond the sometimes avaricious grasp of mayors, aldermen, and party bosses. But it also had the less than desirable consequence of sealing off the whole arena of educational policy from the self-correcting mechanisms of democratic politics. Today, public schools may not be at the mercy of patronage machines; instead, they operate in an anaerobic environment in which interest-group self-dealing flourishes away from the public eye.

Even as civil-service reforms were insulating the schools from politics, the disciples of Frederick Winslow Taylor (1856-1915), the "management-science" genius, began to apply his precepts to education. A heightened emphasis came to be placed on credentialed expertise, disinterested professionalism, orderly management, and uniformity. By the thousands, small school systems were consolidated into districts in which only state-licensed teachers could be hired; principals and superintendents were also required to obtain professional certificates. Once again there were some benefits, and once again there were costs: colleges of education gained a stranglehold on entry into the profession; the priorities of parents and communities receded beneath the theories and dogmas of professors of education;

and villages and towns lost direct control of their children's schools.

The most damaging changes to public education came after World War II. One source of trouble was the expanding role of the federal government, which made its debut in this arena in the 1950's in the name of combating racial segregation. In the 60's, Washington came to play an even greater role on behalf of the poor. In the 70's and 80's, it extended its reach dramatically to feed a growing army of special interests. Thus dawned the era of 1,000-page federal-education statutes and a 5,000-person federal Education Department managing hundreds of distinct programs, each with its stakeholders, state and local bureaucratic counterparts, and a bewildering array of regulatory requirements.

Of course, many of the problems the federal government stepped in to tackle were genuine, beginning with the shame of the "dual" school system for whites and blacks. But federal expansion also meant that schools came to be enveloped in red tape, and many policy disputes that had formerly been settled at the local level were now drawn into national politics.

The growth of Washington's role was mirrored by a process that had equally far-reaching implications: the conversion of teachers' professional organizations into the two giant unions known today as the National Education Association and the American Federation of Teachers. As with most union movements, workers with bona-fide grievances—wretched pay, bans on marriage for female teachers, race-based hiring—joined together to win better terms of employment. With the passage of time, however, the unions came to push onto the bargaining table all manner of policy and management decisions. By generously funding the political campaigns of friendly candidates at every level of government, they gained immense power within the public-education system. Today the teachers' unions have become a formidable bulwark against any true reform of our educational mess.

As if all this were not enough, perhaps the greatest harm befalling public education over the last few decades has come from the intellectuals. The currents of deconstructionism, relativism, and multiculturalism that gained force in the nation's universities in the 1960's and 70's, and then spilled over into the broader society in the 1980's, have left a deep imprint on the schools. It is visible in the pervasive teaching of revisionist anti-American accounts of history, in separatist literature seemingly designed to heighten ethnic discontent, and in the growth of bilingual education, which has turned the traditional mission of the schools on its head by *slowing* the pace at which immigrants assimilate and thus contributing not to the unity but to the further balkanization of American culture and society.

PARENTS, THEN, are right to be up in arms about the schools; politicians are right to be exercised; and professionals like Albert Shanker are right to be alarmed. Indeed, ever since the United States was labeled a "nation at risk" in 1983 by the National Commission on Excellence in Education, a new movement of school reform has taken root. Today, it is a whole industry, staffed by thousands of experts and funded by generous federal and private grants. In a manner characteristic of all such undertakings, its primary output appears to be innumerable conferences, journals, books, studies, and—considering the magnitude of the problem—precious little else.

The most common approach of the school-reform industry has amounted to piecemeal tinkering with the countless gears and levers of the existing educational machinery: upgrading teacher-training programs, stiffening graduation requirements, installing modern technology, revamping reading programs, shrinking class size, adding a period to the school day, and on and on through hundreds of variations and permutations.

Certainly, many such changes are worth making. Ample evidence exists, for example, that tougher graduation requirements impel students to take more challenging courses, Better training for teachers and modern information technology help, too, in their own ways. But piecemeal reform will not fundamentally alter the working of a system in such serious disarray. As Steve Jobs, the full-time computer pioneer and part-time school reformer, wrote earlier this year:

> I've probably spearheaded giving away more computer equipment to schools than anybody else on the planet. But I've had to come to the inevitable conclusion that the problem is not one that technology can hope to solve. . . . It's a political problem. . . . The problems are unions. . . . The problem is bureaucracy.

And Jobs is hardly alone.

A more ambitious reform strategy concentrates on redesigning individual schools. Its proponents would devolve management authority away from the central bureaucracy; empower principals; and experiment with novel designs for school structure and curriculum. Examples include Chicago's local school councils, the fast-spreading charter-school movement, and the seven competing school-design "teams" underwritten by the New American Schools Development Corporation.

Many of the reforms proposed under this rubric make considerable sense, creating a variety of possibilities within school districts and, ideally, enabling parents to choose among them. But in the end this approach, too, faces severe constraints. Resistance from entrenched interest groups has already been great, and seldom has much real power—like the right to hire and fire teachers—actually devolved. Innovative governance schemes have often deteriorated into bickering or paralysis. Finally, in a country with tens of thousands of schools, the proportion of educators with truly fresh ideas and the courage to take risks is too small for an approach like this to transform the entire enterprise.

A different and still more far-reaching tack concentrates less on structure than on substance, bringing academic standards, curriculum, textbooks, tests, and teacher training into national or statewide alignment. This strategy underlies President Clinton's "Goals 2000" program and the efforts by many educators, business leaders, and governors to impose uniform standards on entire states and communities. Here, again, there is much to praise in principle, but there are also pitfalls that have been well-illustrated in practice. For one thing, standards are easily hijacked by ideologues—as we have seen in the recent uproar over national history standards.* For another thing, to press so many divergent elements into true alignment would require additional layers of bureaucratic regulation and the kind of top-down control that has created so much difficulty in the past.

The most radical solution of all is perhaps better characterized as an abandonment of the public-school model than as a reform. Voucher and privatization advocates believe that the market (fueled by the tax revenues that are now directed exclusively to public schools) can rise to the challenge of educating America's young. A number of Wall Street firms have already sponsored conferences for investors eager to play in this emerging field, and several corporations are launching ambitious schemes for privatizing education much in the way security services have been privatized in neighborhoods weary of inadequate police protection. The grass-roots have also become engaged: at one end of the movement, a lobbying organization called the Separation of School and State Alliance is agitating for a complete halt to "government-compelled attendance, financing, curriculum, testing, credentialing, and accreditation."

Would thrusting the schools into the rigors of the marketplace have a beneficial effect? It would certainly free some youngsters from the dead hand of awful schools, and it would crack the government's near-monopoly that now protects such schools. It should also foster innovation and flexibility, although only if a degree of entrepreneurship arises which private schools have thus far rarely demonstrated. But such a move faces intense political opposition from the entire public-education establishment. Nor is it yet clear that for-profit schools can yield a competitive return to investors while providing solid skills to those children whom the public schools have failed most notoriously: namely, the inner-city poor. Finally, the scheme might well conduce to the creation of separatist schools that reject the American civic culture even more forcefully than conventional public schools, under the spell of multiculturalism, are already doing.

Do I have a better way? What I have is not so much a blueprint as a set of principles, and a concept of what an American public education worthy of the name should look like. In other words, I would start from where we want to get to, not from where we are, although I would also borrow freely from the reform strategies already being tried.

To begin with structure: public education ought not mean government-run schools. Society's obligation is to see that instruction is provided and that learning occurs. It is not to operate a bureaucratic system of uniform institutions staffed by government employees.

As I envision them, public schools would continue to be open to all, financed by tax revenues, and accountable to elected authorities; they would thus remain public in every significant sense of the term. But—and here I side with the advocates of vouchers and other such schemes—the schools would also be accountable to their clients, who would be free to select among them and never forced against their will to attend a bad one. The management of the schools would be "out-sourced," "chartered," "contracted," or "co-oped." This means that schools might be directed by groups of parents or teachers, by private firms, or by nonprofit organizations.

Ending the present monopoly of the bureaucrats and unions would liberate thousands of capable educators to break with the orthodoxies of their profession and create schools that parents might actually want their children to attend. Such schools would differ in dozens of ways: style of pedagogy, form of organization and governance, mode of discipline, hours of operations, and so on, up to and including curriculum, at least those elements of curriculum that fall outside the common core.

Where would that core come from? We have had enough sorry experience with standards to have learned that academic experts are not to be

---

* See Walter A. McDougall's "Whose History? Whose Standards?" in COMMENTARY, May 1995 and "What Johnny Still Won't Know About History," July 1996.

trusted with writing a curriculum designed either to inculcate basic skills or to foster civic culture; nor can this task be left entirely to individual teachers and schools, much less to the federal government. Although I am as allergic as the next person to Rockefeller-type commissions, they are not all bad—the one that produced the *Nation at Risk* report in 1983 had much to commend it—and a panel of respected civic, business, and intellectual leaders, *privately* constituted, might be able to frame a core "curriculum of national unity" for states and communities (and textbook writers and test-makers) to adopt if they find it meritorious, and to suggest the performance standards that would denote student mastery of such a curriculum. That, at least, would be a beginning.

Most critically, the schools I have in mind would be built not around intentions, expenditures, or credentials, but around academic results. Examinations, devised and administered from without, would assess whether and how well the curriculum is learned. Students, teachers, and principals would be held accountable for performance, and real consequences would follow from failure and success.

IN SHORT, what I would do is to encourage freedom and rigor at once: freedom of structure, rigor of substance. Above all, however, by means of that substance I would restore to the schools their vital role in fostering, reinforcing, and transmitting the sense of a common civic culture.

Here perhaps is where I differ most fundamentally with recent reform strategies, virtually all of which focus on useful skills and enhanced productivity. If that were all we wanted from our schools, we might well stick with the efforts of the privatizers, standard-setters, and tinkerers. They have shown signs of modest progress and, particularly with greater political support, would likely make more.

But the narrowly utilitarian function of public education was only part of Horace Mann's vision, and should be the lesser part of our own. Simply put, the great project of public education in America should not be the creation of skilled workers but the formation of Americans. If our children do not assimilate a common body of knowledge and values, a shared set of customs and institutional arrangements, there will be little to prevent us from falling even farther apart as a nation.

Yet there, precisely, is the rub. National cultural transmission may be the most pressing task before public education today, but it is also the one task rejected out of hand both by many of the staunchest defenders of the public schools and by many of their most ardent reformers. Worse: although it is a task which seems to have fallen by default to the schools alone, they can hardly be expected to perform it alone.

For reasons having little to do with education *per se*, the base on which the old model of public education was originally constructed—namely, the assumption of an already existing common culture, sustained by home, church, media, and national political institutions, not to mention a wide network of consensual ideas and social habits—has broken down. It would be foolish to suggest that the schools, which all too faithfully reflect the condition in which we find ourselves, can single-handedly mend it. Although I believe that a full-fledged revolution from within could indeed recreate a worthy form of public education, even the best of schools cannot recreate a sound civic culture—and that, alas, not only is the more urgent, but may be the prior, task.

Off course

Women's studies has empowered women to speak up in class. The problem is what they're often talking about.

Karen Lehrman

Karen Lehrman is writing a book on postideological feminism.

It's eight o'clock on a balmy Wednesday morning at the University of California at Berkeley, and Women's Studies 39, "Literature and the Question of Pornography," is about to begin. The atmosphere of the small class is relaxed. The students call the youngish professor by her first name; the banter focuses on finding a man for her to date. She puts on the board: "Write 'grade' or 'no grade' on your paper before turning it in." Students—nine women and one man—amble in sporadically for the first twenty minutes.

Today's discussion involves a previous guest speaker, feminist-socialist porn star Nina Hartley. The professor asks what insights the students gained from Hartley's talk. They respond: "She's free with her sexuality.... I liked when she said, 'I like to... my friends.'... No body-image problems.... She's dependent in that relationship...." The professor tries to move the discussion onto a more serious question: have traditional feminists, in their antiporn stance, defined women out of their sexuality? After a few minutes, though, the discussion fixes on orgasms—how they're not the be-all and end-all of sexual activity, how easy it is to fake one. The lone male stares intently at a spot on the floor; occasionally he squirms.

I never took a porn class when I went to college ten years ago. In fact, I never took a women's studies class and don't even know if the universities I attended offered any. Women's studies was about a decade old at the time, but it hadn't yet become institutionalized (there are now more than six hundred programs), nor gained notoriety through debates over the canon and multiculturalism. But even if I had been aware of a program, I'm certain I would have stayed far away from it. It's not that I wasn't a feminist: I fully supported equal rights and equal opportunities for women.

But I was feminist like I was Jewish—it was a part of my identity that didn't depend on external affirmation.

Perhaps more important, as a first-generation career-woman, I felt a constant need to prove my equality. I took as many "male" courses—economics, political science, intellectual history—as I could; I wanted to be seen as a good student who happened to be a woman. There were a couple of problems, though: I didn't learn much about women or the history of feminism, and like most of my female peers, I rarely spoke in class.

Last spring I toured the world of women's studies, visiting Berkeley, the University of Iowa, Smith College, and Dartmouth College. I sat in on about twenty classes, talked to students and professors at these and other schools, amassed syllabi, and waded through the more popular reading materials. I admit to having begun with a nagging skepticism. But I was also intrigued: rumor had it that in these classes, women talked.

And they do. The problem, as I see it, is what they're often talking about. In many classes discussions alternate between the personal and the political, with mere pit stops at the academic. Sometimes they are filled with unintelligible post-structuralist jargon; sometimes they consist of consciousness-raising psychobabble, with the students' feelings and experiences valued as much as anything the professor or texts have to offer. Regardless, the guiding principle of most of the classes is oppression, and problems are almost inevitably reduced to relationships of power. "Diversity" is the mantra of both students and professors, but it doesn't apply to political opinions.

Not every women's studies course suffers from these flaws. In fact, the rigor and perspective of individual programs and classes vary widely and feminist academics have debated nearly every aspect of the field. But it seems that the vast majority of women's studies professors rely to a greater or lesser extent, on a common set of feminist theories. Put into practice, these theories have the potential to undermine the goals not only of a liberal education, but of feminism itself.

This doesn't mean, as some critics have suggested, that these programs should simply be abolished. Women's studies has played a valuable role in forcing universities to include in the curriculum women other than "witches or Ethel Rosenberg," as Iowa's Linda K. Kerber puts it. The field has generated a considerable amount of first-rate scholarship on women, breaking the age-old practice of viewing male subjects and experience as the norm and the ideal. And it has produced interdisciplinary courses that creatively tie together research from several fields.

Whether all this could have been accomplished without the creation of women's studies programs separate from the traditional departments is a moot question, especially since these programs have become so well entrenched in the academy. The present challenge is to make women's studies as good as it can be. Although the problems are significant, they're not insurmountable. And perhaps more than anything else, women's studies prides itself on its capacity for self-examination and renewal.

Berkeley was the only stop on the tour with an actual women's studies department. It is one of the largest, most established and respected programs in the country. Overall, it impressed me the least. At the other extreme was Smith, where the classes tended to be more rigorous and substantive and there was a greater awareness of the pitfalls of the field. (The students were also far more articulate, though that may have little to do with women's studies.) I found the most thoughtful professors in Iowa's program, which doesn't even offer a major. The program at Dartmouth, perhaps compensating for the school's macho image, seemed the most prone to succumbing to the latest ideological fads.

Discussions run from the personal to the political and back again, with mere pit stops at the academic.

CLASSROOM THERAPY

"Women's studies" is something of a misnomer. Most of the courses are designed not merely to study women, but also to improve the lives of women, both the individual students (the vast majority of whom are female) and women in general. Since professors believe that women have been effectively silenced throughout history, they often consider a pedagogy that "nurtures voice" just as, if not more, important than the curriculum.

Women's studies professors tend to be overtly warm, encouraging, maternal. You want to tell these women your problems—and many students do. To foster a "safe environment" where women feel comfortable talking, many teachers try to divest the classroom of power relations. They abandon their role as experts, lecturing very little and sometimes allowing decisions to be made by the group and papers to be graded by other students. An overriding value is placed on student participation and collaboration: students make joint presentations, cowrite papers, and use group journals for "exploring ideas they can't say in class" and "fostering a sense of community." Because chairs are usually arranged in a circle, in a couple of classes taught by graduate students I couldn't figure out who the teacher was until the end.

Most of the women's studies students I met were quite bright, and many argued certain points very articulately.

To give women voice, many professors encourage all discourse—no matter how personal or trivial. Indeed, since it is widely believed that knowledge is constructed and most texts have been influenced by "the patriarchy," many in women's studies consider personal experience the only real source of truth. Some professors and texts even claim that women have a way of thinking that is different from the abstract rationality of men, one based on context, emotion, and intuition. Fully "validating" women, therefore, means celebrating subjectivity over objectivity, feelings over facts, instinct over logic.

The day I sat in on Berkeley's "Contemporary Global Issues for Women" (all women except for one "occasional" male), we watched a film about women organizing in Ahmadabad, India. The film was tedious, but it seemed like grist for a good political/economic/sociological discussion about the problems of women in underdeveloped countries. After the film ended, though, the professor promptly asked the class: "How do you *feel* about the film? Do you find it more sad or courageous?" Students responded to her question until the end of class, at which point she suggested, "You might think about the film in terms of your own life and the life of your mother. Women are not totally free in this culture. It just might come in more subtle ways."

A previous discussion was apparently not much better. "We had to read an enormous amount of interesting material on reproductive rights, which I was very excited to discuss," Pam Wilson, a women's studies sophomore, told me. "But all she did in class was ask each of us, 'What forms of birth control have you used, and what problems have you had?' We never got to the assigned readings."

Self-revelation is not uncommon to women's studies classes. Students discover that they're lesbian or bisexual, for example, and then share it with the class. In a group journal (titled "The Fleshgoddesses") from last year's porn class, B. wrote: "There is still something about a [man] eating a [woman] out . . . that freaks me out! I guess I'm such a dyke that it seems abnormal." G. recalled that her father used to kiss her on the mouth "real hard" when she was eight or nine.

Of course, self-discovery and female bonding are important for young women, and so, one might argue, are group therapy and consciousness-raising. Indeed, I wish I had had some when I was that age; it might have given me the courage to talk in class and to deal with abusive bosses later in life. But does it belong in a university classroom?

Many of the professors I talked with (including the chair of Berkeley's women's studies department) viewed the more touchy-feely classes as just as problematic as I did. I saw a couple of teachers who were able to use personal experience, either of historical figures or students, to buttress the discussion, not as an end in itself. But even these classes were always on the verge of slipping into confession mode.

This pedagogy does get women talking. But they could do much of this type of talking in support groups at their schools' women's centers. Young women have many needs, and the college classroom can effectively address only one of them: building their intellects. As Ruth Rosen, who helped start the women's studies program at the University of California at Davis, puts it, "Students go to college to be academically challenged, not cared for."

But the problem with a therapeutic pedagogy is more than just allowing students to discuss their periods or sex lives in class. Using the emotional and subjective to "validate" women risks validating precisely the stereotypes that feminism was supposed to eviscerate: women are irrational, women must ground all knowledge in their own experiences, etc. A hundred years ago, women were fighting for the right to learn math, science, Latin—to be educated like men; today, many women are content to get their feelings heard, their personal problems aired, their instincts and intuition respected.

POLITICS, AS USUAL

"Don't worry. We've done nothing here since she forgot her notes a couple of weeks ago," Michael Williams reassures another male student. "We'll probably talk about Anita Hill again." We're waiting for Berkeley's "Gender Politics: Theory and Comparative Study" to begin. When the professor finally arrives and indicates that, yes, we'll be talking about Anita Hill again, the second male student packs up and bolts. Williams tells me that during the first week or two, whenever a male student would comment on something, the professor would say, "What you really mean is . . ." Most men stopped speaking and then dropped out. "Other classes I walk out with eight pages of notes," says Williams. "Here, everybody just says the same thing in a different way" (He stays, though, for the "easy credits.")

Most women's studies professors seem to adhere to the following principles in formulating classes: women were and are oppressed; oppression is endemic to our patriarchal social system; men, capitalism, and Western values are responsible for women's problems. The reading material is similarly bounded in political scope (Andrea Dworkin, Catharine MacKinnon, bell hooks, Adrienne Rich, and Audre Lorde turn up a lot), and opposing viewpoints are usually presented only through a feminist critique of them. *Feminist Frontiers III*, a book widely used in intro courses, purports to show readers "how gender has shaped your life," and invites them to join in the struggle "to reform the structure and culture of male dominance."

Says one student, "The way to get A's was to write papers full of guilt and angst about how I'd bought into society's definition of womanhood."

Although most of the classes I attended stopped short of outright advocacy of specific political positions, virtually all carried strong political undercurrents. Jill Harvey, a women's studies senior at Smith, recalls a feminist anthropology course in which she "quickly discovered that the way to get A's was to write papers full of guilt and angst about how I'd bought into society's definition of womanhood and now I'm enlightened and free."

Sometimes the politicization is more subtle. "I'm not into consciousness-raising," says Linda K. Kerber, a history professor at Iowa. "Students can feel I'm grading them on their competence and not on their politics." Yet in the final project of "Gender and Society in the United States," she asked students: "Reconsider a term paper you have written for another class. How would you revise it now to ensure that it offers an analysis sensitive to gender as well as to race and class?"

Politicization is also apparent in the meager amount of time the classes devote to women who have achieved anything of note in the public sphere. Instead, students scrutinize the diaries and letters of unremarkable women who are of interest primarily because the patriarchy victimized them in one way or another.

According to professors and students, studying "women worthies" doesn't teach you much about oppression. Moreover some added, these women succeeded by male, capitalist standards. It's time for women's traditional roles and forms of expression to be valued.

This may be true, but you don't need to elevate victimized women to the status of heroes to do that. It should also be noted that over the past twenty-five years feminists have been among those who have devalued women's traditional roles most vigorously. I bet not many women's studies majors would encourage a peer's decision to forgo a career in order to stay home and raise children. More important, examples of women who succeeded in the public sphere, possibly even while caring for a family, could be quite inspiring for young women. Instead, the classes implicitly downplay individual merit and focus on the systematic forces that are undermining everything women do.

In general, "core" women's studies courses are more overtly political and less academically rigorous than those cross-listed with a department. The syllabus of Iowa's "Introduction to Women's Studies" course declares: "As we make our collective and individual journeys during this course, we will consider how to integrate our theoretical knowledge with personal and practical action in the world." "Practicums," which typically entail working in a women's organization, are a key part of many courses, often requiring thirty or more hours of a student's time.

Volunteering in a battered-women's shelter or rape crisis center may be deeply significant for both students and society. But should this be part of an undergraduate education? Students have only four years to learn the things a liberal education can offer—and the rest of their lives to put that knowledge to use.

Courses on women don't have to be taught from an orthodox feminist perspective. Smith offers a biology course that's cross-listed with women's studies. It deals with women's bodies and medical issues; feminist theory is not included. Compare that to the course description of Berkeley's "Health and Sex in America": "From sterilization to AIDS; from incest to date rape; from anorexia to breast implants: who controls women's health?" Which course would you trust to be more objective?

Many women's studies professors acknowledge their field's bias, but point out that all disciplines are biased. Still, there's a huge difference between conceding that education has political elements and intentionally politicizing, between, as Women's Studies Professor Daphne Patai puts it, "recognizing and minimizing deep biases and proclaiming and endorsing them." Patai, whose unorthodox views got her in hot water at the University of Massachusetts, is now coauthoring a book on the contradictions of women's studies. "Do they really want fundamentalist studies, in which teachers are not just studying fundamentalism but supporting it?"

A still larger problem is the degree to which politics has infected women's studies scholarship. "Feminist theory guarantees that researchers will discover male bias and oppression in every civilization and time," says Mary Lefkowitz, a classics professor at Wellesley. "A distinction has to be made between historical interpretation of the past and political reinterpretation." And, I would add, between reading novels with an awareness of racism and sexism, and reducing them entirely to constructs of race and gender.

Apparently there has always been a tug of war within the women's studies community between those who most value scholarship and those who most value ideology. Some professors feel obligated to present the work of all women scholars who call themselves feminists, no matter how questionable their methodology or conclusions.

Unfortunately women's studies students may not be as well equipped to see through shoddy feminist scholarship as they are through patriarchal myths and constructs. One reason may be the interdisciplinary nature of the programs, which offers students minimal grounding in any of the traditional disciplines. According to Mary Lefkowitz, women's studies majors who take her class exhibit an inability to amass factual material or remember details; instead of using evidence to support an argument, they use it as a remedy for their personal problems.

But teaching students how to "think critically" is one of the primary goals of women's studies, and both students and professors say that women's studies courses are more challenging than those in other departments. "Women's studies gives us tools to analyze," says Torrey Shanks, a senior women's studies and political science major at Berkeley. "We learn theories about how to look at women and men; we don't just come away with facts."

Women's studies has generated first-rate scholarship on women, but professors often consider a pedagogy that "nurtures voice" just as, if not more, important than the curriculum.

Most of the women's studies students I met were quite bright, and many argued certain points very articulately. But they seemed to have learned to think critically through only one lens. When I asked some of the sharpest students about the most basic criticisms of women's studies, they appeared not to have thought about them or gave me some of the stock women's studies rap. It seemed that they couldn't fit these questions into their way of viewing the world.

For instance, when I expressed the view that an at-times explicit anticapitalist and anti-Western bias pervades the field, a couple of majors told me they thought that being anticapitalist was part of being a feminist. When I asked whether, in the final analysis, women weren't still most free in Western capitalist societies, the seemingly programmed responses ran from "I wouldn't feel free under a glass ceiling" to "Pressures on Iranian women to wear the veil are no different from pressures on women in this country to wear heels and miniskirts."

THE STUDENT PARTY LINE

Despite the womb-like atmosphere of the classrooms, I didn't see much student questioning of the professors or the texts. Although I rarely saw teachers present or solicit divergent points of view the students' reluctance to voice alternative opinions seemed to stem more from political intolerance and conformity on the part of fellow students.

In Smith's "Gender and Politics" class, several students spoke against the ban on gays in the military before Erin O'Connor, her voice shaking, ventured: "I think there is something to the argument of keeping gays out of the military because of how people feel about it."

After several students said things like, "The military should reflect society" O'Connor rebounded: "I'm sick and tired of feeling that if I have a moral problem with something, all of a sudden it's: 'You're homophobic, you're wrong, you're behind the times, go home.' There must be someone else in this classroom who believes as I do."

Professor: "No one is saying that support of the ban is homophobic."

"I would make that assertion," offered a student.

Professor: "But you can argue against the ban from a nonhomophobic perspective."

Another student: "It's homophobic."

When class ended, another woman approached O'Connor and said: "You're absolutely right, and I'm sure there are others who felt the same way but just didn't say anything. You went out on a limb."

No one used the word homophobic until O'Connor did. Still, students, especially in this ostensible "safe environment," shouldn't have to overcome a pounding heart to voice a dissident opinion. "Women's studies creates a safe space for p.c. individuals, but doesn't maintain any space for white Christians," says O'Connor, an English and government major and member of the College Republican Club.

In a study by the Association of American Colleges, 30 percent of students taking women's studies courses at Wellesley said they felt uneasy expressing unpopular opinions; only 14 percent of non-women's studies students felt that way.

Smith's Jill Harvey told me about a "Medical Anthropology" class filled with women's studies students. The professor presented an author's view that one difference between men and women when paralyzed is that men are rendered incapable of getting an erection. "The students jumped down his throat, believing he was insinuating that all women have to do is lie back and enjoy sex," says Harvey. "It was absurd, but I didn't feel like I could speak up. I sometimes feel the other students' attitude is: if you don't agree with me, you're too stupid to understand how oppressed you are."

The pressures on professors to toe the correct feminist line can be even stronger. History Professor Elizabeth Fox-Genovese says she stepped down as chair of Emory's women's studies program because of complaints from students and faculty that she wasn't radical enough. Political theorist Jean Bethke Elshtain left the University of Massachusetts after being attacked for including men on her reading list, allowing men in class, and presenting an array of different feminist positions. She now teaches at Vanderbilt. "Most teachers of women's studies presume that if you don't see yourself as a victim, you're in a state of false consciousness, you're 'male-identified.' The professors here [at Vanderbilt] recognize that feminism is in part an argument."

Women's studies professors take little responsibility for turning female students into Angry Young Women. Yet the effect of these classes, one after another, can be quite intoxicating. (After just a few days, I found myself noticing that the sign on the women's bathroom door in the University of Iowa's library was smaller than the one on the men's room door.) The irony is not only that these students (who, at the schools I visited at least, were overwhelmingly white and upper-middle-class) probably have not come into contact with much oppression, but that they are the first generation of women who have grown up with so many options open to them.

POST-STRUCTURALISM AND MULTICULTURALISM

Perhaps the most troubling influence on women's studies in the past decade has been the collection of theories known as post-structuralism, which essentially implies that all texts are arbitrary all knowledge is biased, all standards are illegitimate, all morality is subjective. I talked to numerous women's studies professors who don't buy any of this (it's typically more popular in the humanities than in the social sciences), but nevertheless it has permeated women's studies to a significant extent, albeit in the most reductive, simplistic way.

According to Delo Mook, a Dartmouth physics professor who is part of a team teaching "Ways of Knowing: Physics, Literature, Feminism," "You can't filter other cultures through our stencil. Nothing is right or wrong."

What about cannibalism? Clitorectomies? "Nope. I can only say 'I believe it's wrong.'"

But post-structuralism is applied inconsistently in women's studies. I've yet to come across a feminist tract that "contextualizes" sexism in this country as it does in others, or acknowledges that feminism is itself a product of Western culture based on moral reasoning and the premise that some things are objectively wrong. Do feminist theorists really want the few young men who take these classes to formulate personal rationales for rape? There's a huge difference between questioning authority, truth, and knowledge and saying none of these exist, a difference between rejecting male standards and rejecting the whole concept of standards.

Like post-structuralism, the concept of multiculturalism has had a deep influence on women's studies. Professors seem under a constant burden to prove that they are presenting the requisite number of books or articles by women of color or lesbians. Issues of race came up in nearly every class I sat through. I wasn't allowed to sit in on a seminar at Dartmouth on "Racism and Feminism" because of a contract made with the students that barred outside visitors.

Terms like sexism, racism, and homophobia have bloated beyond all recognition, and the more politicized the campus, the more frequently they're thrown around. I heard both professors and students call Berkeley's women's studies department homophobic and racist, despite the fact that courses dealing with homosexuality and multiculturalism fill the catalog and quite a number of women of color and lesbians are affiliated with the department.

Although many professors try to work against it, in the prevailing ethos of women's studies, historical figures, writers, and the students themselves are viewed foremost as women, as lesbians, as white or black or Hispanic, and those with the most "oppressed" identities are the most respected. Feminist theorists now generally admit that they can't speak for all women, but some still presume to speak for all black women or all Jewish women or all lesbians. There's still little acknowledgment not only of the individuality of each woman, but of the universal, gender-blind bond shared by all human beings.

THE ROAD NOT TAKEN

Women's studies programs have clearly succeeded with at least one of their goals: whether because of the mostly female classes, the nurturing professors, or the subject matter, they have gotten women students talking.

But getting women to speak doesn't help much if they're all saying the same thing. Women's studies students may make good polemicists, but do they really learn to think independently and critically?

Elizabeth Fox-Genovese says she had envisioned Emory's women's studies program as a mini-women's college: "I thought it should be a special environment that took women seriously and asked them to be the best that they could be by the standards of a good, liberal arts education." Young women—and men—would be steeped in sound scholarship on women, but they would also be offered a variety of theories and viewpoints, feminist and otherwise.

Unfortunately, this hasn't been the perspective of most women's studies professors. Women's studies was conceived with a political purpose—to be the intellectual arm of the women's movement—and its sense of purpose has only gotten stronger through the years. The result is that the field's narrow politics have constricted the audience for nonideological feminism instead of widening it, and have reinforced the sexist notion that there is a women's viewpoint. There's a legitimate reason why two-thirds of college women don't call themselves feminists. "When I got here I thought I was a feminist," Erin O'Connor from Smith told me. "I don't want to call myself that now."

Clearly the first step is for women's studies to reopen itself to internal and external criticism. The intimidation in the field is so great that I had trouble finding dissident voices willing to talk to me on the record. The women's movement has come a long way in the past twenty-five years-feminists should feel secure enough now to take any and all lumps.

Young women should also no longer feel it necessary to shun classes devoted to women, as my friends and I did. Women today still have to work for their equality, but they don't have to prove it every second. And as the status of women in this country evolves, so should the goals of women's studies. It's for its own sake that women's studies should stop treating women as an ensemble of victimized identities. Only when the mind of each woman is considered on its own unique terms will the minds of all women be respected.

A Society of Suspects:
THE WAR ON DRUGS AND CIVIL LIBERTIES

Property seized in drug raids, including large amounts of money, may be forfeited to the government without proof of the owner's guilt.

A decade after Pres. Reagan launched the War on Drugs, all we have to show for it are city streets ruled by gangs, a doubled prison population, and a substantial erosion of constitutional protections.

Steven Wisotsky

Mr. Wisotsky, professor of law, Nova University, Ft. Lauderdale, Fla., is a member of the advisory board of the Drug Policy Foundation, Washington, D.C., and author of Beyond the War on Drugs. *This article is based on a Cato Institute Policy Analysis.*

ON DEC. 15, 1991, America celebrated the 200th anniversary of the Bill of Rights. On Oct. 2, 1992, it marked the 10th anniversary of an antithetical undertaking—the War on Drugs, declared by Pres. Ronald Reagan in 1982 and aggressively escalated by Pres. George Bush in 1989. The nation's Founders would be disappointed with what has been done to their legacy of liberty. The War on Drugs, by its very nature, is a war on the Bill of Rights.

In their shortsighted zeal to create a drug-free America, political leaders—state and Federal, elected and appointed—have acted as though the end justifies the means. They have repudiated the heritage of limited government and individual freedoms while endowing the bureaucratic state with unprecedented powers.

That the danger to freedom is real and not just a case of crying wolf is confirmed by the warnings of a few judges, liberals and conservatives alike, who, insulated from elective politics, have the independence to be critical. Supreme Court Justice Antonin Scalia, for example, denounced compulsory urinalysis of Customs Service employees "in the front line" of the War on Drugs as an "invasion of their privacy and an affront to their dignity." In another case, Justice John Paul Stevens lamented that "this Court has become a loyal foot soldier" in the War on Drugs. The late Justice Thurgood Marshall was moved to remind the Court that there is "no drug exception" to the Constitution.

In 1991, the Court of Appeals for the Ninth Circuit declared that "The drug crisis does not license the aggrandizement of governmental power in lieu of civil liberties. Despite the devastation wrought by drug trafficking in communities nationwide, we cannot suspend the precious rights guaranteed by the Constitution in an effort to fight the 'War on Drugs.'" In that observation, the court echoed a 1990 ringing dissent by the chief justice of the Florida Supreme Court: "If the zeal to eliminate drugs leads this state and nation to forsake its ancient heritage of constitutional liberty, then we will have suffered a far greater injury than drugs ever inflict upon us. Drugs injure some of us. The loss of liberty injures us all."

Those warnings are cries in the wilderness, however, unable to stop the relentless buildup of law enforcement authority at every level of government. In fact, the trend toward greater police powers has accelerated. One summary of the Supreme Court's 1990-91 term observed that its criminal law decisions "mark the beginning of significant change in the relationship between the citizens of this country and its police."

Despite such warnings, most Americans have yet to appreciate that the War on Drugs is a war on the rights of all of us. It could not be otherwise, for it is directed not against inanimate drugs, but against people—those who are suspected of using, dealing in, or otherwise being involved with illegal substances. Because the drug industry arises from the voluntary transactions of tens of millions of individuals—all of whom try to keep their actions secret—the aggressive law enforcement schemes that constitute the war must aim at penetrating their private lives. Because nearly anyone may be a drug user or seller of drugs or an aider and abettor of the drug industry, virtually everyone has become a suspect. All must be observed, checked, screened, tested, and admonished—the guilty and innocent alike.

The tragic irony is that, while the War on Drugs has failed completely to halt the influx of cocaine and heroin—which are cheaper, purer, and more abundant than ever—the one success it can claim is in curtailing the liberty and privacy of the American people. In little over a decade, Americans have suffered a marked reduction in their freedoms in ways both obvious and subtle.

Among the grossest of indicators is that the war leads to the arrest of an estimated 1,200,000 suspected drug offenders each year, most for simple possession or petty sale. Because arrest and incarceration rates rose for drug offenders throughout the 1980s, the war has succeeded dramatically in increasing the full-time prison population. That has doubled since 1982 to more than 800,000, giving the U.S. the highest rate of incarceration in the industrialized world.

It has been established that law enforcement officials—joined by U.S. military forces—have the power, with few limits, to snoop, sniff, survey, and detain, without warrant or probable cause, in the war against drug trafficking. Property may be seized on slight evidence and forfeited to the state or Federal government without proof of the personal guilt of the owner.

Finally, to leverage its power, an increasingly imperial Federal government has applied intimidating pressures to shop owners and others in the private sector to help implement its drug policy.

Ironically, just as the winds of freedom are blowing throughout central and eastern Europe, most Americans and the nation's politicians maintain that the solution to the drug problem is more repression—and the Bill of Rights be damned. As Peter Rodino, former chairman of the House Judiciary Committee, said in expressing his anger at the excesses of the Anti-Drug Abuse Act of 1986, "We have been fighting the war on drugs, but now it seems to me the attack is on the Constitution of the United States."

In the beginning, the War on Drugs focused primarily on supplies and suppliers. Control at the source was the first thrust of anti-drug policy—destruction of coca and marijuana plants in South America, crop substitution programs, and aid to law enforcement agencies in Colombia, Peru, Bolivia, and Mexico.

Because this had no discernible, lasting success, a second initiative aimed to improve the efficiency of border interdiction of drug shipments that had escaped control at the source. There, too, success was elusive. Record numbers of drug seizures—up to 22 tons of cocaine in a single raid on a Los Angeles warehouse, for instance—seemed only to mirror a record volume of shipments to the U.S. By 1991, the amount of cocaine seized by Federal authorities had risen to 134 metric tons, with an additional amount estimated at between 263 and 443 tons escaping into the American market per year.

A reasonable search and seizure in the War on Drugs is interpreted very broadly and favors local police and Federal drug agents.

As source control and border interdiction proved futile, a third prong of the attack was undertaken: long-term, proactive conspiracy investigations targeted at suspected high-level drug traffickers and their adjuncts in the professional and financial worlds—lawyers, accountants, bankers, and currency exchange operators. This has involved repeated and systematic attacks by the Federal government on the criminal defense bar, raising dark implications for the integrity of the adversarial system of justice. Defense lawyers have been subjected to grand jury subpoenas, under threat of criminal contempt, to compel disclosures about their clients. Informants have been placed in the defense camp to obtain confidential information. In each instance, the effect has been to undermine the protections traditionally afforded by the attorney/client relationship. This demonstrates the anything-goes-in-the-War-on-Drugs attitude of the Department of Justice, which publicly defended using lawyers as informants as "a perfectly valid" law enforcement tool.

As these expanding efforts yielded only marginal results, the war was widened to the general populace. In effect, the government opened up a domestic front in the War on Drugs, invading the privacy of people through the use of investigative techniques such as urine testing, roadblocks, bus boardings, and helicopter overflights. Those are dragnet methods; to catch the guilty, everyone has to be watched and screened.

Invading privacy

Drug testing in the workplace. Perhaps the most widespread intrusion on privacy arises from pre- or post-employment drug screening, practiced by 80% of Fortune 500 companies and 43% of firms employing 1,000 people or more. Strictly speaking, drug testing by a private employer does not violate the Fourth Amendment, which protects only against government action. Nevertheless, much of the private drug testing has come about through government example and pressure. The 1988 Anti-Drug Abuse Act, for instance, prohibits the award of a Federal grant or contract to an employer who does not take specified steps to provide a drug-free workplace. As a result of these and other pressures, tens of millions of job applicants and employees are subjected to the indignities of urinating into a bottle, sometimes under the eyes of a monitor watching to ensure that clean urine is not smuggled surreptitiously into the toilet.

In the arena of public employment, where Fourth Amendment protections apply, the courts largely have rejected constitutional challenges to drug testing programs. In two cases to reach the U.S. Supreme Court, the testing programs substantially were upheld despite, as Justice Scalia wrote in dissent in one of them, a complete absence of "real evidence of a real problem that will be solved by urine testing of customs service employees." In that case, the Customs Service had implemented a drug testing program to screen all job applicants and employees engaged in drug interdiction activities, carrying firearms, or handling classified material. The Court held that the testing of such applicants and employees is "reasonable" even without probable cause or individualized suspicion against any particular person, the Fourth Amendment standard.

For Scalia, the testing of Customs Service employees was quite different from that of railroad employees involved in train accidents, which had been found constitutional. In that case, there was substantial evidence over the course of many years that the use of alcohol had been implicated in causing railroad accidents, including a 1979 study finding that 23% of the operating personnel were problem drinkers. Commenting on the Customs case, Scalia maintained that "What is absent in the government's justifications—notably absent, revealingly absent, and as far as I am concerned dispositively absent—is the recitation of even a single instance in which any of the speculated horribles actually occurred: an instance, that is, in which the cause of bribe-taking, or of poor aim, or of unsympathetic law enforcement, or of compromise of classified information, was drug use."

Searches and seizures. Other dragnet techniques that invade the privacy of the innocent as well as the guilty have been upheld by the Supreme Court. In the tug-of-war between the government's search and seizure powers and the privacy rights of individuals, the Court throughout the 1980s almost always upheld the government's assertion of the right of drug agents to use the airport drug courier profile to stop, detain, and question people without warrant or probable cause; subject a traveler's luggage to a sniffing examination by a drug-detecting dog without warrant or probable cause; search without warrant or probable cause the purse of a public school student; and search at will ships in inland waterways.

The right of privacy in the home seriously was curtailed in decisions permitting police to obtain a search warrant of a home based on an anonymous informant's tip; use illegally seized evidence under a "good faith exception" to the exclusionary rule (for searches of a home made pursuant to a defective warrant issued without probable cause); make a trespassory search, without a warrant, in "open fields" surrounded by fences and no trespassing signs and of a barn adjacent to a residence; and conduct a warrantless search of a motor home occupied as a residence, a home on the consent of an occasional visitor lacking legal authority over the premises, and the foreign residence of a person held for trial in the U.S. The Court also validated warrantless aerial surveillance over private property—by fixed-wing aircraft at an altitude of 1,000 feet and by helicopter at 400 feet.

Similarly, it significantly enlarged the powers of police to stop, question, and detain drivers of vehicles on the highways on suspicion with less than probable cause or with no suspicion at all at fixed checkpoints or roadblocks; make warrantless searches

6. CITIES, URBAN GROWTH: Drugs

of automobiles and of closed containers therein; and conduct surveillance of suspects by placing transmitters or beepers on vehicles or in containers therein.

The foregoing list is by no means comprehensive, but it does indicate the sweeping expansions the Court has permitted in the investigative powers of government. Indeed, from 1982 through the end of the 1991 term, the Supreme Court upheld government search and seizure authority in approximately 90% of the cases. The message is unmistakable—the Fourth Amendment prohibits only "unreasonable" searches and seizures, and what is reasonable in the milieu of a War on Drugs is construed very broadly in favor of local police and Federal drug agents.

Surveillance of U.S. mail. Another casualty of the War on Drugs is the privacy of the mail. With the Anti-Drug Abuse Act of 1988, the Postal Service was given broad law enforcement authority. Using a profile, investigators identify what they deem to be suspicious packages and place them before drug-sniffing dogs. A dog alert is deemed probable cause to apply for a Federal search warrant. If an opened package does not contain drugs, it is resealed and sent to its destination with a copy of the search warrant. Since January, 1990, using this technique, the Postal Service has arrested more than 2,500 persons for sending drugs through the mail. The number of innocent packages opened has not been reported.

Wiretapping. As a result of the War on Drugs, Americans increasingly are being overheard. Although human monitors are supposed to minimize the interception of calls unrelated to the purpose of their investigation by listening only long enough to determine the relevance of the conversation, wiretaps open all conversations on the wiretapped line to scrutiny.

Court-authorized wiretaps doubtless are necessary in some criminal cases. In drug cases, though, they are made necessary because the "crimes" arise from voluntary transactions, in which there are no complainants to assist detection. The potential is great, therefore, for abuse and illegal overuse.

Stopping cars on public highways. It is commonplace for police patrols to stop "suspicious" vehicles on the highway in the hope that interrogation of the driver or passengers will turn up enough to escalate the initial detention into a full-blown search. Because the required "articulable suspicion" rarely can be achieved by observation on the road, police often rely on a minor traffic violation—a burned-out taillight, a tire touching the white line—to supply a pretext for the initial stop. In the Alice-in-Wonderland world of roving drug patrols, however, even lawful behavior can be used to justify a stop. The Florida Highway Patrol Drug Courier Profile, for example, cautioned troopers to be suspicious of "scrupulous obedience to traffic laws."

Another tactic sometimes used is the roadblock. Police set up a barrier, stop every vehicle at a given location, and check each driver's license and registration. While one checks the paperwork, another walks around the car with a trained drug-detector dog. The law does not regard the dog's sniffing as the equivalent of a search on the theory that there is no legitimate expectation of privacy in the odor of contraband, an exterior olfactory clue in the public domain. As a result, no right of privacy is invaded by the sniff, so the police do not need a search warrant or even probable cause to use the dog on a citizen. Moreover, if the dog "alerts," that supplies the cause requirement for further investigation of the driver or vehicle for drugs.

Monitoring and stigmatizing. In the world of anti-drug investigations, a large role is played by rumors, tips, and suspicions. The Drug Enforcement Administration (DEA) keeps computer files on U.S. Congressmen, entertainers, clergymen, industry leaders, and foreign dignitaries. Many persons named in the computerized Narcotics and Dangerous Drug Information System (NADDIS) are the subject of "unsubstantiated allegations of illegal activity." Of the 1,500,000 persons whose names have been added to NADDIS since 1974, less than five percent, or 7,500, are under investigation by DEA as suspected narcotic traffickers. Nevertheless, NADDIS maintains data from all such informants, surveillance, and intelligence reports compiled by DEA and other agencies.

The information on NADDIS is available to Federal drug enforcement officials in other agencies, such as the Federal Bureau of Investigation, the Customs Service, and the Internal Revenue Service. State law enforcement officials probably also can gain access on request. Obviously, this method of oversight has troubling implications for privacy and good reputation, especially for the 95% named who are not under active investigation.

Another creative enforcement tactic sought to bring about public embarrassment by publishing a list of people caught bringing small amounts of drugs into the U.S. The punish-by-publishing list, supplied to news organizations, included only small-scale smugglers who neither were arrested nor prosecuted for their alleged crimes.

Military surveillance. Further surveillance of the citizenry comes from the increasing militarization of drug law enforcement. The process began in 1981, when Congress relaxed the Civil War-era restrictions of the Posse Comitatus Act on the use of the armed forces as a police agency. The military "support" role for the Coast Guard, Customs Service, and other anti-drug agencies created by the 1981 amendments expanded throughout the 1980s to the point that the U.S. Navy was using large military vessels—including, in one case, a nuclear-powered aircraft carrier—to interdict suspected drug smuggling ships on the high seas.

By 1989, Congress designated the Department of Defense (DOD) as the single lead agency of the Federal government for the detection and monitoring of aerial and maritime smuggling into the U.S. DOD employs its vast radar network in an attempt to identify drug smugglers among the 300,000,000 people who enter the country each year in 94,000,000 vehicles and 600,000 aircraft. Joint task forces of military and civilian personnel were established and equipped with high-tech computer systems that provide instantaneous communication among all Federal agencies tracking or apprehending drug traffickers.

The enlarged anti-drug mission of the military sets a dangerous precedent. The point of the Posse Comitatus Act was to make clear that the military and police are very different institutions with distinct roles to play. The purpose of the military is to prevent or defend against attack by a foreign power and to wage war where necessary. The Constitution makes the president commander-in-chief, thus centralizing control of all the armed forces in one person. Police, by contrast, are supposed to enforce the law, primarily against domestic threats at the city, county, and state levels. They thus are subject to local control by the tens of thousands of communities throughout the nation.

Since the 1987 enactment of the Uniform Sentencing Guidelines, the penalties for drug crimes have become extreme and mandatory.

To the extent that the drug enforcement role of the armed forces is expanded, there is a direct increase in the concentration of political power in the president who commands them and the Congress that authorizes and funds their police activities. This arrangement is a severe injury to the Federal structure of our democratic institutions. Indeed, the deployment of national military forces as domestic police embarrasses the U.S. in the international arena by likening it to a Third World country, whose soldiers stand guard in city streets, rifles at the ready, for ordinary security purposes.

37. Society of Suspects

The dual military/policing role also is a danger to the liberties of all citizens. A likely military approach to the drug problem would be to set up roadblocks, checkpoints, and roving patrols on the highways, railroads, and coastal waters, and to carry out search-and-destroy missions of domestic drug agriculture or laboratory production. What could be more destructive to the people's sense of personal privacy and mobility than to see such deployments by Big Brother?

Excessive punishment

These are some of the many ways the War on Drugs has cut deeply—and threatens to cut deeper still—into Americans' privacy, eroding what Justice Louis D. Brandeis described as "the right to be let alone—the most comprehensive of rights and the right most valued by civilized men." Working hand-in-hand with the political branches, the courts have diminished constitutional restraints on the exercise of law enforcement power. In addition to expanded powers of surveillance, investigation, and prosecution, punishment has been loosed with a vengeance, against enemy and bystander alike.

Punishments have become draconian in part because of permission conferred by Justice William Rehnquist's 1981 circular dictum: "the question of what punishments are constitutionally permissible is not different from the question of what punishments the Legislative Branch intended to be imposed." The penalties have become so extreme, especially since the 1987 enactment of the Uniform Sentencing Guidelines, that many Federal judges have begun to recoil. U.S. district court Judge J. Lawrence Irving of San Diego, a Reagan appointee, announced his resignation in protest over the excessive mandatory penalties he was required to mete out to low-level offenders, most of them poor young minorities. Complaining of "unconscionable" sentences, the judge said that "Congress has dehumanized the sentencing process. I can't in good conscience sit on the bench and mete out sentences that are unfair."

Judge Harold Greene of the District of Columbia went so far as to refuse to impose the minimum guideline sentence of 17.5 years on a defendant convicted of the street sale of a single Dilaudid tablet, pointing to the "enormous disparity" between the crime and the penalty. In the judge's view, the minimum was "cruel and unusual" and "barbaric." Fourth circuit Judge William W. Wilkins objected to mandatory penalties because "they do not permit consideration of an offender's possibly limited peripheral role in the offense." Agreeing with that thinking, the judicial conferences of the District of Columbia, Second, Third, Seventh, Eighth, Ninth, and Tenth circuits have adopted resolutions opposing mandatory minimums.

As drug control policymakers came to realize that the drug dealers were, in an economic sense, merely entrepreneurs responding to market opportunities, they learned that attacks on dealers and their supplies never could succeed as long as there was demand for the products. Thus, they would have to focus on consumers as well as on suppliers. Pres. Reagan's 1986 Executive Order encouraging or requiring widespread urine testing marked a step in that direction. By 1988, Administration policy was being conducted under the rubric of "zero tolerance." In that spirit, Attorney General Edwin Meese sent a memorandum to all U.S. Attorneys on March 30, 1988, encouraging the selective prosecution of "middle and upper class users" in order to "send the message that there is no such thing as 'recreational' drug use...."

Because of the volume of more serious trafficking cases, however, it was not remotely realistic, as the Attorney General must have known, to implement such a policy. Indeed, in the offices of many U.S. Attorneys, there were minimum weight or money-volume standards for prosecution, and the possession and small-scale drug cases routinely were shunted off to state authorities. In fact, in many districts, the crush of drug cases was so great that the adjudication of ordinary civil cases virtually had ceased. The courthouse doors were all but closed to civil litigants.

In the name of zero tolerance, Congress purposely began enacting legislation that did not have to meet the constitutional standard of proof beyond a reasonable doubt in criminal proceedings. In 1988, it authorized a system of fines of up to $10,000, imposed administratively under the authority of the Attorney General, without the necessity of a trial, although the individual may request an administrative hearing. To soften the blow to due process, judicial review of an adverse administrative finding is permitted, but the individual bears the burden of retaining counsel and paying court filing fees. For those unable to finance a court challenge, this system will amount to punishment without trial. Moreover, it has been augmented by a provision in the Anti-Drug Abuse Act of 1988 that may suspend for one year an offender's Federal benefits, contracts, grants, student loans, mortgage guarantees, and licenses upon conviction for a first offense.

Both sanctions are a form of legal piling on. The legislative intent is to punish the minor offender more severely than is authorized by the criminal law alone. Thus, the maximum penalty under Federal criminal law for a first offense of simple possession of a controlled substance is one year in prison and a $5,000 fine, with a minimum fine of $1,000. Fines up to $10,000 plus loss of Federal benefits obviously exceed those guidelines.

The most recent innovation of this kind is a form of greenmail, a law that cuts off highway funds to states that do not suspend the driver's licenses of those convicted of possession of illegal drugs. The potential loss of work for those so punished and the adverse consequences on their families are not considered. The suspension is mandatory.

Seizure and forfeiture

The War on Drugs not only punishes drug users, it also penalizes those who are innocent and others who are on the periphery of wrongdoing. The most notable example is the widespread and accelerating practice, Federal and state, of seizing and forfeiting cars, planes, boats, houses, money, or property of any other kind carrying even minute amounts of illegal drugs, used to facilitate a transaction in narcotics, or representing the proceeds of drugs. Forfeiture is authorized, and enforced, without regard to the personal guilt of the owner. It matters not whether a person is tried and acquitted; the owner need not even be arrested. The property nonetheless is forfeitable because of a centuries-old legal fiction that says the property itself is "guilty." Relying on it, in March, 1988, the Federal government initiated highly publicized zero tolerance seizures of property that included the following:

- On April 30, 1988, the Coast Guard boarded and seized the motor yacht *Ark Royal*, valued at $2,500,000, because 10 marijuana seeds and two stems were found on board. Public criticism prompted a return of the boat, but not before payment of $1,600 in fines and fees by the owner.
- The 52-foot *Mindy* was impounded for a week because cocaine dust in a rolled up dollar bill was found on board.
- The $80,000,000 oceanographic research ship *Atlantis II* was seized in San Diego when the Coast Guard found 0.01 ounce of marijuana in a crewman's shaving kit. The vessel eventually was returned.
- A Michigan couple returning from a Canadian vacation lost a 1987 Mercury Cougar when customs agents found two marijuana cigarettes in one of their pockets. No criminal charges were filed, but the car was kept by the government.
- In Key West, Fla., a shrimp fisherman lost his boat to the Coast Guard, which found three grams of cannabis seeds and stems on board. Under the law, the craft was forfeitable whether or not he had any responsibility for the drugs.

Not surprisingly, cases like the foregoing generated a public backlash—perhaps the only significant one since the War on Drugs was declared in 1982. It pressured Congress into creating what is known as the "innocent owner defense" to such *in rem*

205

6. CITIES, URBAN GROWTH: Drugs

forfeitures, but even that gesture of reasonableness is largely illusory.

First, the defense does not redress the gross imbalance between the value of property forfeited and the personal culpability of the owner. For example, a Vermont man was found guilty of growing six marijuana plants. He received a suspended sentence, but he and his family lost their 49-acre farm. Similarly, a New York man forfeited his $145,000 condominium because he sold cocaine to an informant for $250. The law provides no limit to the value of property subject to forfeiture, even for very minor drug offenses.

Second, the innocent owner defense places the burden on the property claimant to demonstrate that he or she acted or failed to act without "knowledge, consent or willful blindness" of the drug activities of the offender. Thus, the Federal government instituted forfeiture proceedings in the Delray Beach, Fla., area against numerous properties containing convenience stores or other businesses where drug transactions took place, claiming that the owners "made insufficient efforts to prevent drug dealings."

Placing the burden on the claimant imposes expense and inconvenience because the claimant must hire a lawyer to mount a challenge to the seizure. Moreover, many cases involve the family house or car, and it often is difficult to prove that one family member had no knowledge of or did not consent to the illegal activities of another. For instance, a Florida court held that a claimant did not use reasonable care to prevent her husband from using her automobile in criminal activity; thus, she was not entitled to the innocent owner defense.

A particularly cruel application of this kind of vicarious responsibility for the wrongs of another is seen in the government's policy of evicting impoverished families from public housing because of the drug activities of one unruly child. The Anti-Drug Abuse Act of 1988 specifically states that a tenant's lease is a forfeitable property interest and that public housing agencies have the authority to hire investigators to determine whether drug laws are being broken. The act authorizes eviction if a tenant, member of his or her household, guest, or other person under his or her control is engaged in drug-related activity on or near public housing premises.

To carry out these provisions, the act funded a pilot enforcement program. In 1990, the Departments of Justice and Housing and Urban Development announced a Public Housing Asset Forfeiture Demonstration Project in 23 states. The project pursued lease forfeitures and generated considerable publicity.

In passing this law, it must have been obvious to Congress that many innocent family members would suffer along with the guilty. Perhaps it was thought vital, nonetheless, as a way of protecting other families from drugs in public housing projects. As experience proves, however, even evicted dealers continue to deal in and around the projects. It is hard to take public housing lease forfeitures very seriously, therefore, other than as a symbolic statement of the government's tough stand against illegal drugs.

Destructive consequences

A policy that destroys families, takes property from the innocent, and tramples the basic criminal law principles of personal responsibility, proportionality, and fairness has spillover effects into other public policy domains. One area in which the fanaticism of the drug warriors perhaps is most evident is public health. Drugs such as marijuana and heroin have well-known medical applications. Yet, so zealous are the anti-drug forces that even these therapeutic uses effectively have been banned.

Marijuana, for instance, has many applications as a safe and effective therapeutic agent. Among them are relief of the intraocular pressure caused by glaucoma and alleviating the nausea caused by chemotherapy. Some AIDS patients also have obtained relief from using cannabis.

Yet, marijuana is classified by the Attorney General of the U.S., not the Surgeon General, as a Schedule I drug—one having a high potential for abuse, no currently accepted medicinal use, and lack of accepted safety for utilization. It thereby is deemed beyond the scope of legitimate medical practice and thus is not generally available to medical practitioners.

The only exception was an extremely limited program of compassionate treatment of the terminally or seriously ill, but even that has been eliminated for political

---

The intensive pursuit of drug offenders has generated an enormous population of convicts held in prison for very long mandatory periods of time; so much so that violent criminals (murders, robbers, and rapists) often serve less time than the drug offenders.

---

reasons. Assistant Secretary James O. Mason of the Department of Health and Human Services announced in 1991 that the Public Health Service's provision of marijuana to patients seriously ill with AIDS would be discontinued because it would create a public perception that "this stuff can't be so bad." After a review caused by protests from AIDS activists, the Public Health Service decided in March, 1992, to stop supplying marijuana to any patients save the 13 then receiving it.

There also are beneficial uses for heroin. Terminal cancer patients suffering from intractable pain generally obtain quicker analgesic relief from heroin than from morphine. Many doctors believe that heroin should be an option in the pharmacopeia. Accordingly, in 1981, the American Medical Association House of Delegates adopted a resolution stating that "the management of pain relief in terminal cancer patients should be a medical decision and should take priority over concerns about drug dependence." Various bills to accomplish that goal were introduced in the 96th, 97th, and 98th Congresses. The Compassionate Pain Relief Act was brought to the House floor for a vote on Sept. 19, 1984, but was defeated by 355 to 55. Although there were some concerns voiced about thefts from hospital pharmacies, the overwhelming concern was political and symbolic—a heroin legalization bill could not be passed in an election year and, in any event, would send the public the "wrong message."

The final and perhaps most outrageous example in this catalog of wrongs against public health care is the nearly universal American refusal to permit established addicts to exchange used needles for sterile ones in order to prevent AIDS transmission among intravenous drug users. In 1991, the National Commission on AIDS recommended the removal of legal barriers to the purchase and possession of intravenous drug injection equipment. It found that 32% of all adult and adolescent AIDS cases were related to intravenous drug use and that 70% of mother-to-child AIDS infections resulted from intravenous drug use by the mother or her sexual partner. Moreover, the commission found no evidence that denial of access to sterile needles reduced drug abuse, but concluded that it did encourage the sharing of contaminated needles and the spread of the AIDS virus. Notwithstanding the commission's criticism of the government's "myopic criminal justice approach" to the drug situation, the prevailing view is that needle exchange programs encourage drug abuse by sending the wrong message.

Public safety is sacrificed when, nationwide, more than 18,000 local, sheriff's, and state police officers, in addition to thousands of Federal agents, are devoted full time to special drug units. As a result, countless hours and dollars are diverted

37. Society of Suspects

from detecting and preventing more serious violent crimes. Thirty percent of an estimated 1,100,000 drug-related arrests made during 1990 were marijuana offenses, nearly four out of five for mere possession. Tax dollars would be spent better if the resources it took to make approximately 264,000 arrests for possession of marijuana were dedicated to protecting the general public from violent crime.

The intensive pursuit of drug offenders has generated an enormous population of convicts held in prison for very long periods of time as a result of excessive and/or mandatory jail terms. It is estimated that the operating cost of maintaining a prisoner ranges from $20,000 to $40,000 per year, depending upon the location and level of security at a particular prison. With more than 800,000 men and women in American correctional facilities today, the nationwide cost approaches $30,000,000,000 per year. This is a major diversion of scarce resources.

These financial burdens are only part of the price incurred as a result of the relentless drive to achieve higher and higher arrest records. More frightening and damaging are the injuries and losses caused by the early release of violent criminals owing to prison overcrowding. Commonly, court orders impose population caps, so prison authorities accelerate release of violent felons serving non-mandatory sentences in order to free up beds for non-violent drug offenders serving mandatory, non-parolable terms.

For example, to stay abreast of its rapidly growing inmate population, Florida launched one of the nation's most ambitious early release programs. However, prisoners serving mandatory terms—most of them drug offenders, who now comprise 36% of the total prison population—are ineligible. As a result, the average length of sentence declined dramatically for violent criminals, while it rose for drug offenders. Murderers, robbers, and rapists often serve less time than a "cocaine mule" carrying a kilo on a bus, who gets a mandatory 15-year term.

A Department of Justice survey showed that 43% of state felons on probation were rearrested for a crime within three years of sentencing. In short, violent criminals are released early to commit more crimes so that their beds can be occupied by non-violent drug offenders. Civil libertarians are not heard often defending a societal right to be secure from violent criminals, much less a right of victims to see just punishment meted out to offenders. In this they are as shortsighted as their law-and-order counterparts. The War on Drugs is a public safety disaster, making victims of us all.

However uncomfortable it may be to admit, the undeniable reality is that drugs always have been and always will be a presence in society. Americans have been paying too high a price for the government's War on Drugs. As Federal judge William Schwarzer has said, "It behooves us to think that it may profit us very little to win the war on drugs if in the process we lose our soul."

It's Drugs, Alcohol and Tobacco, Stupid!

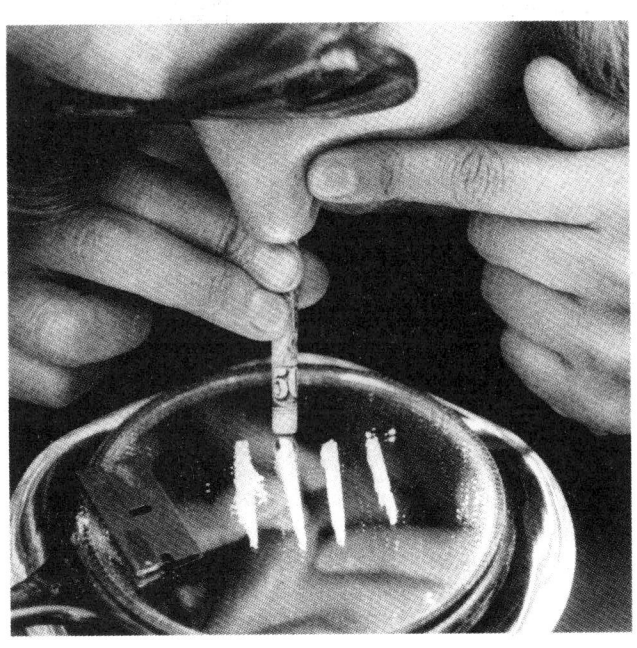

As the new generation of political leaders in Washington, state capitals and city halls grapples with America's collapsing judicial systems, rising medical costs, persistent poverty amid plenty and the defiant federal budget deficit that looms over future generations, they confront the same 800-pound gorilla: drug, alcohol and tobacco abuse and addiction. The sooner these leaders see how substance abuse has fundamentally changed the nature of the pressing social and economic problems they face, the sooner they'll deal with them effectively.

For 30 years, we've tried to curb crime and renew our ailing court system with tougher punishments, bigger prisons, and more cops and judges; rein in health costs by manipulating payments to doctors and hospitals for delivering sick care; wage war on poverty with a welfare system that encourages dependence and drives families apart; and reduce the deficit by cooking the federal books. Trying to reform our court and criminal justice systems, restrain health care spending, reduce welfare rolls, trim the deficit and nourish the American family without confronting, front and center, substance abuse and addiction is like trying to clean coastal waters without stopping the flow of oil from the ruptured offshore well. It can't be done.

Criminals and Courts

Congress and state legislatures have been passing laws designed for celluloid gangsters and inmates played in classic 1930s movies by James Cagney and Humphrey Bogart. But today's prisons are wall to wall with drug dealers, addicts, alcohol abusers and the mentally ill (often related to drug and alcohol abuse).

In 1960, less than 30,000 Americans were arrested for drug offenses; by 1991, the number had soared to more than a million. Since 1989, more individuals have been incarcerated for drug offenses than for all violent crimes—and most violent crime is committed by drug and alcohol abusers. Alcohol and drug abuse are implicated in

38. Drugs, Alcohol, and Tobacco

three-quarters of all spouse abuse, rapes, child molestations, suicides and homicides.

In 1994, the number of Americans in prison broke the one million barrier and, on its current trajectory, will double soon after the turn of the century. The United States is second only to Russia in the rate of citizens it imprisons: 519 per 100,000, compared to 558 in Russia, 368 in third-place South Africa, 116 in Canada and 36 in Japan.

Probation and parole are sick jokes in most American cities. With so many parolees needing drug treatment and aftercare as essential first steps to rehabilitation, they demand far more monitoring than their drug-free predecessors of a generation ago. Yet in Los Angeles, for example, probation officers must handle as many as 1,000 cases at a time. With most offenders committing drug- or alcohol-related crimes, it's no wonder so many of them go right back to jail: 80 percent of prisoners have prior convictions and more than 60 percent have served time before.

Drugs have turned the private security industry from a less than four-billion-dollar weakling in 1970 into a 70-billion-dollar behemoth in 1994, as office

buildings and homes install sophisticated protection systems and commercial properties post guards around the clock.

Judges and prosecutors are demoralized as they juggle caseloads more than double the recommended maximums. The rush of drug-related criminal cases has created intolerable delays for civil litigants: four years in Newark, five in Philadelphia, up to ten in Cook County, Illinois. In many jurisdictions, divorce and separation cases languish for years, as splitting parents and their children struggle to survive in a limbo of nasty uncertainty.

The safety and civility of urban life have been shattered by alcohol- and drug-related crimes. Children kill children and innocents are downed by random gunfire from warring drug gangs. Elementary and high school students are required to pass through metal detectors in order to check for weapons, the deadly companions of the drug trade, and teachers are locked in classrooms for their own protection.

City dwellers can no longer buy out of the mess. Individuals walking Wisconsin Avenue in Washington D.C.'s Georgetown, Madison Avenue in New York, Newbury Street in Boston and the Miracle Mile in Chicago are accosted by angry, aggressive panhandlers, many seeking money for their next fix. The ugly scrawls of graffiti on city buildings mark not only the arrival of spray paint, but also the widespread abuse of drugs and alcohol.

Substance abuse is an equal opportunity killer, snaring addicts in every social and economic class. Store owners lock their doors during daytime business hours in fear of robbery by alcohol- and drug-crazed criminals. Office managers bolt computers to desks to prevent theft. Customers and employees warily read headlines about murders and assaults, often committed under the influence of alcohol and drugs, that have torn apart the comfortable routines where America works, eats and shops—post offices, fast-food restaurants, banks and supermarkets. Two-thirds of illegal drug users are employed, adding an element of Russian roulette to going to work each day.

6. CITIES, URBAN GROWTH: Drugs

Health care Costs

In 1995, drugs, alcohol and tobacco will trigger some $200 billion in health care costs.

Hospital emergency rooms are piled high with the debris of drug use on city streets. From Boston to Baton Rouge, hospitals teem with the victims of gunshot wounds and other violence caused by alcohol abusers, drug addicts and dealers, and of a variety of medical conditions, such as cancer, emphysema and cardiac arrest, caused by alcohol, tobacco, cocaine and other drugs.

AIDS and tuberculosis spread rapidly and not just among intravenous drug users and crack addicts. Beyond sharing dirty needles and trading sex for drugs, individuals high on beer, other alcohol and pot are far more likely to have sex and to have it without a condom.

The more than 500,000 newborns exposed each year to drugs and/or alcohol during pregnancy is a slaughter of innocents of biblical proportions. Crack babies, a rarity a decade ago, crowd $2,000-a-day neonatal wards. Many die. Each survivor can cost one million dollars to bring to adulthood. Fetal alcohol syndrome is a top cause of birth defects.

Even where prenatal care is available, women on drugs and alcohol are not likely to take advantage of it. Those who do seek help must often wait in line for scarce treatment slots. Mothers abusing drugs during pregnancy account for most of the $3 billion that Medicaid spent in 1994 on inpatient hospital care for illness and injury due to drug abuse.

Poverty in History's Most Affluent Society

Drugs have changed the nature of poverty in America. Nowhere is this more striking than in the persistent problem of welfare dependency.

At least 20 percent of Chicago and Maryland's Montgomery County adults on welfare have drug problems. And that may be low compared to other urban areas. Many of the million teenagers who get pregnant each year are high on alcohol or drugs at the time they conceive, and one of the surest ways to get locked in poverty is to become an unwed mother before graduating from high school. At least half the homeless men and women—some say 80 to 90 percent—are alcohol and drug abusers.

The American electorate is hell-bent on putting welfare mothers to work. But all the financial sticks and carrots and all the job training in the world will do precious little to make employable the hundreds of thousands of welfare recipients who are drug and alcohol abusers. For too long, reformers have had

38. Drugs, Alcohol, and Tobacco

their heads in the sand about this unpleasant reality. Liberals fear that acknowledging the extent of alcohol and drug use among welfare recipients will incite even more punitive reactions than those currently in fashion. Conservatives don't want to face up to the cost of treatment.

This political denial ensures failure. Any reform that will move individuals from welfare to work must provide funds to treat drug and alcohol abuse.

Supplemental Security Income, the welfare program that provides monthly checks to blind, disabled and poor adults, reveals the grim and expensive consequence of the alternative. Of 90,000 individuals receiving SSI primarily because of substance abuse, fewer than ten percent are in treatment. Not surprisingly, the U.S. Department of Health and Human Services found that thousands of these addicts and alcoholics receive benefits until they die.

Illegal drugs have added a vicious strain of intractability to urban poverty. Drugs are the greatest threat to family stability, decent housing, public schools and even minimal social amenities. Widespread drug use derails the emotional, social and intellectual development not only of the children who abuse them, but also of their peers and neighbors who must grapple with the violent consequences of rampant drugs in housing projects and schools. It becomes difficult—sometimes

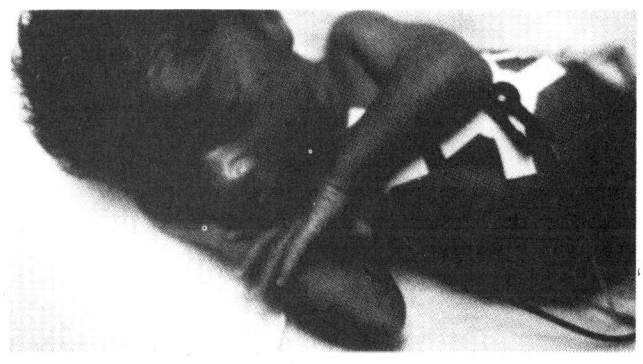

impossible—for children in this sordid environment to acquire the basic educational and social skills they need to get out of poverty.

The Federal Budget Deficit

In fiscal 1995, tobacco, alcohol and drug abuse will account for at least $77.6 billion in entitlement expenditures, an amount equivalent to 40 percent of the 1995 federal budget deficit.

Of that amount, $66.4 billion are costs to health and disability programs, such as Medicare, Medicaid and veterans' health and disability. Cigarette smoking is by far the biggest culprit. Two-thirds of the $66.4 billion—$44 billion—is attributable to tobacco. Alcohol accounts for 18 percent and drugs for 16 percent.

6. CITIES, URBAN GROWTH: Drugs

Substance abuse takes its biggest slice from the veterans' health care program. Nearly 30 percent of the dollars spent on veterans' health is due to substance abuse, more than half of that as a result of alcohol and drug abuse. Welfare payments to illegal drug addicts and alcoholics draw the rhetorical fire of legislators. But American taxpayers fork over $4.6 billion a year to individuals on Social Security disability as a result of smoking cigarettes.

Of the $77.6 billion, the remaining $11.2 billion is spent on welfare, food stamp and Supplemental Security Income recipients who regularly use alcohol and drugs and are unlikely to get off the rolls without treatment and aftercare.

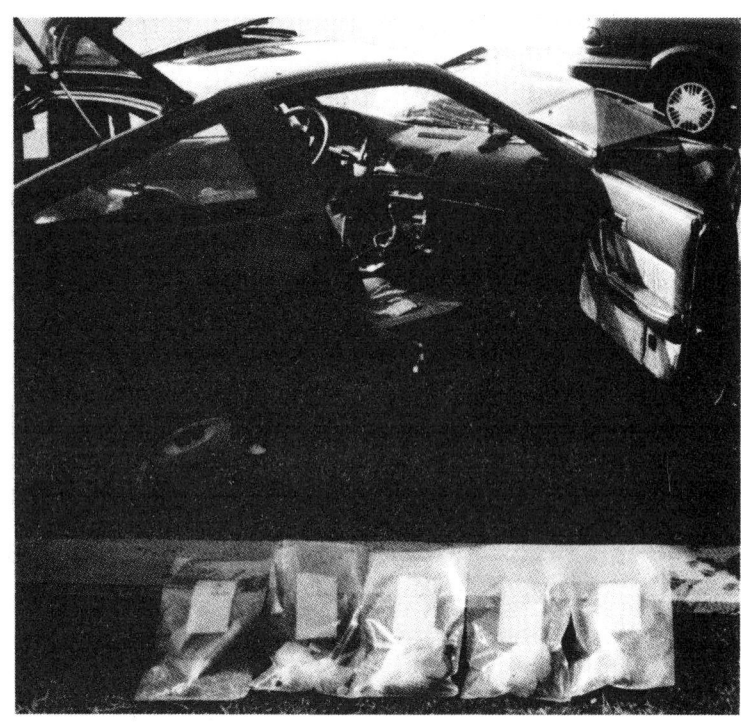

Any honest attack by the President and the Congress on entitlement programs—from Medicare and veterans' health and disability to Medicaid and welfare—has to confront substance abuse and addiction. That means a significant investment in prevention and treatment of all abuse. Simply removing individuals who abuse alcohol and drugs from disability, welfare and health care programs will only shift costs to the states, cities and counties, which will then have to deal with the resulting illness, hunger, homelessness and crime. Indeed, a wholesale denial of benefits to alcoholics and drug addicts without providing treatment and aftercare will push up the crime rate and scatter thousands more homeless individuals on America's streets.

Is There Anything We Can Do?

We can begin by ending our national and personal denial of the tough truth that the common denominator of the nation's hot buttons—crime and violence, health care costs, welfare reform and the budget deficit—is substance abuse. Our denial keeps our sights on the wrong targets. Indeed, 92 percent of federal health entitlement program costs attributable to substance abuse is spent to treat the *consequences* of tobacco, alcohol and drugs; only eight percent is spent to treat the tobacco, alcohol or drug dependence itself.

Our leaders and citizens focus on the top killers: heart disease (720,000 deaths in 1990), cancer (505,000), stroke (144,000), accidents (92,000), emphysema (87,000), pneumonia and influenza (80,000), diabetes (48,000), suicide (31,000), chronic liver disease and cirrhosis (26,000), and AIDS (25,000). But they give scant attention to the *causes* of these killers, which, according to a 1993 *Journal of the American Medical Association* study, include tobacco (435,000 deaths), alcohol (100,000) and illicit drug use (20,000).

Our obsession with the consequences and neglect of the causes is not limited to health care. We pump billions into combatting crime—cops, courts, prisons and punishment—and pennies into preventing the drug and alcohol abuse and addiction that spawn so much criminal activity. We pour resources into shoveling up city slums—rebuilding gutted housing, putting more cops on unsafe streets and barbed wire around housing projects—and little into curbing drug and alcohol abuse. And we often use our hefty budget-cutting axes to chop down prevention and treatment programs, which are most likely to reduce the deficit over the long run.

Dealing effectively with the causes requires up-front investments—the kind that corporations make every day to produce long-term results for their stockholders, the kind that parents make to give their children the best education they can get. It also requires that we scrub the stigma off drug and alcohol abuse and devote the kind of energy and resources to research on addiction and its prevention that we have committed to cancer and heart disease. And it requires common sense.

Here are a few starter suggestions:

• Provide federal funds to state and federal prison systems only if they provide drug and alcohol treatment and aftercare for all inmates who need such care.

• Instead of across-the-board mandatory sentences, keep inmates in jails, boot camps or halfway houses until they demonstrate at least one year of sobriety after treatment.

• Require drug and alcohol addicts to go regularly to treatment and aftercare, like Alcoholics Anonymous, while on parole or probation.

• Provide federal funds for police only to cities that agree to enforce drug laws throughout their jurisdictions. End acceptance of drug bazaars in Harlem, southeast Washington, D.C., and south-central Los Angeles, which would not be tolerated on the streets of New York's Upper East Side, Georgetown or Beverly Hills.

• Encourage judges with lots of drug cases to employ public health professionals, just as they hire economists to assist with anti-trust cases. Drug cases present far more complex human and medical problems than the economic issues posed by commercial litigation.

• Charge higher Medicare premiums to individuals who smoke.

• Cut off welfare payments to drug addicts and alcoholics who refuse to seek treatment and pursue aftercare. As employers and health professionals know, addicts from CEOs to chambermaids need lots of carrots and sticks, including the threat of losing their jobs and incomes, to get the monkey off their backs.

• Subject inmates, parolees and welfare recipients with a history of drug or alcohol abuse to random tests and fund the treatment they need. Conservatives who preach an end to recidivism and welfare dependency must recognize that reincarceration and removal from the welfare rolls for those who test positive is a cruel catch-22 unless treatment is available. Liberals must recognize that getting off drugs is the only chance these individuals (and their babies) have to enjoy their civil rights.

• Identify parents who abuse their children by their own drug and alcohol abuse and place those children in decent orphanages and foster care until the parents go into treatment and shape up.

These are only a few suggestions. The overriding point is that addiction and abuse—involving heroin, cocaine, hallucinogens, amphetamines, inhalants, marijuana, alcohol and tobacco—have fundamentally changed the nature of America's pressing social and economic challenges, and we must rethink how we address them. If a mainstream disease like diabetes or cancer affected as many individuals and families as drug, alcohol and tobacco abuse and addiction do, this nation would mount an effort on the scale of the Manhattan Project to deal with it.

Joseph A. Califano Jr.

Global Issues

Many of the social problems facing Americans today are shared by people worldwide, such as the environment, pollution, disease, health, and inflation. What Americans do impacts on the world, and what happens around the world impacts on the United States. Some problems facing the entire world are fueled by the consumerism of Americans, and some problems facing the United States are the product of other nations' improvements in production and their desire to improve their economic conditions. The world is no longer the exclusive marketplace for U.S. goods. In addition, health problems facing one nation or group of people are spreading quickly around the globe, with devastating results. In this unit, several major problems with global implications are examined and discussed.

In "U.S. Competitiveness: 'Resurgence' versus Reality," Robert Hayes contrasts what is happening in America with what is happening in other nations, particularly Japan, and raises some major concerns. He argues that unless America changes its arrogant attitude of superiority, increases its competitive edge, and revitalizes its infrastructure, it may be on the verge of economic disaster. In "The Civil Rights Issue of the '90s," Nancie Marzulla examines the social factors encouraging people to band together to protect their property rights. Legislation to protect the environment, provide flood control, and protect endangered species is often implemented at the expense of private property owners.

Rensselaer Lee, in "Global Reach: The Threat of International Drug Trafficking," documents how the flow of drugs between nations has become a critical issue in global security. While in the short run the economies of some developing nations are benefited, thus explaining their reluctance to eliminate the problem, their long-range stability is jeopardized. Resistance to multinational cooperation is compromised by concerns over national sovereignty.

In "A Growing Global Crisis," Ann Marie Kimball argues that AIDS is by far the most significant pandemic disease in the world in the latter half of the twentieth century. How it became so and its implications for the future are explored.

Lester Brown, in "Earth Is Running Out of Room," points out scientists' concern that the continuation of destructive human activities may interrupt and limit the world's ability to sustain life as we know it.

Looking Ahead: Challenge Questions

Which is the most effective strategy to use in protecting the world's environment, public or private ownership? Explain your choice.

Can the United States continue to be the police force and/or savior of the world? Can technological innovation continue to meet the world's ever-increasing demand for food, fuel, and security? Defend your answer.

Just how far should the government go in creating and implementing laws designed to protect the environment?

What should be done to nations that not only do not discourage drug production within their borders but sometimes even actively promote it?

What strategies should national leaders consider in attempting to cope with the AIDS pandemic? What are the implications for these strategies?

What are the possible implications for world peace of increasing populations and shrinking resources?

In what significantly different ways would the three major sociological theoretical perspectives argue that we should study global issues?

What are the major values, rights, obligations, and harms associated with each of the issues covered in this unit?

UNIT 7

U.S. Competitiveness: "Resurgence" versus Reality

Robert H. Hayes

This experienced business professor and consultant has been on the front lines of the battle to increase America's productivity. He thinks reports of progress among U.S. businesses are exaggerated. And he says don't count out the Japanese.

The business press these days is full of glowing stories about the resurgence of U.S. industry during the 1990s. *Challenge* examined this issue at length (November–December 1995), yet I remain struck by the outpouring of articles that have supported this notion. *Business Week* (October 9, 1995) opined that it is "Riding High," a year after it proclaimed that the United States had entered "The New Golden Age of Productivity" (September 26, 1994). The *Wall Street Journal* ran a series of articles under the banner "America Ascendent—U.S. Companies' New Competitiveness" (August 9, 1994). *Fortune* breathlessly reported that the recent increase in U.S. business investment is "The Most Important Economic Event of the Decade" (April 3, 1995). And from across the ocean, *The Economist* rhetorically asked, "American Business: Back on Top?" (September 16, 1995)—and answers with a guarded affirmative.

As a long-time practitioner in this field, I believe the truth is much less reassuring. Based on both aggregate statistics and my own observations of practice within companies, I will argue that America's current international competitiveness is still shaky. Moreover, we are poorly positioned for the future. Not only are we investing insufficiently in our future, but the real challenge we are facing today—both from Japan and from such developing countries as China, India, Malaysia, and Mexico—is not low-cost labor but low-cost technology.

All the articles quoted above base their analyses on similar "facts": our GDP has been rising vigorously for the past three years; nonfarm productivity is "soaring" (*Fortune*'s term), after growing less than 1 percent a year for two decades; investment in new equipment is "booming" (*Fortune*, again); corporate profits are way up and stock prices are up even further. Meanwhile, unemployment is down close to 5 percent, and the average wages of our workers are finally beginning to increase again. This new vitality is reflected in our increasing success in world trade, particularly in high-tech products. Not only are American exports flourishing, but we appear to have blunted the Japa-

ROBERT H. HAYES is Professor of Business Administration, Harvard University.

nese invasion of several key markets, such as autos, computers, and integrated circuits. We are even—at last—beginning to have some success invading *their* markets.

And all agree that these improvements are based on real, sustainable accomplishments. Over the past fifteen years U.S. companies have engaged in a flurry of improvement activities that are finally bearing fruit. They have been downsizing (or, more euphemistically, "right-sizing"), delayering their organizations,

> *Japanese companies continue to invest two to three times as much per employee as does the United States—even though they have been experiencing a prolonged recession, while we have enjoyed strong years throughout most of the 1990s.*

benchmarking the "best in class," reengineering their business processes, outsourcing globally, getting close to their customers, and pursuing "world class manufacturing" status through a shotgun blast of TLAs (three-letter acronyms), such as TQM (total quality management), JIT (just-in-time production scheduling), DFM (design for manufacturability), QFD (quality function deployment), QPD (quick product/process development), CIM (computer-integrated manufacturing), etc.

UNDERNEATH THE EMPEROR'S CLOTHES

In the midst of the euphoria, fueled by skyrocketing stock prices and this barrage of feel-good journalism, one feels awkward raising a few contrary facts. Consider that in 1995 (although the data are not yet complete) we may well have run the largest merchandise trade deficit in our history—well in excess of $150 billion. This level is comparable to the deficits we were running in the mid-1980s, with or without the effect of imported oil. A decade ago, however, those deficits were attributed largely to the strength of the U.S. dollar during that period. Since then, the dollar has fallen between 40 and 60 percent against the currencies of our major competitor nations. Yet still our deficits persist. And our exports, although up from the mid-1980s, have returned only to about the same percentage of GDP that they represented in 1975.

Even worse, one-third of our total trade deficit is in such high-tech products as computers, communications equipment, and office machines. For example, although U.S. producers dominate the world computer industry (their share of global shipments now exceeds 70 percent), we still ran a trade deficit of over $15 billion this past year because those producers have become so dependent on foreign sources of computer parts and peripheral equipment. And in motor vehicles—that most familiar and important example of our supposed industrial resurgence—we continue to run an annual trade deficit approaching $50 billion (about as large as the trade surplus we show for all services, not including the repatriation of dividends from foreign investments).

A closer examination of our "soaring" productivity and "booming" investment is similarly disappointing. The growth in our GDP (and hence our national productivity) does not reflect adequately the plummeting costs of computers over the past decade. When adjusted to incorporate changes in the prices of computers and peripheral equipment, the recent apparent growth in U.S. nonfarm productivity essentially disappears (*Challenge*, November–December 1995). Similarly, although U.S. investment in new plant and equipment has risen over 50 percent during the past four years, it still represents only about 2 percent of corporate revenues—a third of its level in 1980 (see Figure 1, next page). Moreover, the investment in new plant and equipment per worker (which, along with worker education and skill levels, is the key to future productivity growth) has also dropped to about a third of its levels in the late 1970s. More ominous, Japanese companies continue to invest two to three times as much per employee as does the United States—even though they have been experiencing a prolonged recession, while we have enjoyed strong years throughout most of the 1990s.

DON'T PITY "POOR" JAPAN

Speaking of Japan, Americans have taken some comfort over the past five years in watching as it gets buffeted by a collapse in land values, a persistent recession, a banking crisis, political scandals, a collapsing stock market (that fell, in percentage terms, farther than did the U.S. stock market between 1929 and 1932), and an ever-strengthening currency that keeps devaluing its foreign investments and forcing its export prices up. As a result, Japan's assaults on U.S. markets have faltered, and it is increasingly receptive to imported goods. Many Americans therefore appear

7. GLOBAL ISSUES

Figure 1

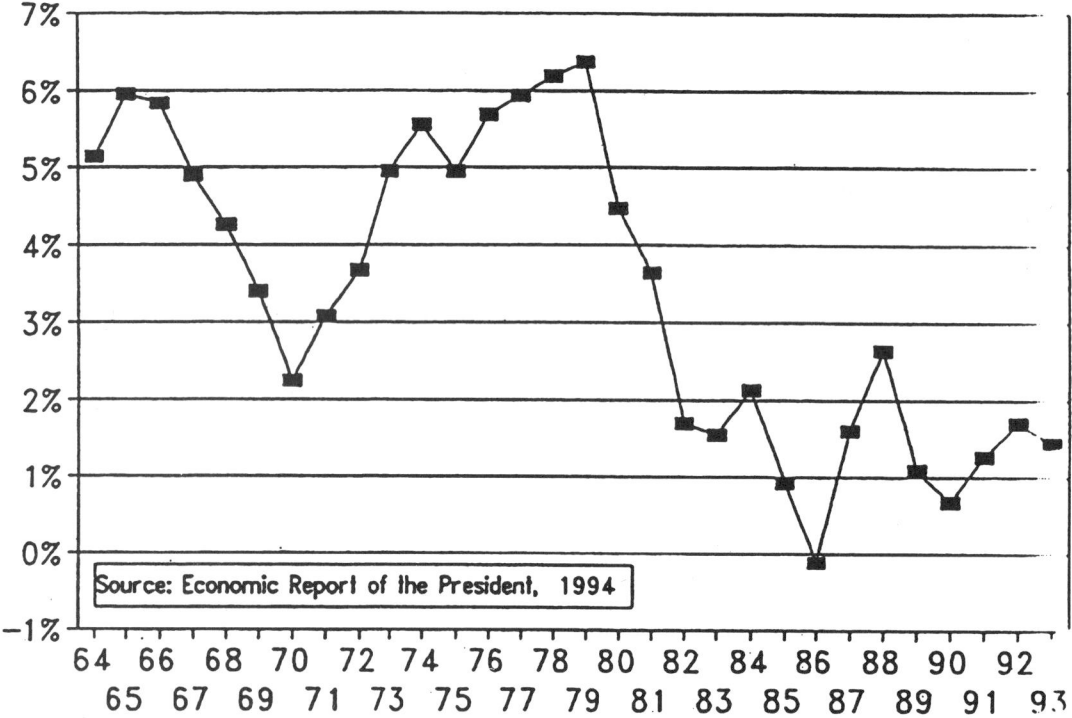

to feel that Japan no longer poses a serious threat to us. Nothing could be further from the truth.

Japanese industry is stronger than ever. It has continued to invest in advanced equipment and new methods and has steadily improved the design, cost, and quality of its products. It has responded to slow domestic growth by increasing its exports, primarily to the fast-growing economies of Southeast Asia—particularly China. It has been preparing for a strong yen for years. I remember touring a number of Japanese factories in the late 1980s and seeing on their walls banners that featured the number "100" in the midst of several Japanese characters. I learned that all expressed variations of the same theme: "We must continually seek improvement, so that we can still be competitive when 100 yen equals one dollar." Seven years later, they are.

A recent study by Goldman Sachs suggests that export-oriented Japanese industries can operate profitably today as long as the yen is above 93 to the dollar ("Yen Appreciation's Effects on the Japanese Economy," March 29, 1995). Another study by Deloitte & Touche, comparing various Japanese and American industries, found that Japanese companies still possessed both a quality and a flexibility advantage over their U.S. counterparts. Moreover, they had much greater experience with advanced process technologies and were receiving greater benefits from their investments in these technologies. The study concluded, rather gloomily, that this "technology experience gap" is likely to widen in the future ("Comparison of U.S. and Japanese Manufacturers," 1992).

Supporting evidence is provided by Professor Lawrence Franko, who has been analyzing trends in world trade for over a decade. He reports that the percentage of world sales accounted for by Japanese companies continues to increase in such key industries as autos (from 23 percent in 1990 to 31 percent in 1993), computers and office equipment (23 percent to 32 percent), electrical/electronic equipment (47 percent to 52 percent), and nonelectrical machinery (23 percent to 41 percent) (see "Global Corporate Competition in the 1990s: The Japanese Juggernaut Rolls On," University of Massachusetts working paper, June 1995). These, not surprisingly, are the same industries for which Japan runs large trade surpluses with the United States. Meanwhile, such high-tech products as fish, wood in the rough, and unmilled corn account for three of the four largest product categories that Japan imports from the United States (to be fair, the largest category is aircraft).

We do not mean to downplay the seriousness of Japan's banking crisis or its current political instability. Both are at least as disruptive as America's Savings & Loan crisis of the 1980s, which we eventually overcame, and our current political gridlock. But neither create

> *A growing chorus of business executives and observers is expressing alarm about corporate downsizing and restructuring. They find little evidence that it has led to increased competitiveness and is in fact more likely to result in "corporate anorexia"—a long-term downward spiral of lower sales and profits, followed by further cutbacks.*

serious impediments to a vigorous, innovative industrial sector. And given the choice between a country with a strong industrial base but a weak banking sector and one with strong banks but weak industry (think of Switzerland), most people would choose the former.

BUT WHAT ABOUT ALL THAT REENGINEERING?

How can we reconcile these more gloomy statistics with all the stories about companies that have become "lean and mean" through right-sizing, delayering, and reengineering? The fact is, the great majority of those efforts have been failures—by the companies' own admissions. A two-million-dollar study of 584 companies conducted jointly by Ernst & Young and the American Quality Foundation in 1991 found that most of these programs had achieved "shoddy results." In 1992 studies by two highly respected consulting companies—McKinsey & Company (of thirty quality programs) and Arthur D. Little (of 500 manufacturing and service companies)—both found that only about a third of these programs had had a significant impact on their companies' market success. The following year both the Electric Power Institute (in a study of more than 300 large companies) and another management consulting firm, Sibson & Co., found the overall success rate to be even lower (based on 4,000 focus groups, consisting of both employees and customers). And another recent study of over 100 reengineering projects by McKinsey & Co. reported that "most process reengineering efforts have . . . had little measurable impact on the overall business."

These results square with my own observations. I am hardly unbiased; I have been arguing in a number of books and articles over the last fifteen years that U.S. companies need to improve their productivity, quality, and response times. Yet in company after company—many of which had proclaimed their improvement efforts to be highly successful and, in some cases, had even invited me in to see what they had done—I found contradictory evidence and deep cynicism down in the ranks. Most employees felt that they had been fragmenting their time trying to implement a "blizzard of buzzwords," which was augmented each time one of their senior managers attended a conference, read the latest "secrets of success" management book, or hired a consultant. Worse, many of these initiatives turned out to be in conflict with one another (either because they adopted contrary goals or methods or because they fought for common resources) and most had little discernible strategic value.

Meanwhile, a growing chorus of business executives and observers is expressing alarm about corporate downsizing and restructuring. They find little evidence that it has led to increased competitiveness and is in fact more likely to result in "corporate anorexia"—a long-term downward spiral of lower sales and profits, followed by further cutbacks. A recent American Management Association survey of corporate downsizing efforts undertaken between 1989 and 1994 found, for example, that only about half led to higher profits and only a third to increased productivity. But 86 percent reported lower employee morale. Nor has "delayering" had the impact that anecdotal evidence would suggest. In fact, a recent *Wall Street Journal* study of Fortune 500 and other U.S. companies discovered that the total number of managers they reported has remained essentially unchanged since its high point in 1990. And a Bureau of Labor Statistics study estimated that the category "executives, managers, and administrative personnel" for U.S. companies as a whole grew almost 30 percent between 1983 and 1993.

THE PLIGHT OF THE AMERICAN WAGE-EARNER

In 1985, President Reagan's Commission on Industrial Competitiveness, composed of distinguished managers and academics, reported that "A close look at U.S. performance during the past two decades reveals a declining ability to compete—a trend that if not reversed, will . . . [undermine] our ability to provide for our people the standard of living and opportunity to which they aspire." Since then, not much has changed.

Despite the recent uptick, the average U.S. worker today earns about the same wages, in terms of buying power, as she or he did in 1985. In fact, average wages and benefits today are, if anything, slightly less then they were in 1973 (the increases in employee-contributed benefits that have taken place have been essentially offset by higher taxes). More alarming, income inequality is increasing as the wages of lower-skilled workers fall sharply in real terms while those of highly educated "knowledge workers" grow briskly. I say "alarming" because both trends portend increasing social divisiveness and instability into the future.

To understand why, it is helpful to know "the rule of 69"—that is, the number of years required for something to double in value is equal to 69 divided by the compound growth rate of that "something." Hence, during the twenty-five years after World War II, when our national productivity growth and inflation-adjusted average wages were both growing at about 3 percent a year, the typical worker could expect his or her real annual income to double over the next twenty-three years. In addition, those wages would be further bolstered by the increasing experience, skills, and responsibilities that come with the passage of time. In other words, the average wage-earner could expect his or her wages to at least quadruple in buying power over the course of a working career. In such an environment people tend to trust in the future, to place confidence in the institutions of their society (e.g., political parties, the media, the justice system), and to share their good fortune with those less fortunate.

But in an environment where wages are not growing at all (or are growing less than 1 percent a year, in line with national productivity growth since 1973), workers see little hope for a better life in the future through honest work alone. This undermines their confidence in society and encourages them to seek improvements in their standard of living through gambling (which includes speculation and lotteries) and illicit activities, and by attacking those who have more than they do. They also become increasingly hostile to those who have less than they do, and who are similarly seeking a larger portion of the common pie.

WHAT DOES THE FUTURE HOLD?

Tomorrow's standard of living is largely the fruit of the investments made today—in new equipment, R&D, human skills, and public infrastructure. Unfortunately, there is little evidence to suggest that the future will be much better than the present, either for American industry or the American worker. Not only are we underinvesting in our future, but tectonic shifts in the world around us are increasingly devaluing the skills of our people. Moreover, two Faustian pacts we made in the past to support our current standard of living are now coming back to haunt us.

Despite the widely heralded increase of investment over the past three years (not an unusual phenomenon when economies emerge from the bottom of a business cycle), as Figure 2 shows (see next page), American industry is still reinvesting a substantially smaller percentage of its revenues and profits than it did fifteen years ago. In addition, government spending on public infrastructure has steadily declined to less than half (as a percentage of GNP) its level of thirty years ago. And with the cutbacks in government-sponsored R&D, total R&D has also fallen. Even though nondefense R&D has risen somewhat to compensate, the total is still about the same percentage of GDP as it was twenty-five years ago—and a third less than Germany's and Japan's. (See Figure 2.) Adding together all forms of investment—private plant and equipment, public infrastructure, public and private R&D, and even public and private education—comes to about 18 percent of U.S. GDP, down 25 percent from its level two decades ago.

Why are we underinvesting? The reason lies partly in U.S. companies' preoccupation with high returns on investment; historically their ROIs have been far higher than those achieved by their German and Japanese competitors. That is not necessarily a mark of strength; the most profitable time in a company's history is often right after it stops investing—and before it goes out of business. But the major reason runs deeper: Americans save less than 5 percent of

> *A close look at U.S. performance during the past two decades reveals a declining ability to compete—a trend that if not reversed, will . . . [undermine] our ability to provide for our people the standard of living and opportunity to which they aspire.*

Figure 2.

Nondefense R&D Spending as a percentage of GDP

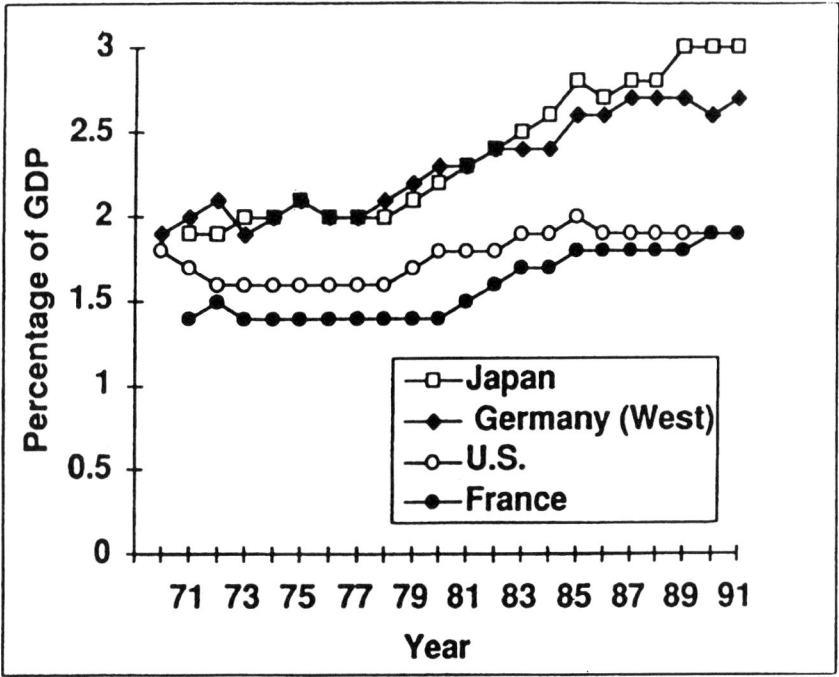

Source: National Science Foundation, "Science & Engineering Indicators," 1993.

their disposable income. If one takes into account all forms of saving, including pension funds and home purchases, we save about 15 percent of GDP. Both figures are at least 50 percent below where they were fifteen years ago, and below those of most other developed nations.

This level is insufficient to fund the amount of investment we are making, which brings us to our first Faustian pact: since we cannot save enough, we pay dollars to foreigners for their goods and depend on them to save for us—and then loan back to us the money we need. The huge trade deficits that result debase our currency, which lowers our standard of living by feeding inflation, and have driven us in ten years from being the world's biggest creditor nation to being its biggest debtor. And as people everywhere discover, the more debt one takes on, the more onerous the terms of future borrowing become.

The reduced level of R&D spending by the large companies that, along with universities, conduct most of the basic research in this country is having the impact on patents that one might expect. Six of the top ten recipients of U.S. patents in 1993 were Japanese companies, and a recent *Business Week* analysis indicated that the three firms with the strongest overall patent portfolios were all Japanese.

This brings us to our second Faustian pact. In an effort to maintain, even increase, their standard of living in the face of stagnant wages, over the past twenty-five years American families have been steadily increasing the number of hours they work. They do this both through having additional family members join the workforce and by increasing the number of hours all of them work. The average employed American today works about 8 percent more hours in a year—the equivalent of an additional month—than he or she did in 1969, more even than the supposedly workaholic Japanese and about 15 percent more than the stereotypical "hardworking" German. How much longer can "making up in volume what we lost in margin" continue to be effective? The percentage of the civilian population holding jobs now seems stuck at about 63 percent, with about 55 percent of all women being employed, and there is increasing concern that long working hours are leading to employee health problems and "burnout." As families approach a limit in the annual number of working hours they can sustain, any improvement in their welfare becomes

7. GLOBAL ISSUES

totally dependent on the growth in their wage rate. But today, more than ever before, the wages an American worker can command are driven by the forces of world supply and demand.

Large American retailers, such as The Gap, are able to achieve turnaround times of just a few weeks on orders for customized garments from Chinese suppliers—about as quickly as from their best domestic suppliers.

A WORLD GLUT OF WORKERS

The world has always been awash with unskilled workers—most of whom reside in the so-called "third world" of developing nations. Until about ten years ago, however, this excess supply had relatively little impact on wage rates or the availability of jobs in the developed countries for two reasons. First, these third-world workers were largely inaccessible to companies in the developed world, for a variety of reasons ranging from the politics of the cold war to inadequate methods of communication to problems of bridging different cultures. Second, when companies in the developed countries were able to perfect and simplify procedures and processes to the point where they could be transferred to lower-cost workers in remote locations, they tended to retain the higher-skilled and higher-value-added work in their home countries. This included activities that required close interaction (e.g., customized products and services, as well as fast response) with domestic customers, the invention, development, and introduction of new products and processes, and corporate management.

This exchange (in effect, of mature products and processes for innovative ones, and of low-value-added work for higher-value-added work) usually led to layoffs of unskilled workers in the developed countries in the short term, but it also set in motion a sequence of longer-term benefits. By reducing production costs, it enabled the remaining skilled people to produce greater profits than was possible previously. This increased the value of these skilled (and, usually, more-educated) people and allowed them to command higher wages than before. This in turn expanded the wage differential between skilled/educated and unskilled workers, which intensified the motivation of younger people to seek more education. Over time this "virtuous cycle" both fueled the productivity growth of the developed countries and encouraged their populations to become more educated, thereby enabling them to command ever higher wages.

Over the past decade or so, however, changes have been taking place that are undermining this virtuous cycle and therefore will have profound implications for future wage growth and job stability in the developed countries. First, the end of the cold war—and, more important, the failure of socialism and of centrally planned economies around the world—has opened up countries and regions that were previously closed in much of the third world: China, India, Latin America, and Russia/Eastern Europe. Moreover, improvements in transportation and telecommunications have made it easier than ever before to communicate with companies in those countries and to integrate them into a worldwide supply network. As an example, product specifications and shipping schedules can now be transmitted instantaneously between the United States and China. These plans can be interactively clarified and modified to meet the constraints of local conditions, and components and products can be shipped back and forth by overnight delivery firms. Large American retailers, such as The Gap, are thereby able to achieve turnaround times of just a few weeks on orders for customized garments from Chinese suppliers—about as quickly as from their best domestic suppliers.

These two changes have effectively added more than a billion unskilled workers to the pool of unskilled workers that is available to companies in the United States and other developed countries. Most of them are willing to work for a fraction of the wages that unskilled workers in the United States require.

Second, these workers are increasingly literate, trainable, and motivated. Television and movies have given them a vision of "the good life" and an insatiable hunger to achieve it for themselves (about 95 percent of those living in China's coastal provinces have access to color television sets; a large percentage of them are able to receive programs transmitted from Japan, Korea, Taiwan, Hong Kong, and Macao). Five years ago I met a young university student in Guilin, China, who told me that his awareness of the standard of living enjoyed by the "outside world," and therefore his expectations for his own future, had been profoundly changed by a James Bond movie! The material success achieved by the developed countries makes young people like him hungry to learn the latest management techniques. Moreover, an increasing percentage of them speak English (as did my university student). It has become the second language of choice throughout Asia and Latin America; when a Korean businessman communicates with his Taiwanese, Chinese, or Mexican counterpart, their common language is usually English.

Moreover, it is much easier to train workers who have little previous industrial experience to accept the demands and discipline required by TQM or JIT than it is to get older, "experienced" workers to adopt these same techniques. Again and again in my visits to factories in Mexico and Southeast Asia, I have been struck by their enthusiastic adoption of the most advanced approaches to production management. A couple of years ago, the Cummins Engine Company judged its factory in San Luis Potosí, Mexico, to be the most successful adopter—in a worldwide network of over fifty plants—of the new production system that the company had been encouraging all its plants to utilize. Similarly, most of the top-rated General Motors factories in North America (according to its own rating system, which incorporates several criteria) are in Mexico. And one of the top-rated factories in Hewlett-Packard's worldwide network is in Penang, Malaysia.

Finally, and perhaps most ominously, these developing countries have evolved into sources of skilled as well as unskilled workers. Whereas the percentage of college students who major in engineering is decreasing in the United States, as well as in most European countries, China, Eastern Europe, India, Latin America, and Russia are churning out increasing numbers. Six Asian countries (not including Japan), for example, graduate about six times as many scientists and engineers each year as does the United States, and over half those who receive doctorates in science and engineering from U.S. universities have student visas. The second largest pool of English-speaking, scientifically trained people in the world is in India. As a result of this expanded supply, the wage differential between U.S. engineers and technicians and their third-world counterparts is now about the same as that for unskilled workers. Good computer circuit board designers in California, for example, cost their companies well over $100,000 a year, if one adds support costs to their salaries and benefits. The comparable cost of an English-speaking Indian engineer having similar experience is well under $30,000. The greatest threat to the U.S. standard of living posed by the newly open economies of the third world, therefore, is that they have evolved into viable, low-cost sources of technology.

Rather than using U.S. engineers to develop new products and production processes and then transfer them in their entirety to third-world locations as they used to do, American companies are increasingly encouraging engineers and other skilled people in those countries to take on a greater share of the creative work. Several years ago, for example, Hewlett-Packard established a factory in Guadalajara, Mexico, to take advantage of the low-cost assembly labor there. Over time, that facility evolved into an international purchasing office, coordinating the worldwide acquisition of component parts for printer production. Now it increasingly is being seen as a center for new product development. Similarly, H-P's facility in Singapore has been designated as the global R&D and production center for its portable ink-jet printers. Several years ago, I visited an advanced product design facility in Monterrey, Mexico. Using the latest CAD (computer-aided-design) technology, it was in direct satellite communication with customers in Europe. Once a design had been agreed upon, a three-dimensional model was automatically produced using what was then a cutting-edge process called stereo lithography. I estimate that this facility used more technically advanced design and prototype approaches than did 95 percent of its counterparts in the United States at that time.

This trend is encouraged by the increasing availability in those countries of skilled managers and technical personnel who have been educated in the developed world and then worked for several years for foreign companies. For example, for many years Taiwanese lamented their "brain drain" to the United States—young Taiwanese who came here for their education and then stayed on to work for U.S. companies, forsaking the land of their birth. Now, as Taiwan's prosperity and income levels rise rapidly, as its public infrastructure and living amenities approach those of the developed countries, and as those expatriates begin to perceive constraints on—and increasing

---

It is much easier to train workers who have little previous industrial experience to accept the demands and discipline required by TQM or JIT than it is to get older, "experienced" workers to adopt these same techniques.

7. GLOBAL ISSUES

uncertainty about—their personal development and prosperity in the United States, many of them are returning to Taiwan. The "brain drain" is now perceived as a "brain bank"—even a "brain incubator," since these repatriates bring with them all the skills and experience they have gained in the States. Moreover, like similar returnees from Hong Kong, Malaysia, and Singapore, their knowledge of Chinese language and culture has prepared them to manage operations not only in their own countries, but also in China and other Southeast Asian countries whose populations include a substantial proportion of people of Chinese descent.

This shift is not confined to manufacturing. Around the world, skilled people who speak English are being utilized by service companies as well. The friendly and helpful customer service person on the other end of your telephone is increasingly likely to be located in the Caribbean, Ireland, or even India. Accounting firms are using the services of experienced, low-cost accountants in the Philippines, and software development is becoming a world commodity. India alone is estimated to have over 100,000 software engineers, and China has at least three times that many. India's exports of software products surpassed $700 million this year, thirty times its level in 1987. No matter where they are, these programmers can communicate by satellite directly with their clients in other countries. Although they are not yet as efficient as U.S. programmers, Indian and Philippine programmers are still able to produce comparable software products at about half the cost. As a result, an increasing number of U.S. companies, such as Sea-Land Service, Inc., are laying off domestic programmers and outsourcing their work to other countries.

CONCLUSION

Some economists discount the importance—even disparage the very notion—of "national competitiveness," but the idea is very simple. A competitive nation, at the very least, ought to be able to pay its own bills and create a social and economic environment within which future generations can at least maintain their level of prosperity. Moreover, it ought to be able to compete effectively for the world's business, capital, and work without undermining its citizens' standard of living through devaluations of its currency. The United States is uncomfortably close to failing on all these counts. We import more of the world's goods and services than we export to them. We are increasingly captive to the financial scrutiny and political whims of the "gnomes of Zurich" and Tokyo. Foreign workers, both skilled and unskilled, are able to perform equivalent jobs at lower cost. And all the while the U.S. dollar has been weakening.

The virtuous cycle of economic and educational development is losing its momentum. As skill levels in the developing countries increase, and as these skills become more and more accessible to us, the value of additional education in the United States is undermined. I do not refer here to higher education, such as that leading to undergraduate and graduate degrees, but to such modest efforts at self-improvement as simply finishing high school and perhaps seeking some additional technical skills—the kind of thing most unskilled workers in the United States might aspire to. And as the reward for self-improvement decreases, so does the motivation to do so. In fact, for several years we have been seeing declines in the average literacy and education levels of the new entrants to our work force, as both the high school dropout rate increases and the literacy of those who do graduate declines. At the same time, companies are finding it increasingly difficult to find people who can take on "high-skill" jobs (i.e., requiring at least a bachelor's degree); for example, a recent study predicted that by 2010 the state of Massachusetts will not be able to fill about 400,000 such jobs.

It is all too possible for the virtuous cycle to reverse and become a downward spiral. The United States is not "riding high," as the business press would have us believe. It is riding perilously close to a slippery slope.

The Civil Rights Issue of the '90s

Nancie G. Marzulla

Nancie G. Marzulla is president and chief legal counsel of Defenders of Property Rights in Washington, D.C.

When people think of civil rights, they tend to conjure up images of segregated schools in the 1950s or the fight we had in the 1960s (indeed, are still having) in this country to ensure that all people—regardless of their skin color or nationality—can enjoy the same privileges of living under the rule of law in a free society. That right to equality regardless of race, color, or creed is guaranteed by the Fourteenth Amendment to the U.S. Constitution.

Today, however, another civil rights revolution is under way. At issue is another guaranteed right, that of obtaining payment of just compensation whenever one's private property is taken by the government for public use. As specified in the Fifth Amendment to the Constitution, "No person shall be . . . deprived of life, liberty or property, without due process of law; nor shall private property be taken for public use, without just compensation."

Just as segregation led to the racially based civil rights movement, so it is that incursions on property rights, largely in the name of protecting the environment, have sparked the property rights movement now, some 30 years later. Starting in the 1960s, federal, state, and local governments increasingly began to regulate property rights through environmental protection policies.

Today, environmental regulations touch every conceivable aspect of property use and ownership, often infringing directly upon private-property rights protected by the Constitution. Through its ability to regulate, the government has steadily and increasingly tended to "take" whatever uses and benefits of property it wishes rather than condemning the property outright and paying for what it has taken.

The property rights movement comprises thousands upon thousands of individuals across America who are being singled out by the government to bear the unfair burden of implementing land use and environmental policies the government itself is simply not prepared to pay for. The complaint of the property owner is not so much what the government is trying to achieve through its policies but the means by which it is achieving them.

THE COURT TAKES A STAND

A case in point is last June's decision by the U.S. Supreme Court in *Dolan v. City of Tigard*. At issue was whether the city could demand from Florence Dolan approximately 10 percent of her land—which the city planned to use for a bicycle/pedestrian pathway and public greenway—in exchange for being granted permission to enlarge her plumbing-supply store. No one, including the Dolans, objected to the city's plans for a bike path or greenway. Clearly, the fair thing for the city to do—the constitutionally mandated way of achieving its objectives—was to buy the land necessary to complete the project rather than hold Dolan's planned use of her property hostage, as the city did.

Such infringements have provoked people to unite to form the

7. GLOBAL ISSUES

property rights movement. These people are objecting to an environmental regulatory regime that by 1993, according to a report from the American Enterprise Institute, was costing Americans $140 billion a year in compliance costs and that ranks even the most insignificant snail's welfare as more important than human interests.

They are objecting to a wetlands program that delineates one to two hundred million acres of land—75 percent of which is privately owned—as "wet" and stipulates that it must be maintained untouched by human hands. They are objecting to a criminal enforcement program that puts innocent people in jail for simply placing dry sand on existing dry sand, all on land that is privately owned, because it violates the government's policy of no net loss of wetlands.

Until recently, the Supreme Court showed little interest in property rights law. In 1922, Justice Oliver Wendell Holmes declared the bedrock principle of takings law: "The general rule at least is, that while property may be regulated to a certain extent, if regulation goes too far it will be recognized as a taking." After this ruling, however, the Court did not select any cases that would advance this doctrine. This was apparent in 1978, when Justice William Brennan expressed his dismay over the Court's inability to "develop any 'set formula' for determining when 'justice and fairness' require that economic injuries caused by public action be compensated by the government rather than remain disproportionately concentrated on a few persons."

In 1987, however, the Court showed a renewed interest in property rights, ruling on three cases. In *First English Evangelical Church v. County of Los Angeles*, the Court held the county could be required to compensate a church barred by a flood-control ordinance from reconstructing summer camp buildings destroyed during a 1978 flood. In *Nollan v. California Coastal Commission*, the justices ruled the public could not require that the owner of a home next to a beach donate a third of his land to the state in order to obtain a permit to rebuild the house without paying just compensation. *Hodel v. Irving* invalidated a regulation eliminating Native Americans' right to devise reservation lands to their heirs.

In 1992, the Supreme Court again ruled for the plaintiff in *Lucas v. South Carolina Coastal Council*. The case involved two beachfront lots the owner planned to develop that were rendered useless by a beachfront protection statute. The central holding of *Lucas* is that

> regulations that deny the property owner of all 'economically viable use of his land' constitute one of the discrete categories of regulatory deprivations that require compensation without the usual case-specific inquiry into the public interest advanced in support of the restraint.

The state was forced to purchase the property from Lucas. The state then put the property up for sale to potential developers: the very activity from which Lucas had been barred. This shows how the government's attitude changes when it is forced to bear the cost of its regulations, at least when the Fifth Amendment is applied.

In his opinion in the *Dolan* case, Chief Justice William Rehnquist noted, "We see no reason why the Takings Clause of the Fifth Amendment, as much a part of the Bill of Rights as the First Amendment, should be relegated to the status of a poor relation in these comparable circumstances." The high court reversed the decision of the lower court, sending it back to be retried. This time, the burden is on the city to prove why it needs to take the Dolan property.

CONGRESS GOING IN THE RIGHT DIRECTION

The courts, however, are not the only place where the muscle of the property rights movement is being felt. The 103rd Congress has been active on three fronts—unfunded mandates, risk-assessment analyses of regulations, and property rights bills—which environmentalists collectively refer to as the "unholy trinity." These three reforms have become

---

Property Rights in America

- They are protected under the Fifth Amendment.
- They have been under assault from local, state, and federal governments for at least 30 years.
- They have been aided by the Supreme Court's proproperty decision in the Dolan case, handed down this June.

40. Civil Rights Issue of the '90s

dreaded amendments to regulatory reauthorizations in the eyes of the environmentalist movement.

In a memo to environmental leaders, a Natural Resources Defense Council lobbyist warned that regulations long overdue for reauthorization could not be passed during this session of Congress without one or more of the trinity amendments attached. The strategy the lobbyist offered was to keep most of the reauthorization bills off the table, allowing the environmental movement to focus on just one or two pieces of legislation. Regulations that have not been reauthorized are in no danger of expiring because they remain valid until reauthorized.

Property rights were the focus early in the session, when Interior Secretary Bruce Babbitt's National Biological Survey—a plan to map the entire

■

Today, environmental regulations touch every conceivable aspect of property use and ownership, often infringing directly upon private property rights protected by the Constitution.

■

nation by ecosystems—was saddled with property rights amendments when it passed the House. The bill was never introduced in the Senate.

The importance of risk-assessment analyses led to the downfall of the bill that would have elevated the EPA to cabinet-level status. When the bill arrived on the House floor for debate, the leadership would not allow consideration of a risk-assessment amendment that had already passed by an overwhelming margin in the Senate. Angered over this move, congressmen sent it back to committee, effectively stopping the bill.

Lawmakers with political ideologies as divergent as Sens. Dirk Kempthorne (R-Idaho) and Carol Moseley-Braun (D-Illinois) have come to agreement on the issue of curbing unfunded mandates, with each sponsoring bills to reduce or even eliminate them completely. An unfunded mandate is federal legislation that seeks to achieve its goal by compelling states and localities to pick up the bill for enforcement.

These mandates are the bane of mayors and governors

■ *The current justices:* The Court's decision in *Dolan v. City of Tigard* has given a boost to private property rights.

7. GLOBAL ISSUES

across America, and the Clinton administration has been forced to take notice of the problem. Environmentalists are particularly concerned about this because so many environmental regulations passed at the federal level compel the states to pay for their compliance and fiscal outlay. In some cases, they are forced to set up their own programs.

The property rights issue has gained such prominence that the Senate has established a property rights caucus led by Sens. Bob Dole (R-Kansas) and Howell Heflin (D-Alabama), a former chief justice of the Alabama Supreme Court. In the House, Louisiana Democrat Billy Tauzin is finding himself at odds with his party colleagues in the House leadership as he circulates a discharge petition to bring his "Property Owner's Bill of Rights" to the floor for debate. Sen. Phil Gramm (R-Texas), an oft-mentioned Republican candidate for the White House in 1996, has also introduced property rights legislation, and he has expressed interest in using it as a campaign issue in this fall's elections. (Gramm is the head of the National Republican Senatorial Campaign Committee.)

Property rights are also becoming a major issue in the states. Over the past legislative session, 37 state legislatures introduced almost 100 bills to protect property rights. Ten of these states have turned the bills into law. At the American Legislative Exchange Council's 1994 conference, a whole morning was set aside to discuss the issue of property rights. Many states are eagerly awaiting the results of a vote to take place in Arizona in November, where voters will be asked to decide the fate of property rights legislation.

After a petition campaign rife with deceitful allegations that property rights legislation would ruin Arizona's economy and ecology, environmentalists were able to force a referendum on property rights legislation already passed by the legislature and signed by the governor. Proposition 300, as it is now known, "would require state agencies, before a taking results, to examine their activities, including rules and other regulatory actions that effect the use of property, to determine if an action requires compensation from the state." In addition to this important election battle, Massachusetts and Florida are also putting property rights to a vote in referendums this fall.

From the very beginning of our republic, property rights have been considered sacred. The Founding Fathers envisioned a strong system of property rights as a means of protecting individual liberty. John Adams, in his *Defense of the Constitutions of Government*, said, "The moment the idea is admitted into society that property is not as sacred as the laws of God, and there is not force of law and public justice to protect it, anarchy and tyranny commence." The founders believed property rights to be of overreaching importance. Indeed, the Constitution provides that money be awarded to citizens for incurred damages in only one case. As stipulated in the Fifth Amendment, "Nor shall private property be taken for public use, without just compensation."

Property rights are the true linchpin of all the rights we hold dear. If the government possesses the ability to take away our land at any given time for any given reason, our rights to speak freely, associate with whom we choose, and other individual freedoms are at risk. It has taken the American people over two decades to take serious notice of this erosion, but they are making up for lost time. Property rights, like any civil rights movement, is strong at the grass roots.

■

John Adams said: "The moment the idea is admitted into society that property is not as sacred as the laws of God, and there is not force of law and public justice to protect it, anarchy and tyranny commence."

■

Facing the equally strong and well-funded environmental establishment, it is shaping up as one of the biggest battles of this decade—and beyond.

ADDITIONAL READING

Terry Anderson and Donald Leal, *Free Market Environmentalism*, Pacific Research Institute, San Francisco, Calif., 1991.

Richard Epstein, *Takings: Property Rights and the Power of Eminent Domain*, Harvard University Press, Cambridge, Mass., 1985.

"The Pocket Guide to Your Property Rights," Defenders of Property Rights, Washington, D.C., 1994.

Dixy Lee Ray, *Trashing the Planet*, Regnery Gateway, Washington, D.C., 1991.

———, *Environmental Overkill*, Regnery Gateway, Washington, D.C., 1993.

The international trade in drugs has become an increasingly important issue in global security. It is a problem, however, that falls outside traditional national security concerns, even though it threatens the political stability of many states.

Global Reach: The Threat of International Drug Trafficking

RENSSELAER W. LEE III

RENSSELAER W. LEE III, *president of Global Advisory Services in Alexandria, Virginia, and associate scholar at the Foreign Policy Research Institute in Philadelphia, is the author of* The White Labyrinth: Cocaine and Political Power *(New Brunswick, N.J.: Transaction Press, 1989).*

Narcotics industries rank as the world's most successful illegal enterprises, generating annual profits of roughly $200 billion to $300 billion. Major production and trafficking complexes in the Andes, Southwest Asia, and the Golden Triangle of Southeast Asia thrive, impervious to international enforcement programs. Indeed, between 1982 and 1994, worldwide opium production more than doubled, and the global output of coca leaves rose by 300 percent. Opium cultivation is expanding rapidly in several communist or former communist states, primarily China, Vietnam, Uzbekistan, Tajikistan, and Turkmenistan. (In China, the revolutionary government claimed four decades ago that the once pervasive problem of opiate addiction was all but obliterated, but poppy plantings have been reported in at least 17 of the nation's 30 provinces, and drug use is escalating rapidly.)

China currently serves as an important transit country for Burmese heroin, and entrepreneurial North Koreans are entering the heroin business, possibly with the backing of their government. Colombia, which produces an estimated 70 to 80 percent of the world's refined cocaine, is the world's third largest opium producer, with an estimated 20,000 hectares under poppy cultivation. Moreover, the distinctive signature of Colombian-refined heroin is increasingly appearing in United States retail markets. A similar trend toward diversification in the industry is also seen in Peru, the world's largest producer of coca leaf, where low levels of poppy cultivation are being recorded. In addition, several communist or former communist countries—Poland, China, Russia, Azerbaijan, and the Baltic states—are emerging as important producers and exporters of sophisticated amphetamines. And narcotics such as LSD, Ecstasy, and trimethylphentanil (a powerful synthetic opiate) are marketed with increasing frequency in former Soviet states.

Narcotics industries are becoming larger, more powerful, and more entrenched in the global economy and in the economies and societies of individual producing states, although they differ significantly in organizational sophistication and systematic effects. For example, trafficking organizations in China are at a rudimentary stage of development, with small staffs that often disband after one or two deals. Furthermore, the profits generated by smuggling opium or heroin out of China flow primarily to overseas Chinese, not mainlanders. Consequently, Chinese drug organizations exert little influence on China's economic system and are generally not viewed as a threat to state authority. (Virtually no reports of high-level drug corruption in China have surfaced.)

At the other extreme, Colombia's highly developed trafficking enterprises employ hundreds of specialized personnel—pilots, shippers, chemists, accountants, lawyers, financial managers, and assassins—directly or on contract, and earn an estimated $4 billion to $7 billion annually, mainly from cocaine sales in the United States and Europe. These revenues endow the narcotraffickers with a significant capability to bribe or otherwise influence the behavior of key Colombian officials and political leaders. (United States law enforcement officials estimate that the Colombian cartels spend more than $100 million annually on bribes in the country.) Colombia constitutes a model of advanced or mature narcotics enterprises that have the economic and political problems associated with a large criminal sector.

Some worrisome trends have arisen in narcotic industry strategies: widening economic influence, that is, the impact of the illicit drug trade on illegal economic structures and processes in major producing or transit countries; the increasing political corruption in such countries; the growing intrusion of narcocriminal enterprises into the realm of the state and the law, a process that some scholars associate with the delegitimation of government; the successes of narcotics businesses in innovation, avoiding detection, and increasing operating efficiency; and, especially apparent since the late 1980s, the growing transnational cooperation among criminal empires that deal in drugs and other black market items. All these trends suggest that narcotics industries are enhancing their power and reach, developing new and advanced capabilities, and establishing new bases of support. At the same time, the leaders and citizens of some trafficking countries are exhibiting clear signs of drug war fatigue. Much to the dismay of the United States, support is growing for peaceful resolutions of the drug trade issue, ranging from negotiated surrenders that treat drug kingpins leniently to the outright legalization of narcotics.

Drug Economics

The economic effects of the drug trade stem mainly from the processes of legitimizing narcotics earnings in the country or countries of origin. Different nations display different patterns. For example, officials in the Chinese Ministry of Public Security believe that opium and heroin smugglers invest few of their earnings in ventures that benefit the Chinese economy. Most drug proceeds are banked in noncommunist Asian countries and the little money that does return to the country tends to be used to buy luxury housing, furniture, electronic equipment, and gold jewelry. In Russia the sales of illicit drugs total an estimated $800 million each year, and law enforcement officials contend that much of the startup capital for small, legitimate businesses, such as stores, restaurants, and fruit stands, is supplied by the narcotics trade. (However, the economic effect of drugs in Russia is difficult to separate from the effect of organized crime groups in general, which operate many profitable illegal enterprises.) Colombia is probably suffering from the most advanced case of narcoeconomic penetration; traffickers annually repatriate an estimated $2 billion to $5 billion from narcotics exports, or approximately 4 to 9 percent of Colombia's GDP of $55 billion. One Colombian economist, Francesco Thoumi, has calculated that accumulated trafficker assets in Colombia and abroad reached anywhere from $39 billion to $66 billion between 1989 and 1990, a scale of narcowealth so immense that it could easily alter Colombia's economic and political status quo.

Indeed, drug money pervades the Colombian economy. For example, according to the Colombian Institute of Agrarian Reform and the Colombian Farmers Association, drug dealers expanded their direct or intermediary ownership of agricultural land from 1 million hectares in the late 1980s to an estimated 4 million hectares in 1994. Today, traffickers own or control between 8 and 11 percent of agriculturally usable land in at least 250 of 1,060 Colombian municipalities, making them a powerful force in the rural Colombian economy. Technological improvements in cattle raising and commercial agriculture sometimes accompanied narco land investments, strengthening traffickers as rural leaders in some areas. Furthermore, as the so-called Cali cartel gained ascendancy, the infiltration of legitimate businesses by drug dealers reached significant new levels. Earlier generations of traffickers, such as the leaders of the Medellín cartel of the 1980s (then the dominant trafficking group in Colombia), were relatively unsophisticated economic actors who were more concerned with laundering drug earnings than realizing adequate returns on their investments. However, these earlier traffickers clearly spent to enhance their status. Conspicuous examples abound. Jose Gonzalo Rodriguez Gacha accumulated 140 country estates, collectively worth an estimated $100 million; some of these residences were lavishly furnished with items such as pillows stuffed with ostrich feathers, gold-plated bathroom fixtures, and imported Italian toilet paper stamped with likenesses of Botticelli's *The Birth of Venus*.

In contrast, the Cali group cultivated an image of business respectability by investing in a wide range of economic activities. According to a recent report by the Colombian Department of Administrative Security, Cali drug money has "infiltrated the construction industry, drugstore chains, radio stations, automobile dealerships, department stores, factories, banks, sports clubs, and investment firms." Agribusiness enterprises such as cut flowers, tropical fruit production, and poultry farms can be added to this list. "In what sectors of the economy has the Cali cartel not invested?" asked Gabriel de Vega Pinzon, the head of the Colombian National Drug Directorate, in a December 1994 interview with this writer.

Drug Politics

Drug trafficking also has wide-ranging effects on political and administrative systems in developing countries. Narcotics industries in countries such as Myanmar, Afghanistan, and Colombia (especially between 1989 and 1991) are associated with extreme antistate violence or with the disintegration of national authority. However, most drug dealers are not pursuing independent political initiatives, preferring to coexist with and manipulate the state authority. "We don't kill judges or ministers, we buy them," remarked Cali cartel leader Gilberto Rodriguez Orejuela on one occasion. Indeed, corruption has assumed outlandish proportions in Colombia. In 1994 police and judicial investigations detected evidence of trafficker payoffs to: a former president of the Colombian national congress, a former comptroller general, a recently elected congressman, 12 retired army officers (communication and security specialists "decorated for their outstanding service to the army"), more than 150 Cali police officers, almost the entire contingent of Cali airport police, employees of the El Valle telephone system, the Cali regional prosecutor, 6 of 22 Cali city councillors, and the mayors of 4 Colombian cities, among them Medellín.

The pattern of corruption in Latin America also includes attempts by traffickers to purchase influence at the highest political levels. During the 1980s, narcocorruption involving top national leaders or their closest associates was documented in Bolivia, Panama, the Bahamas, and the Turks and Caicos Islands. One drug informant claims that Fidel Castro's brother Raúl personally authorized the shipment of 6 tons of cocaine through Cuba between 1987 and 1989. Traffickers have also indirectly sought political leverage by contributing to presidential election campaigns. Trafficker support of the 1989 Bolivian campaign of President Jaime Paz Zamora prompted Paz in 1991 to appoint a known drug dealer, Faustino Rico Toro, to head the Bolivian Special Narcotics Force, although pressure from the United States subsequently forced Paz to dismiss Rico Toro from that post; in early 1995 the Bolivian Supreme Court authorized the extradition of Rico Toro to the United States on drug charges. In Colombia a major scandal erupted in June 1994 when a taped telephone conversation leaked to the press showed Gilberto Rodriguez Orejuela and his brother discussing a possible donation of $3.8 million to the presidential campaign of Ernesto Samper Pizano. Some Colombian and United States observers—among them Jose Toft, the United States Drug Enforcement Agency (DEA) representative in Colombia at the time—contend that the Samper campaign did in fact receive millions of dollars from the Cali cartel. Toft, who resigned from the DEA last September, summarized a widely held belief about the state of Colombian politics when he commented to a Colombian television news station that, "I cannot think of a single political or judicial institution that has not been penetrated by the narcotraffickers—I know that people don't like to hear the term 'narcodemocracy,' but the truth is it's very real and it's here."

Modern narcotics enterprises have also helped criminal authority grow at the expense of legitimate state authority. In Latin America this encroachment spans issues such as social welfare, counterinsurgency, and (ironically) the maintenance of law and order. For example, in Mexico, Colombia, and Bolivia, traffickers have cultivated a Robin Hood image by devoting vast resources to community development projects such as roads, schools, airport repairs, and housing, or by donating money and gifts to the poor. Such activities cemented political support for the drug capos among marginalized social groups such as Medellín slum dwellers and poor farmers in the Bolivian Beni— populations that governments and legitimate nongovernment organizations cannot serve. In Colombia a weak government presence in the countryside, an ongoing rural insurgency, and the acquisition of landed estates by drug lords in the 1980s created new political opportunities and roles for narcotics dealers. For example, paramilitary organizations financed by trafficking interests emerged, supplanting an impotent Colombian state by furnishing local security against predatory guerrilla groups. Curbed somewhat by the Colombian government's 1990 crackdown on the Medellín cartel, narco-backed paramilitaries nonetheless pursue their mission in the middle Magdalena Valley, Cordoba, Uraba, and other guerrilla-infested regions. (Of course, paramilitary operations to root out and exterminate leftist guerrilla sympathizers pose serious human rights challenges for Colombia.) Legitimate private groups conducting business in the Colombian hinterlands— coffee growers, cattlemen associations, and foreign oil companies, for example—admittedly provide public welfare and security protection functions. However, the assumption of such roles by the narcotraffickers generates particularly ominous overtones for the Colombian political process.

Traffickers tend to support their local police on law and order issues such as the defense of property rights and maintenance of basic community services; police who spearhead government narcotics crackdowns or work for rival trafficking organizations, however, stand a good chance of being murdered. The Cali cartel supported a perverted and socially regressive form of law enforcement, the so-called social cleansing groups, which targeted marginal urban dwellers such as prostitutes, thieves, beggars and drug addicts. In some regions of Latin America and Asia, trafficking interests for all practical purposes are the law, since the government does not exercise real sovereignty in those areas. (Drug trafficker Khun Sa's Shan state enclave in Myanmar represents perhaps the most egregious modern example of narcowarlordism.) Yet, opportunistic traffickers also assist or form alliances with governments that persecute rival criminal organizations. For example, in Myanmar, the Wa insurgent trafficking groups are enlisting government help to fight Khun Sa's Shan United Army—and managing to broaden their territorial base in the process.

In Colombia the Cali cartel found it politically and commercially expedient to furnish "valuable information" to the government for its ultimately successful manhunt for Pablo Escobar and some of his lieutenants, a contribution recently acknowledged by Colombian prosecutor general Alfonso Valdivieso (according to the Cali regional prosecutor's office, the Cali group hired Japanese communication experts to track Escobar's movements in the months before his demise). Moreover, in a June 1994 interview with this writer, Valdivieso's predecessor, Gustavo de Grieff, referred to a report that a special government search force in Medellín received a $10 million payment from Cali traffickers shortly after Escobar was killed last December and allegedly distributed the funds among ranking members of the force. "Apparently, [the force] was an instrument of Escobar's enemies, not of the government," de Grieff commented.

Such scattered examples confirm the ability of traffickers, who command enormous power and resources, to pirate government functions or inherit them by default. In surrender negotiations with the Colombian government, the Cali traffickers surprised no one by wielding their contributions to the anti-Escobar campaign as leverage against the government. In a letter to President Cesar Gaviria in January 1994, Gilberto Rodriguez Orejuela petitioned for house arrest rather than a jail cell, in part on grounds of his "collaboration with the prosecutor general's office and the search group to achieve the well-known results."

Finally, new patterns of domestic and international cooperation have spawned among criminal empires that deal in illicit drugs. Such cooperation connects criminal groups such as the Colombian cartels, Mexican smuggling organizations, Japanese *yakuza,* Hong Kong Chinese syndicates, Sicilian mafia, and Russian organized crime. Central issues of common concern include the organization of markets, trade deals (for example, exchanges of drugs for weapons, drugs for cash, and drugs for drugs), smuggling logistics, and laundering or repatriation of trafficking proceeds.

Cooperation between Colombian traffickers and Italian organized crime groups to sell cocaine in Italy and the rest of Europe apparently stands at a particularly advanced stage. The Cali cartel and the Sicilian mafia are experimenting with franchise arrangements that would allow the mafia to distribute large consignments of Cali cocaine to European buyers outside Italy. The Cali group has also established working relationships with organized crime figures in Poland, the Czech Republic, and Russia. The Cali traffickers' strategic design uses these countries as a back door to deliver cocaine to western Europe. Such relationships are underscored by Russian government seizures of 1.1 tons of cocaine in Vyborg in February 1993, and 400 kilograms of the drug in St. Petersburg in April 1994; both shipments could be traced to Cali trafficking organizations. In general, international narcocooperation opens new markets for narcotics and other illegal products, exploits economies of scale for selling in those markets, enhances organized crime's penetration of legal economic and financial systems, and generally increases the power of criminal formations relative to national governments.

Treating the Drug Problem

Confronting powerful narcotics lobbies and publics weary of drug wars, government commitment to suppress narcotrafficking is waning perceptibly in some source countries. One manifestation of this trend is rising political support to legalize drugs. Bolivia's president, Gonzalo Sanchez de Losada, openly favors this; he declared to a Spanish newspaper in 1993 that "The antidrug fight is the politician's tomb—prohibition has achieved nothing but making vices extremely profitable for traffickers. It is terrible to say it, but some tax on drugs should be created." The Bolivian government promised coca farmers in September 1994 that it would mount an international campaign to decriminalize the coca leaf (but not the products derived from it). Colombian President Ernesto Samper recently canceled plans for a popular referendum to overturn a May 1994 decision by the Colombian constitutional court that legalized personal drug use.

The legalization or selective decriminalization of drugs is gaining ground elsewhere in the world. Poland, Russia, and Italy have lifted criminal penalties for personal drug use, and cannabis products are openly sold to adults in coffeehouses in the Netherlands. In China, where drug dealers are routinely executed with great public fanfare, some local cadres advocate removing restrictions on poppy growing to help isolated mountain areas "get rid of poverty"— possibly a sign of the significant proportions of the private opium trade in that country.

In the Andean countries, governments have not legalized drug production—an action that would spur certain retaliation by the United States—but they have attempted to diminish conflicts with the cocaine industry by negotiating with participants and leaders in the trafficking chain. Colombia's negotiations with Medellín cartel leaders date to May 1984, when former President Alfonso Lopez Michelson and Attorney General Carlos Jimenez Gomez held separate meetings with Pablo Escobar and other kingpins in Panama. Since 1990, Colombia has offered reduced sentences and other legal inducements to traffickers who surrender, confess, and turn state's evidence.

Colombian officials see negotiations as a tool of social policy that can subdue the power of individual trafficking organizations. Negotiations doubtless helped reduce narcoterrorist violence in the 1990s, but produced few successes against cocaine trafficking. Important traffickers negotiated relatively short sentences that ranged from 4 to 8 years, but furnished little information on the workings of cocaine enterprises. Ivan Urdinola, for example, refused to name major accomplices, averring that such disclosures would place him in mortal danger, and liberally laced his confessions with fatuous statements. (At one point, he informed a judge, "Aside from being a drug trafficker, I am an admirable person."). Perhaps the late Pablo Escobar abused the surrender policy most notoriously. After negotiating a deal with the government in mid-1991, Escobar was incarcerated in the La Catedral prison near Medellín, where he continued to manage his cocaine business until his escape 13 months later. (Subsequent revelations indicated that Escobar paid $2 million for construction of the facility, which was equipped with cellular telephones, fax machines, and computers.) In Bolivia, the government's repentance program produced similarly disappointing results. Repenters characterized themselves as simple cattle farmers who only dabbled in cocaine or lent money to traffickers; the three most important traffickers who surrendered under the Bolivian program received sentences of only 4 to 6 years.

Cali cartel leaders recently offered to implement a plan that would reduce cocaine exports from Colombia by 60 percent (their estimate of their share of the business) if they spent little or no time in jail. Of course, such an offer invites skepticism, since the Cali dons might not control or directly influence a sufficiently large percentage of Colombian refining and exporting capacity to fulfill such a commitment. Recent information suggests that the Colombian cocaine industry is more decentralized and balkanized than during the 1980s. Gilberto Rodriguez Orejuela himself noted in a November 1994 letter to *El Tiempo* that "there are many cartels"; moreover, the industry depends on a multitude of subcontractors and freelancers. Of course, drug kingpins possess considerable leverage over lower level operators; they can stop purchasing products and services or simply withhold protection from laboratories, transport companies, distribution cells, laundering operations, and other key trafficking entities. But, in putting forth their offer, the Cali traffickers provided no blueprint or timetable for dismantling their multibillion-dollar enterprises. Also, a number of factors—the size of the illicit drug industry, the prevalence of official corruption, and the weakness of the Colombian criminal justice and judicial institutions—indicate that Colombia could not successfully implement such a deal.

Debates over legalization, democratization, and negotiated accords with traffickers in key source countries have produced consternation in Washington. Yet, disillusionment with overseas narcotics control and with drug prohibition in general is also widespread in the United States. Many Americans favor scrapping supply-side programs altogether, shifting resources to education and prevention programs, or even legalizing the production and use of some drugs. United States international initiatives, including the roughly $1 billion allocated to counternarcotics operations in the Andes since 1989, certainly have had few long-term effects on the availability or purity of drugs in America's major urban markets. Some United States policies are wasteful, counterproductive, or worse. For example, the Bolivian government spent $48.1 million in American aid between 1987 and 1993 to pay farmers to eradicate 26,000 hectares of coca. Farmers, however, planted more than 35,000 new hectares of coca during the same period. The planned compensation for eradication transformed into little more than a coca support program.

At least in the short term, the objective of restricting internationally the suppliers of illegal drugs is probably not attainable. The number of potential drug suppliers is virtually unlimited; few geographical, organizational, or technological barriers obstruct entry into narcotics industries, and crops, laboratories, drug shipments, planes, money, chemicals, and routes can be easily replaced if destroyed. Of course, the value of international drug control does not necessarily lie solely in controlling narcotics. The war on narcotrafficking can be justified as a moral imperative even if it is a practical failure. Furthermore, the United States has staked its prestige and predicated its diplomatic relations with several countries on combating the drug scourge.

More important, however, is the fact that enterprises such as the Colombian cartels and counterpart groups in Europe and Asia dangerously aggregate power that can destabilize governments and facilitate global breakdowns in law and order. (For example, some United States intelligence officials believe that drug-trafficking networks and routes can be easily reconfigured to smuggle chemical weapons, plutonium, or tactical nuclear weapons to terrorist nations and groups.) Demolishing such power can stand alone as a worthwhile objective. Similarly, United States policy expresses legitimate concerns when helping governments curb the political and economic reach of the drug lords, contain narcoterrorist violence, and in general cope with the divisive effects of the drug business. Between 1989 and 1993, the United States supported a crackdown on the Medellín cartel that decimated the group's leadership (all the Medellín founding fathers are either dead or in jail) and removed a lethal threat to the Colombian political order. American pressure or intervention prompted the ouster of narcotics-linked military regimes in Bolivia in 1980 and 1981, and in Panama in 1989, two countries where narcotics trafficking interests had built cozy relationships with the military, giving them de facto control of the national government apparatus for controlling drug crime. In Bolivia, United States pressure on the Paz administration in 1991 prevented the appointment of Bolivians apparently linked to the cocaine trade to head the Ministry of Interior, the National Police, and the Special Narcotics Force.

In a number of countries—such as Bolivia, Thailand, and Laos—United States foreign assistance has fostered positive economic growth, widened income opportunities for farmers who cultivate drugs, and weakened the relative economic clout of narcotics industries. Perhaps international drug policy cannot substantially control entrenched drug trafficking, but supply-side programs can be reconfigured to target criminal organizations, promote stability and growth in drug-torn countries, and enhance positive United States influence.

A Growing Global Crisis

AIDS represents the most significant pandemic of disease in the latter half of the twentieth century.

ANN MARIE KIMBALL

The magistrate of Bambari District in the Central African Republic passed away last month, leaving a household of four wives and 17 children. The loss of a leader, husband, and head of household is always tragic; this loss was unusually so. The magistrate died of AIDS, and at least one of his wives is already suffering symptoms of the illness.

As I walked the dusty streets of this *carrefour*—a crossroads of trucking between Chad, the Central African Republic, and Zaire—my colleagues told me of other losses, other funerals, and many more who are ill. A decade ago, 10 young women were trained here to be *animatrices*—health promoters carrying the HIV prevention message to the people of the town. Today, just one survives. The rest have all died of AIDS.

This scenario is being enacted over and over again across the breadth of Central Africa as the death toll from AIDS rises. AIDS is taking Africa's leadership, the intelligentsia, and losses are extremely high.

AIDS represents the most significant global pandemic of disease in the latter half of the twentieth century. It is an illness caused by HIV—a RNA virus that was not identified until the mid 1980s. It is not a casually communicated virus. It is relatively noninfectious. In fact, to catch HIV, intimate contact is required: sexual contact, the injection or infusion of blood or blood products, or transmission from mother to child at the time of birth or in breast-feeding.

Before the virus was known to scientists, it had circulated broadly in sub-Saharan Africa and North America. The virus has a long latent period of infection: Only half of those infected will experience any clinical illness in the first 10 years of their infection, according to studies done in the United States.

The crisis that HIV has caused is only now beginning to severely affect our global community. Unhappily, the predictions made in the past five years are proving themselves. In fact, they are proving conservative. This article will explore a number of issues related to the current crisis, putting forward some lessons learned that can serve the human community as we navigate our rocky future of living with HIV and AIDS.

How did it start?

The origin of HIV has been a popular area for scientific and nonscientific speculation. Early in the epidemic, cases were occurring in two widely distant populations on the globe: homosexual men in the United States and urban populations in Africa. In the last decade, virologic techniques have taken a giant step forward, and it is now possible to identify the genetic similarity of viruses.

This is important because, as with any other family, looking at the HIV family allows us to examine the relationship between different clades of HIV. Such studies have shown that HIV-1, the most common cause of AIDS, is centuries away from simian (monkey) retroviruses, whereas HIV-2, an agent found in West Africa that is somewhat less virulent, is closely related to the simian viruses. The earliest sample from which HIV-1 virus has been isolated was drawn during the Ebola virus investigations in Zaire, in 1976, and stored on dry ice until testing was carried out recently.

Viral studies have shown that HIV-1 is highly variable genetically and has undergone significant changes as it traveled around the world's populations. HIV strains from Africa, North America, and Southeast Asia have all been detected and probably contributed to the epidemic now under way in India.

We are beginning to appreciate that HIV is not the only new infection to have arrived recently. It is merely the best studied and possibly the most lethal. As our human activities begin to change the ecology of the world, we are also changing the ecology of the microbial world. New strains of influenza virus arise each year and circulate the globe.

Travel, transportation, and tourism are burgeoning, bringing the peoples of the world into closer and closer contact. Just as we exchange gifts, learn new languages, and try new foods and styles, so we also exchange our microbes. Traveler's diarrhea has become a leading cause of morbidity among airline personnel, and all of us who travel know how frequently the unwanted cold interrupts our work or vacation abroad.

Other human activity is also important to this new paradigm of disease emergence. In 1992, the prestigious Institute of Medicine called attention to this global phenomenon in a report, *Emerging Infections*. Factors that contribute to the resurgence of infections include the following:

● Increasing human population and changes in human behavior;
● Travel and transportation;
● Land use;
● Technology;
● Microbial adaptation and change; and
● The breakdown of public health.

Many of these factors were important to the widespread dissemination of HIV. Certainly, the increasing urbanization of the world's people contributed to our vulnerability to HIV. The loss of traditional mores among people moving to the city contributes to increasing numbers of sexual partners and the rise of commercial sex. Travel disseminated the virus around the world, along with technological advances in blood products.

While HIV was in the U.S. blood supply, we were the major global exporters of factor VIII. Factor VIII is a lifesaving product for hemophiliacs. It frees them from the painful bleeding crises that otherwise can cripple them.

42. Growing Global Crisis

Unfortunately for thousands of hemophiliacs in our hemisphere, factor VIII was contaminated and exported from the United States in the years before we were able to test and assure its safety. In a number of Central American countries, this was a major route of introduction.

Where is it going?

The global pandemic of HIV/AIDS is best viewed as a series of regional epidemics. Scientific discoveries have occurred during the pandemic, not before. This is important to keep in mind as we map out where we are and where we are going with this ongoing crisis. The oldest epidemics are those in the United States and sub-Saharan Africa. These are distinct.

In the United States, transmission occurred in the mid to late 1970s but was not detected until 1982, when a cluster of a new illness was reported among homosexual men in San Francisco. The illness caused swollen lymph nodes and serious drops in lymphocyte counts. Patients would succumb to infections that were previously not considered to be serious in humans.

As scientists put the picture together, it became clear that a virus was at work—a virus that actually knocked out the human immune system. As this virus worked, infections like Pneumocystis carinii pneumonia or tuberculosis would take advantage of the lack of immune response. HIV was opening the door to other lethal infections.

Meanwhile, in Africa, a new disease was being picked up by workers in Uganda. It caused weight loss and diarrhea. Indeed, it was dubbed "slims disease," because of the profound wasting it caused. After a short course, the disease was fatal.

At the same time, cryptococcal meningitis, a rare type of

Lessons from the AIDS Pandemic

■ Valuable time was lost early in prevention because of denial, scapegoating, and stigmatization.

■ A broad societal response beyond the formal health sector is necessary.

■ Adequate investment in prevention must be made.

7. GLOBAL ISSUES

severe neurological infection, was changing its pattern. It was more frequent than before, and the kinds of organism involved also shifted. This, again, was HIV at work. Heterosexually transmitted, it wreaked havoc with the already tenuous balance between man and microbe that characterizes the tropics. By the time testing was available in 1985, high levels of seroprevalence of up to 20 to 30 percent among some adult population groups were found.

Latin America and Asia represent more recent epidemics. In these geographic areas, the virus and means of prevention were known prior to the onset of widespread transmission. In Latin America, testing shows high levels of infection among homosexual men and intravenous drug users. In the Caribbean states, high levels of infection have been reported, especially on the island of Hispaniola, which includes both Haiti and the Dominican Republic. These levels of infection remain about half as high as those seen in Africa among similar groups.

As in many areas, Cuba took an exceptional approach to HIV. The first infection in that country was detected in a returnee from the Angolan conflict. The veteran was promptly put in involuntary confinement. Saying that "Cuba cannot afford an AIDS epidemic," policymakers in that country tested the entire adult population of the island and quarantined those who tested positive. At least four such sanatoriums were set up. When I visited the largest of these, Las Cocas, in 1990, the strategy was being rethought.

The government was finding it costly, undesirable, and unsustainable to keep the internees locked up. However, when it tried to reintegrate internees into their workplaces, difficulties arose. Because testing had been funded over education, the population remained fearful of contagion from these individuals. In addition, the life in the sanatoriums separated families, with wives sometimes deliberately becoming infected to join their spouses.

Did the Cuban strategy work? Certainly, it generated a great deal of controversy about ethics and human rights. When it came to epidemic control, however, it is not clear that the policy was effective. Nicaragua, another country off the trade and travel route in the 1970s and early '80s, also exhibited very low seroprevalence. Thus, the role of the sanatoriums in preventing infection is inconclusive. The sustainability of this strategy is being questioned by the Cuban authorities themselves.

In many other parts of the region, infection has also remained low. Particularly in Chile, Argentina, and the Andean region, the rate of infection in the general population may reflect some success of the more conventional information, education, and condom-promotion activities. In addition, the blood supply is routinely screened for HIV in these countries.

The Asian epidemic is the newest. As in all other regions, the epidemic was greeted with denial. In Thailand, however, the risk was clear. With a thriving sex trade that had been fueled by many years as an R and R stopover for U.S. troops during the Vietnam War, the stakes were high. The government moved aggressively to set up surveillance and then to address the bad news. In the north, infection was high in the brothels. In the army, infection rates were high among new recruits coming from the north.

The government initiated a vigorous condom-promotion campaign. The "100 percent condom use" approach was unique. The police could be called if a client refused to use a condom at a brothel, and the client, not the brothel, would be the target of discussion. This and other campaigns seem to have leveled off the rate of new infections. However, the population-based estimated number of infections in Thailand now exceeds that of the United States.

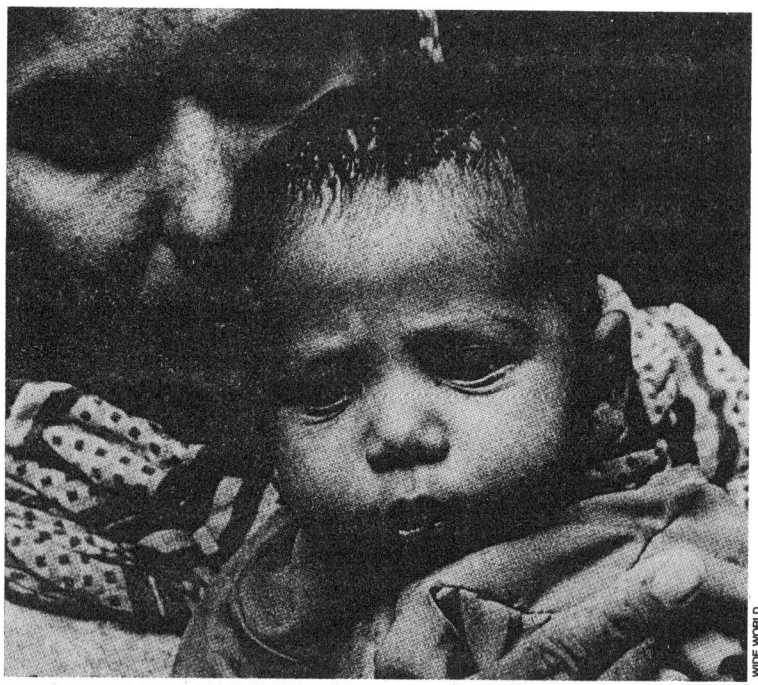

■ *HIV positive:* A two-month-old baby contracted AIDS after a blood transfusion in New Delhi. The government has no program for screening blood samples.

What can be done?

It is not too late to prevent HIV/AIDS, despite the terrific toll of mortality we are already experiencing. It is certainly not too late in Asia to prevent widespread disease in countries such as Korea, Japan, Hong Kong, the Philippines, China, and Singapore, where large segments of the population remain HIV free. Increasingly, the potential of involving the powerful corporate sector in the work of AIDS prevention has been advocated as a strategy well suited to the region.

In fact, companies such as Matsushita and the Saison Group in Japan have begun to put prevention education into their employee training programs. Multinational corporations working in Asia (Digital Equipment, Boeing, Levi Strauss) are also beginning to transfer the policies from their U.S. sites into their Asian subsidiaries. In addition, a number of effective community groups have been active. In Indonesia, Malaysia, and the Philippines, nongovernmental organizations have been taking a leading role in providing education and promoting prevention of HIV.

In the older epidemic areas, prevention must remain a priority. Each new generation of young adults presents a fresh opportunity for HIV to take hold and an imperative for HIV prevention. Recent studies in Uganda suggest that even in highly infected societies, the provision of anonymous counseling and testing can provide the tools needed to reduce the rate of infection. These services are not now available in most of sub-Saharan Africa and should be. Hopefully, as policymakers learn of the success in Uganda, new resources can be found to make similar intensive efforts elsewhere in the region.

A final area of hope is new progress made in preventing the transmission from mother to child. Recently, an important trial (the "076 trial") of AZT therapy in mothers demonstrated a 60 percent efficacy in reducing transmission from infected mothers to their newborns. Other studies have suggested that timely cesarean section may diminish transmission at birth.

Without any intervention, about one-third of infants born to infected mothers will be HIV positive; the majority of these will die in infancy and childhood. Having new tools to interrupt such transmission is exciting. The majority of mother-to-infant transmission now occurs in the developing world, so another challenge is to identify interventions that are practical in these resource-poor settings.

What have we learned?

With this crisis, the human community has learned some lessons. Such lessons will be important to our collective survival in the face of HIV/AIDS and new infections of the future. First, we have learned that solidarity is crucial in the face of such a serious pandemic. Valuable time was lost early in prevention because of denial, scapegoating, and stigmatization. The United States, with the Reagan-era policy, presents a classic example of this.

If the United States had vigorously addressed HIV in 1985 with the highest-level political commitment to prevention, thousands of lives would have been saved. The international "ripple" of our benighted characterization of the epidemic as a homosexual disease contributed to denial in other countries.

Second, we have learned that a broad societal response is necessary to effect prevention. We are beginning to see the nongovernmental social and corporate sector take charge in Asia. A response beyond the formal health sector appears absolutely key.

Finally, the experience in Uganda is demonstrating that, in every setting, prevention is possible if the primary direction is from within the society. However, adequate investment in prevention must be made.

The future of the HIV/AIDS pandemic stretches before us. There will be much greater mortality to bear. In sub-Saharan Africa, an estimated 16 million people are infected. In Asia, up to 3 million new infections are occurring each year. At this time, AIDS remains largely incurable.

With current therapies, largely unaffordable in much of the world, life can be prolonged and symptoms diminished, but the increased survival is measured in months to years, not decades. Vaccine-development efforts continue, but current vaccines are still in testing. Having a useful product on the market that can be deployed throughout broad regions of the globe remains out of reach at this time. Prevention efforts that focus on promoting safe sex, discouraging injection drug use, and assuring a safe blood supply will remain the mainstays of the worldwide effort for the foreseeable future.

Additional Reading

R.M. Anderson and R.M. May, "Understanding the AIDS Pandemic," *Scientific American* 266, 1992.

R.J. Biggar, "When Ideals Meet Reality: The Global Challenge of HIV/AIDS," *American Journal of Public Health* 83, 1993.

A.M. Kimball et al., "International Aspects of the AIDS/HIV Epidemic," *Annual Review of Public Health* 16, 1995.

J. Mann, D. Tarantola, T.W. Netter, eds., *AIDS in the World*, the Global AIDS Policy Coalition, Harvard University Press, Cambridge, 1992.

Ann Marie Kimball is director of community medicine and associate professor of epidemiology and health services at the University of Washington School of Public Health and Community Medicine. She has served as the head of national program support for the Global Program on AIDS in the Americas and director of the Washington AIDS Program.

EARTH is Running Out of Room

Brazilian children searching garbage dump for food.

Food scarcity, not military aggression, is the principal threat to the planet's future.

Lester R. Brown

Mr. Brown is president, Worldwatch Institute, Washington, D.C., and co-author of Full House: Reassessing the Earth's Population Carrying Capacity.

THE WORLD is entering a new era, one in which it is far more difficult to expand food output. Many knew that this time would come eventually; that, at some point, the limits of the Earth's natural systems, cumulative effects of environmental degradation on cropland productivity, and shrinking backlog of yield-raising technologies would slow the record increase in food production of recent decades. Because no one knew exactly when or how this would happen, food prospects were debated widely. Now, several constraints are emerging

43. Earth Is Running Out of Room

simultaneously to slow that growth.

After nearly four decades of unprecedented expansion in both land-based and oceanic food supplies, the world is experiencing a massive loss of momentum. Between 1950 and 1984, grain production expanded 2.6-fold, outstripping population growth by a wide margin and raising the grain harvested per person by 40%. Growth in the fish catch was even more spectacular—a 4.6-fold increase between 1950 and 1989, thereby doubling seafood consumption per person. Together, these developments reduced hunger and malnutrition throughout the world, offering hope that these biblical scourges would be eliminated one day.

In recent years, these trends suddenly have been reversed. After expanding at three percent a year from 1950 to 1984, the growth in grain production has slowed abruptly, rising at scarcely one percent annually from 1984 until 1993. As a result, grain production per person fell 12% during this time.

With fish catch, it is not merely a slowing of growth, but a limit imposed by nature. From a high of 100,000,000 tons, believed to be close to the maximum oceanic fisheries can sustain, the catch has fluctuated between 96,000,000 and 98,000,000 tons. As a result, the 1993 per capita seafood catch was nine percent below that of 1988. Marine biologists at the United Nations Food and Agriculture Organization report that the 17 major oceanic fisheries are being fished at or beyond capacity and that nine are in a state of decline.

Rangelands, a major source of animal protein, also are under excessive pressure, being grazed at or beyond capacity on every continent. This means that rangeland production of beef and mutton may not increase much, if at all, in the future. Here, too, availability per person will decline indefinitely as population expands.

With both fisheries and rangelands being pressed to the limits of their carrying capacity, future growth in food demand can be satisfied only by expanding output from croplands. The increase in demand for food that was satisfied by three food systems must now be satisfied by one.

Until recently, grain output projections for the most part were simple extrapolations of trends. The past was a reliable guide to the future. However, in a world of limits, this is changing. In projecting food supply trends now, at least six new constraints must be taken into account:
- The backlog of unused agricultural technology is shrinking, leaving the more progressive farmers fewer agronomic options for expanding food output.
- Growing human demands are pressing against the limits of fisheries to supply seafood and rangelands to supply beef, mutton, and milk.
- Demands for water are nearing limits of the hydrological cycle to supply irrigation water in key food-growing regions.
- In many countries, the use of additional fertilizer on currently available crop varieties has little or no effect on yields.
- Nations that already are densely populated risk losing cropland when they begin to industrialize at a rate that exceeds the rise in land productivity, initiating a long-term decline in food production.
- Social disintegration by rapid population growth and environmental degradation often is undermining many national governments and their efforts to expand food production.

New technologies are not enough

In terms of agricultural technology, the contrast between the middle of the 20th century and today could not be more striking. When the 1950s began, a great deal of technology was waiting to be used. Except for irrigation, which goes back several thousand years, all the basic advances were made between 1840 and 1940. Chemist Justus von Liebig discovered in 1847 that all nutrients taken from the soil by crops could be replaced in mineral form. Biologist Gregor Mendel's work establishing the basic principles of heredity, which laid the groundwork for future crop breeding advances, was done in the 1860s. Hybrid corn varieties were commercialized in the U.S. during the 1920s, and dwarfing of wheat and rice plants in Japan to boost fertilizer responsiveness dates back a century.

These long-standing technologies have been enhanced and modified for wide use through agricultural research and exploited by farmers during the last four decades. Although new developments continue to appear, none promise to lead to quantum leaps in world food output. The relatively easy gains have been made. Moreover, public funding for international agricultural research has begun to decline. As a result, the more progressive farmers are looking over the shoulders of agricultural scientists seeking new yield-raising technologies, but discovering that they have less and less to offer. The pipeline has not run dry, but the flow has slowed to a trickle.

In Asia, rice crops on maximum-yield experimental plots have not increased for more than two decades. Some countries appear to be "hitting the wall" as their yields approach those on the research plots. Japan reached this point with a rice yield in 1984 at 4.7 tons per hectare (2.47 acres), a level it has been unable to top in nine harvests since then. South Korea, with similar growing conditions, may have run into the same barrier in 1988, when its rice yield stopped rising. Indonesia, with a crop that has increased little since 1988, may be the first tropical rice-growing nation to see its yield rise lose momentum. Other countries could hit the wall before the end of the century.

Farmers and policymakers search in vain for new advances, perhaps from biotechnology, that will lift food output quickly to a new level. However, biotechnology has not produced any yield-raising technologies that will lead to quantum jumps in output, nor do many researchers expect it to. Donald Duvick, for many years the director of research at Iowa-based Pioneer Hi-Bred International, one of the world's largest seed suppliers, makes this point all too clearly: "No breakthroughs are in sight. Biotechnology, while essential to progress, will not produce sharp upward swings in yield potential except for isolated crops in certain situations."

The productivity of oceanic fisheries and rangelands, both natural systems, is determined by nature. It can be reduced by overfishing and overgrazing or other forms of mismanagement, but once sustainable yield limits are reached, the contribution of these systems to world food supply can not be expanded. The decline in fisheries is not limited to developing countries. By early 1994, the U.S. was experiencing precipitous drops in fishery stocks off the coast of New England, off the West Coast, and in the Gulf of Mexico.

With water—the third constraint—the overpumping that is so widespread eventually will be curbed to bring it into balance with aquifer recharge. This reduction, combined with the growing diversion of irrigation water to residential and industrial uses, limits the amount of water available to produce food. Where farmers depend on fossil aquifers for their irrigation water—in the southern U.S. Great Plains, for instance, or the wheat fields of Saudi Arabia—aquifer depletion means an end to irrigated agriculture. In the U.S., where more than one-fourth of irrigated cropland is watered by drawing down underground water tables, the downward adjustment in irrigation pumping will be substantial. Major food-producing regions where overpumping is commonplace include the southern Great Plains, India's Punjab, and the North China Plain. For many farmers, the best hope for more water is from gains in efficiency.

Perhaps the most worrisome emerging constraint on food production is the limited capacity of grain varieties to respond to the use of additional fertilizer. In the U.S., Western Europe, and Japan, usage has increased little if at all during the last decade. Utilizing additional amounts on existing crop varieties has little or no effect on yield in these nations. After a tenfold increase in world fertilizer use from 1950 to 1989—from 14,000,000 to 146,000,000 tons—use declined to the following four years.

A little-recognized threat to the future world food balance is the heavy loss of cropland that occurs when countries that already are densely populated begin to industrialize. The experience in Japan, South Korea, and Taiwan gives a sense of what to expect. The conversion of grainland to non-

7. GLOBAL ISSUES

farm uses and to high-value specialty crops has cost Japan 52% of its grainland; South Korea, 42%; and Taiwan, 35%.

As the loss of land proceeded, it began to override the rise in land productivity, leading to declines in production. From its peak, Japan's grain output has dropped 33%; South Korea's, 31%; and Taiwan's, 19%. These declines occurred at a time when population growth and rapidly rising incomes were driving up the demand for grain. The result is that, by 1993, Japan was importing 77% of its grain; South Korea, 68%; and Taiwan, 74%.

Asia's densely populated giants, China and India, are going through the same stages that led to the extraordinarily heavy dependence on imported grain in the three smaller countries that industrialized earlier. In both, the shrinkage in grainland has begun. It is one thing for Japan, a country of 120,000,000 people, to import 77% of its grain, but quite another if China, with 1,200,000,000, moves in this direction.

Further complicating efforts to achieve an acceptable balance between food and people is social disintegration. In an article in the February 1994 *Atlantic* entitled "The Coming Anarchy," writer and political analyst Robert Kaplan observed that unprecedented population growth and environmental degradation were driving people from the countryside into cities and across national borders at a record rate. This, in turn, was leading to social disintegration and political fragmentation. In parts of Africa, he argues, nation-states no longer exist in any meaningful sense. In their place are fragmented tribal and ethnic groups.

The sequence of events that leads to environmental degradation is all too familiar to environmentalists. It begins when the firewood demands of a growing population exceed the sustainable yield of local forests, leading to deforestation. As firewood become scarce, cow dung and crop residues are burned for fuel, depriving the land of nutrients and organic matter. Livestock numbers expand more or less apace with the human population, eventually exceeding grazing capacity. The combination of deforestation and overgrazing increases rainfall runoff and soil erosion, simultaneously reducing aquifer recharge and soil fertility. No longer able to feed themselves, people become refugees, heading for the nearest city or food relief center.

Crop reports for Africa now regularly cite weather and civil disorder as the key variables affecting harvest prospects. Not only is agricultural progress difficult, even providing food aid can be a challenge under these circumstances. In Somalia, getting food to the starving in late 1992 required a UN peacekeeping force and military expenditures that probably cost 10 times as much as what was distributed.

As political fragmentation and instability spread, national governments no longer can provide the physical and economic infrastructure for development. Countries in this category include Afghanistan, Haiti, Liberia, Sierra Leone, and Somalia. To the extent that nation-states become dysfunctional, the prospects for humanely slowing population growth, reversing environmental degradation, and systematically expanding food production are diminished.

Other negative influences exist, but they have emerged more gradually. Among those that affect food production more directly are soil erosion, the waterlogging and salting of irrigated land, and air pollution. For example, a substantial share of the world's cropland is losing topsoil at a rate that exceeds natural soil formation. On newly cleared land that is sloping steeply, soil losses can lead to cropland abandonment in a matter of years. In other situations, the loss is slow and has a measurable effect on land productivity only over many decades.

Growing pessimism

Until recently, concerns about the Earth's capacity to feed ever-growing numbers of people adequately was confined largely to the environmental and population communities and a few scientists. During the 1990s, however, these issues are arousing the concerns of the mainstream scientific community. In early 1992, the U.S. National Academy of Sciences and the Royal Society of London issued a report that began: "If current predictions of population growth prove accurate and patterns of human activity on the planet remain unchanged, science and technology may not be able to prevent either irreversible degradation of the environment or continued poverty for much of the world."

It was a remarkable statement, an admission that science and technology no longer can ensure a better future unless population growth slows quickly and the economy is restructured. This abandonment of the technological optimism that has permeated so much of the 20th century by two of the world's leading scientific bodies represents a major shift, though perhaps not a surprising one, given the deteriorating state of the planet. That they chose to issue a joint statement, their first ever, reflects the deepening concern about the future within the scientific establishment.

Later in 1992, the Union of Concerned Scientists issued a "World Scientists' Warning to Humanity," signed by some 1,600 of the planet's leading scientists, including 102 Nobel Prize winners. It observes that the continuation of destructive human activities "may so alter the living world that it will be unable to sustain life in the manner that we know." The scientists indicated that "A great change in our stewardship of the earth and the life on it is required, if vast human misery is to be avoided and our global home on this planet is not to be irretrievably mutilated."

In November, 1993, representatives of 56 national science academies convened in New Delhi, India, to discuss population. At the end of their conference, they issued a statement in which they urged zero population growth during the lifetimes of their children.

Between 1950 and 1990, the world added 2,800,000,000 people, an average of 70,000,000 a year. Between 1990 and 2030, it is projected to add 3,600,000,000, or 90,000,000 a year. Even more troubling, nearly all this increase is projected for the developing countries, where life-support systems already are deteriorating. Such population growth in a finite ecosystem raises questions about the Earth's carrying capacity. Will the planet's natural support systems sustain such growth indefinitely? How many people can the Earth support at a given level of consumption?

Underlying this assessment of population carrying capacity is the assumption that the food supply will be the most immediate constraint on population growth. Water scarcity could limit population growth in some locations, but it is unlikely to do so for the world as a whole in the foreseeable future. A buildup of environmental pollutants could interfere with human reproduction, much as DDT reduced the reproductive capacity of bald eagles, peregrine falcons, and other birds at the top of the food chain. In the extreme, accumulating pollutants in the environment could boost death rates to the point where they would exceed birth rates, leading to a gradual decline in human numbers, but this does not seem likely. For now, it appears that the food supply will be the most immediate, and therefore the controlling, determinant of how many people the Earth can support.

Grain supply and demand projections for the 13 most populous countries—accounting for two-thirds of world population and food production—show much slower growth in output than the official projections by the Food and Agriculture Organization and the World Bank. If those projections of relative abundance and a continuing decline of food prices materialize, governments can get by with business as usual. If, on the other hand, the constraints discussed above continue, the world needs to reorder priorities.

The population-driven environmental deterioration/political disintegration scenario described by Robert Kaplan not only is possible, it is likely in a business-as-usual world. However, it is not inevitable. This future can be averted if security is redefined, recognizing that food scarcity, not military aggression, is the principal threat to the future. Government must give immediate attention to filling the family planning gap; attacking the underlying causes of high fertility, such as illiteracy and poverty; protecting soil and water resources; and raising investment in agriculture.

Index

abolitionists, nuclear weapons and, 68–71
abortion, 5, 10, 12, 37–38
Adams, Scott, 118
adoption, 41, 43, 148
advertising subsidies, 143
affirmative action, 5, 112, 146–150, 166–167
African Americans. *See* blacks
aging, of America and Social Security, 74–90
agri-business, subsidies for, 143
agriculture, environment and, 238–140
Aid to Families with Dependent Children (AFDC). *See* welfare
AIDS, 5, 12, 15, 175, 234–237
alcohol abuse, 11, 15, 16, 18, 208–213
Allen, Robert E., 120
Americans with Disabilities Act (ADA), 106
Anderson, Elijah, 163
anti-capitalism, loss of middle class and, 115
Armstrong, Robert, 185, 186
Aryan Nation, 59
Asian Americans, glass ceiling and, 112
asset forfeiture, war on drugs and, 205–206
AT&T, 120
ATF (Alcohol, Tobacco and Firearms), Bureau of, 54, 57, 59
Auletta, Ken, 134
Australia, 76
authority, respect for, and family values, 23, 30

Baby Boomers, retirement of, and Social Security, 74–90
Baby Doe laws, 92, 93
Baraka, Amiri, 157
Big Brothers/Big Sisters of America, 122–123, 124, 125–126
Bing, Leon, 162
black helicopter sightings, domestic terrorism and, 56–60
blacks, 155–165; affirmative action, 146–150; glass ceiling and, 112
Bowl for Kids' Sake, 122
Branch Davidians, 54, 58, 59
Bush, George, 58, 174, 177, 202

Cali Cartel, 229–233
Calvinist theology, social welfare policy and, 136
cartels, drug, 229–233
Casarjian, Robin, 64–67
Central African Republic, 234
CEOs: downsizing and, 119–120; salaries of, 130
cerebal palsy, premature babies and, 91–95
Character Development Act, 126
charities, welfare and, 136, 179, 181
Chicago, Illinois, drug enterprise of Gangster Disciples in, 61–63
child abuse, 5, 15, 17–18, 40–44; forgiveness and, 64–67
churches, social problems and, 179–182
civil rights, 146, 149, 169; property rights and, 225–228
civil-service reform, in public education, 192
Clinton, Bill, 53, 54, 56, 58, 97, 139, 171
Coats, Dan, 126, 182
Coleman, James, 122, 127
Colombia, drug trafficking in, 229–233

community, crisis of, 168–173
competition: health care reform and, 96–99; U.S. productivity and, 216–224
Comprehensive Terrorism Prevention Act of 1995, 51–55
Compton, Dean, 59, 60
conflict theory, 8–10
Constitution, U.S., rights and, 11, 17. *See also* individual amendments
corporate welfare, government subsidies as, 141–143
Coulter, Ernest, 123
crime, 5, 9, 12; forgiveness and, 64–67; medical costs of, and guns, 100–103; poverty and, 131; race and, 24–25
crusaders, social problems and, 14–19
Cuba, 236

Daley, Richard, M., 175
Daro, Deborah, 42, 44
Dawson, Deborah, 29, 30
de Grieff, Gustavo, 231, 232
deduction, mortgage interest, as government subsidy, 143
"description," difference between "claim" and, 107–108
Dilbert, 118
disease, myth of mental illness as, 104–109
divorce, 28–29, 31, 38
Do or Die (Bing), 162
Dolan v. City of Tigard, 225, 227
Dole, Bob, 112, 228
domestic violence, 34
downsizing, 116–121
drug trafficking: by Gangster Disciples in Chicago, 61–63; international, 229–233
drugs: crime and, 208–211; war on, and civil liberties, 202–207
Duncan, Andrea, 184, 185, 186

Earned Income Tax Credit (EITC), 139, 154
economic inequality, welfare reform and, 130
economy: loss of middle class and, 114–115; unemployment and, 116–121
education, 30, 38; affirmative action and, 166–167; family literacy programs and, 45–47; need for reform in, 191–195; women and, in developing countries, 153; women's studies and, 196–201
Eighteenth Amendment, 17
employers, health-care reform and, 96–99
environmental issues, 5, 10, 11, 12, 15, 16; agriculture and, 238–140; property rights and, 225–228
Escobar, Pablo, 231, 232–233
ethics, medical, and treatment of premature babies, 91–95

factor VIII, 235
Faith and Families program, 179–180
families: importance of, 22–34, 36–40; literacy and, 45–47; reunification of, and child abuse, 40–44
Farrakhan, Louis, 157
Farrell, Elaine, 93–94, 95
fathers, importance of, 24–26, 28, 37, 176
FBI (Federal Bureau of Investigation), terrorism and, 55, 57

FDA (Food and Drug Administration), 99
Feagin, Joe, 158
federal government: affirmative action and, 146–150; budget deficit and, 211–212; cities and, 176–177; involvement of, in public education, 193–195; subsidies and, 141–143; terrorism and, 56–60; war on drugs and, 202–207
FEMA (Federal Emergency Management Agency), terrorism and, 56–60
feminists: domestic violence and, 34; women's studies and, 196–201
fertilizer, 239
Fifth Amendment, 226
First English Evangelical Church v. County of Los Angeles, 226
First Wave, Toffler's theory of, 39
food stamps, 143, 212
Fordice, Kirk, 179, 180
forfeiture, asset, and war on drugs, 205–206
forgiveness, 64–67
foster care, 41, 42
Foster Grandparent Program, 126–127
Fourth Amendment, war on drugs and, 203, 204
Fox-Genovese, Elizabeth, 200, 201
Freud, Sigmund, 105
Friendly Visiting movement, 127
Friends of the Children, 126
functionalism, 7–8, 9–10

Gangster Disciples, drug enterprise of, 61–63
Gelles, Richard, 41, 42, 43
Generation X, voluntary simplicity movement and, 190
Gilder, George, 26
Gingrich, Newt, 39
glass ceiling, 112–113
government, blacks and, 159–160. *See also* federal government; state government
Gramm, Phil, 228
Greece, ancient, patriarchy in, 33
Greenstein, Robert, 139
Grossman, Michael, 188–189
gun violence, 18–19; medical costs of crime and, 100–103; terrorism and, 56–60
Gusfield, Joseph, 14, 15, 16

habeas corpus, terrorism and, 50, 52
Harrison, Helen, 92
Hartmann, Heidi, 138, 140
health care costs, of crime and guns, 100–103
health care reform, 81, 96–99, 174, 210
health issues, 12; AIDS and, 5, 12, 15; families and, 28–29; forgiveness and, 65; myth of mental illness and, 104–109; premature babies and, 91–95
health maintenance organizations (HMOs), 97
Hispanics, glass ceiling and, 112
HIV. *See* AIDS
HIV-1, 234–235
HIV-2, 234–235
Holdel v. Irving, 226
homelessness, 175
homosexuality, 5, 11

hospital alliances, health-care reform and, 97–98

illegal immigrants, 170
illegitimacy, 129–130, 135, 162–163
immigration, 170, 175
impairments to the collective welfare, harm and, 12
individualism, 6, 32
information society, 38–39
International Monetary Fund (IMF), 151–152
Iron Law of Oligarchy, 9

Jackson, Jesse, 19
James, Kay, 23–24
Japan, 76, 217–219, 237
job growth, 176
Jobs, Steve, 193
John Birch Society, 58
Juvenile Mentoring Program Act (JUMP), 126

Kenan Trust Family Literacy Project, 96
Kennedy, John F., 14
Kerber, Linda K., 197, 198
King, Martin Luther, Jr., 169, 170
Kinsey, John, 183–184
knowledge problem, social theories and, 15–16
knowledge workers, 220
Koernke, Mark, 57, 59
Kornfield, Jack, 65, 66, 67
Korones, Sheldon, 92–93, 94
Ku Klux Klan, 57, 59

Latin American, AIDS in, 236
Latinos, glass ceiling and, 112
literacy, for families, and poverty, 45–47
Living with Racism (Feagin and Sikes), 158
loopholes, government subsidies and, 141–143
love, family and, 30–33
loyalty, family and, 30
Lucas v. South Carolina Coastal Council, 226

Maddux, William, 41, 44
mail, surveillance of, and war on drugs, 204
mandatory sentencing, war on drugs and, 205, 213
Mann, Horace, 192
mansion subsidy, 143
marginalizers, nuclear weapon and, 68–71
marriage, family and, 26, 28–29
Marshall, Thurgood, 202
Martinez, Arthur C., 120
Marx, Karl, 33, 105
McDonough, William J., 114
Mead, Lawrence, 136–137
Medellin cartel, drug trafficking and, 229–233
media, domestic violence by women and, 34
Medicaid, 97, 174, 211, 212
medical ethics, treatment of premature babies and, 91–95
Medicare, 75, 76, 77, 81, 86, 97, 142, 143, 211, 212, 213
Mendel, Gregor, 239
mental illness, myth of, 104–109
mental retardation, premature babies and, 91–95
"mental-hygiene" movement, in public education, 192
mentoring movement, 122–128
Messenger, Gregory, 92, 93

middle class: blacks, 156, 157–158, 159; loss of, in U.S., 114–115
Miedzian, Myriam, 24, 25–26
military surveillance, war on drugs and, 204–205
militia groups, 56–60
Miller, Walter, 25–26
mining companies, subsidies for, 143
minorities, glass ceiling and, 112
Mississippi, Faith and Families program in, 179–180
Monster (Shakur), 161–162
mortgage interest deduction, as government subsidy, 143
Moynihan, Daniel Patrick, 26, 32
multiculturalism, women's studies and, 200–201
Murphy, Patrick, 41–42, 43
Murray, Charles, 134–135, 137, 168

neonatology, premature babies and, 91–95
New Politics of Poverty, The (Mead), 136–137
Nicaragua, 236
Nollan v. California Coastal Commission, 226
nuclear family, 25, 26, 28, 32, 39
nuclear weapons, 68–71

Office of Management and Budget (OMB): Directive, 15; affirmative action and, 147–148, 149
Oklahoma City, Oklahoma, bombing in, 51, 53, 55, 56, 57, 59, 60
Olasky, Marvin, 135–136, 137, 181

Parallel Time (Staples), 158–159
paranoia, terrorists and, 56–60
parents: family literacy programs and, 45–47; rights of, concerning ill premature babies, 91–95
parole, 209
patriarchy, 33, 36, 37
patriots, domestic terrorism and, 56–60
Patterson, Orlando, 161
Personal Responsibility and Work Opportunity Act, 134
"personal responsibility," welfare reform and, 135
pharmaceutical companies, health-care reform and, 98
Plato, 33
police, powers of, and war on drugs, 203–204
population growth, 240
Posse Comitatus, 55, 58, 59
post-structuralism, women's studies and, 200–201
poverty, 5, 6, 8, 14, 27–28, 38, 208, 209–210; family literacy programs and, 45–47; urban, 177, 184; welfare and, 138–140
premature babies, viability of, 91–95
privacy, invasion of, and war on drugs, 202–207
privatization, school reform and, 194
productivity growth, paying for Social Security and, 80–81
productivity, U.S. competitiveness and, 216–224
progressive plan, for welfare, 131–133
Project for American Renewal, 126
property rights, environmental issues and, 225–228
Proposition 300, 228

psychiatry, myth of mental illness and, 104–109
public education, need for reform in, 191–195
public housing, 183–186
Public/Private Ventures, 124, 125

Quindlen, Anna, 187

race: affirmative action and, 148; crime and, 24–25
racism, 5, 6, 168–169; as excuses, 156–157
Racism: American Style (Travis), 157
Reagan, Ronald, 174, 175, 202, 205, 219–220
Rehnquist, William, 205, 226
Reich, Robert, 115, 153
research and development, competitiveness and, 220–222
residential mobility programs, urban, 183–186
rights, conflict theory and, 10–12
Rohatyn, Felix, 128
romantic progressivism, in public education, 192

savings rate, 221–222; retirement and, 78–80
Scalia, Antonin, 202, 203
school reform, 191–194
search and seizures, war on drugs and, 203–204; 205–206
Sears, Roebuck and Company, 120
Second Amendment, militia groups and, 57
Second Wave, Toffler's theory of, 39
sectarians, social problems and, 14–19
self-isolation, conflict theory and, 9
self-revelation, women's studies and, 198
set-aside programs, affirmative action and, 147
Shakur, Sanyika, 161–162
Sikes, Melvin, 158
single-parent families, 22–34, 175, 176; welfare reform and, 138–140
Smith, Adam, 10
social capital, 122
Social Security, 74–90, 142, 143
Sowell, Thomas, 155
Spalter-Roth, Roberta, 138, 139, 140
Stalin, Josef, 33
Staples, Brent, 158–159
state governments, cities and, 176, 177–178
stress, voluntary simplicity movement and, 187–190
Structural Adjustment programs (SAPs), women and, 151–153
subsidies, 141–143
substance abuse: crime and, 208–213; war on drugs and, 202–207
Supplementary Security Income (SSI), 211, 212
Supreme Court, U.S., property rights and, 225–228
survivalists, 59
suspicious vehicles, search of, and war on drugs, 204
symbolic interaction, 6–7

Takings Clause, of the Fifth Amendment, 226
Tanzania, 152
tax breaks, subsidies and, 141–143
taxes, cities and, 174–176
technology, viability of premature babies and, 91–95
Tenneco Inc., 120
Tenth Amendment, 59

terminal illnesses, euthanasia and, 5
terrorism, 50–55, 56–60
Thailand, 236
Third Wave, Toffler's theory of, 38–39
third-party payments, health insurance and, 96–99
Thompson, Linda, 57, 59
tobacco industry, 208–213
Toffler, Alvin and Heidi, 38–39
Tragedy of American Compassion, The (Olasky), 135–136
Travis, Dempsey, 157
Trochmann, John, 57, 59
twofers, affirmative action and, 166–167

Uganda, 235, 237
unemployment, 116–121
unfunded liabilities, federal pension plan and, 92–95
unfunded mandates, environmental issues and, 226–228
unions: municipal, 175, 178; teacher, and school reform, 193
United Nations, terrorism and, 56–60

unskilled workers, competitiveness and, 222–224

Vale, Lawrence, 184, 186
values, family, 22–33
viability, of premature babies, 91–95
voluntary simplicity movement, stress and, 187–190
volunteerism: mentoring movement and, 122–128; welfare reform and, 136
von Leibig, Justus, 239
vouchers, school reform and, 194

war on drugs, civil liberties and, 202–207
War on Poverty, 174, 177
Washington, Booker T., 156
water pollution, 239
wealthfare, government subsidies as, 141–143
Weaver, Randy, 54, 58
Weber, Max, 5, 10
welfare, 174, 208, 210–211, 212; cuts in, and women, 153–154; government subsidies as corporate, 141–143
welfare reform, 129–133; theories of, 134–137; women and, 153–154

Werner, Emmy E., 123–124
white supremacists, 57, 59
Whiting, Beatrice and John, 24–26
Wildavsky, Aaron, 15–16
Williams, Patricia, 158
Wilson, James, 6, 10
Wilson, William Julius, 156, 164
wiretapping, 55, 204
women: affirmative action and, 166–167; domestic violence by, 34; glass ceiling and, 112–113; Structural Adjustment programs and, 151–153; welfare and, 153–154
women's studies, 196–201
workforce, 153–154
workplace: drug testing in, 202, 203; glass ceiling and, 112
World Bank, Structural Adjustment programs of, and women, 151–153
World Trade Center, terrorist bombing of, 51, 54

zero tolerance, war on drugs and, 205

Credits/Acknowledgments

Cover design by Charles Vitelli

Introduction
Facing introduction—United Nations photo by Paulo Fridman.

1. Parenting and Family Issues
Facing overview—Courtesy of Sandra Nicholas. 45-47—Photos by the National Center for Family Literacy.

2. Crime, Terrorism, and Violence
Facing overview—© Dorothy Littell Greco/Stock•Boston.

3. Aging, Health, and Health Care Issues
Facing overview—United Nations photo by John Isaac.

4. Poverty and Inequality
Facing overview—United Nations photo by P. S. Sudhakaran.

5. Cultural Pluralism and Affirmative Action
Facing overview—Photo by Jim Pickerell.

6. Cities, Urban Growth, and the Quality of Life
Facing overview—Dushkin/McGraw-Hill Companies.

7. Global Issues
Facing overview—United Nations photo by J. Frank.

PHOTOCOPY THIS PAGE!!!*

ANNUAL EDITIONS ARTICLE REVIEW FORM

■ NAME: _____ DATE: _____

■ TITLE AND NUMBER OF ARTICLE: _____

■ BRIEFLY STATE THE MAIN IDEA OF THIS ARTICLE: _____
_____

■ LIST THREE IMPORTANT FACTS THAT THE AUTHOR USES TO SUPPORT THE MAIN IDEA:
_____
_____
_____
_____

■ WHAT INFORMATION OR IDEAS DISCUSSED IN THIS ARTICLE ARE ALSO DISCUSSED IN YOUR TEXTBOOK OR OTHER READINGS THAT YOU HAVE DONE? LIST THE TEXTBOOK CHAPTERS AND PAGE NUMBERS:
_____
_____
_____
_____

■ LIST ANY EXAMPLES OF BIAS OR FAULTY REASONING THAT YOU FOUND IN THE ARTICLE:
_____
_____
_____
_____

■ LIST ANY NEW TERMS/CONCEPTS THAT WERE DISCUSSED IN THE ARTICLE, AND WRITE A SHORT DEFINITION:
_____
_____
_____
_____

*Your instructor may require you to use this ANNUAL EDITIONS Article Review Form in any number of ways: for articles that are assigned, for extra credit, as a tool to assist in developing assigned papers, or simply for your own reference. Even if it is not required, we encourage you to photocopy and use this page; you will find that reflecting on the articles will greatly enhance the information from your text.

We Want Your Advice

ANNUAL EDITIONS revisions depend on two major opinion sources: one is our Advisory Board, listed in the front of this volume, which works with us in scanning the thousands of articles published in the public press each year; the other is you—the person actually using the book. Please help us and the users of the next edition by completing the prepaid article rating form on this page and returning it to us. Thank you for your help!

ANNUAL EDITIONS: SOCIAL PROBLEMS 97/98
Article Rating Form

Here is an opportunity for you to have direct input into the next revision of this volume. We would like you to rate each of the 43 articles listed below, using the following scale:

1. **Excellent: should definitely be retained**
2. **Above average: should probably be retained**
3. **Below average: should probably be deleted**
4. **Poor: should definitely be deleted**

Your ratings will play a vital part in the next revision. So please mail this prepaid form to us just as soon as you complete it.
Thanks for your help!

Rating	Article	Rating	Article
	1. Social Problems: Definitions, Theories, and Analysis		24. It's Not Working: Why Many Single Mothers Can't Work Their Way Out of Poverty
	2. How Social Problems Are Born		25. Aid to Dependent Corporations: Exposing Federal Handouts to the Wealthy
	3. Fount of Virtue, Spring of Wealth: How the Strong Family Sustains a Prosperous Society		26. Reclaiming the Vision: What Should We Do after Affirmative Action?
	4. Things That Go Bump in the Home		27. Balancing Budgets on Women's Backs: The World Bank and the 104th U.S. Congress
	5. Growing Up against the Odds		28. Black America's Moment of Truth
	6. Why Leave Children with Bad Parents?		29. A Twofer's Lament
	7. The Three R's Spell Success		30. Crisis of Community: Make America Work for Americans
	8. Terrorism in America		31. Can We Stop the Decline of Our Cities?
	9. Enemies of the State		32. Can Churches Save America?
	10. How Nation's Largest Gang Runs Its Drug Enterprise		33. The Projects Come Down
	11. Forgiving the Unforgivable		34. Time Out
	12. Fearsome Security: The Role of Nuclear Weapons		35. Can the Schools Be Saved?
	13. Will America Grow Up before It Grows Old?		36. Off Course
	14. The Cruelest Choice		37. A Society of Suspects: The War on Drugs and Civil Liberties
	15. A New Look at Health Care Reform		38. It's Drugs, Alcohol, and Tobacco, Stupid!
	16. Guns, Money & Medicine		39. U.S. Competitiveness: "Resurgence" versus Reality
	17. Mental Illness Is Still a Myth		40. The Civil Rights Issue of the '90s
	18. 'Glass Ceiling' Still Too Hard to Crack, U.S. Panel Finds		41. Global Reach: The Threat of International Drug Trafficking
	19. Death of the Middle Class		42. A Growing Global Crisis
	20. On the Battlefields of Business, Millions of Casualties		43. Earth Is Running Out of Room
	21. Social Change One on One: The New Mentoring Movement		
	22. Dismantling the Welfare State: Is It the Answer to America's Social Problems?		
	23. Welfare Fixers		

(Continued on next page)

ABOUT YOU

Name _____ Date _____

Are you a teacher? ☐ Or a student? ☐

Your school name _____

Department _____

Address _____

City _____ State _____ Zip _____

School telephone # _____

YOUR COMMENTS ARE IMPORTANT TO US!

Please fill in the following information:

For which course did you use this book? _____

Did you use a text with this *ANNUAL EDITION*? ☐ yes ☐ no

What was the title of the text? _____

What are your general reactions to the *Annual Editions* concept?

Have you read any particular articles recently that you think should be included in the next edition?

Are there any articles you feel should be replaced in the next edition? Why?

Are there other areas of study that you feel would utilize an *ANNUAL EDITION*?

May we contact you for editorial input?

May we quote your comments?

ANNUAL EDITIONS: SOCIAL PROBLEMS 97/98

| **BUSINESS REPLY MAIL** |
| First Class   Permit No. 84   Guilford, CT |

Postage will be paid by addressee

Dushkin/McGraw-Hill
Sluice Dock
Guilford, Connecticut 06437